拱棚辣椒栽培技术试验研究

高晶霞 谢 华 编著

GONGPENG LAJIAO ZAIPEI JISHU SHIYAN YANJIU

黄河出版传媒集团
阳光出版社

图书在版编目（CIP）数据

拱棚辣椒栽培技术试验研究 / 高晶霞, 谢华编著. -- 银川：阳光出版社，2020.8
ISBN 978-7-5525-5442-7

Ⅰ.①拱… Ⅱ.①高…②谢… Ⅲ.①辣椒－大棚栽培 Ⅳ.①S626.4

中国版本图书馆 CIP 数据核字（2020）第 153739 号

拱棚辣椒栽培技术试验研究   高晶霞 谢 华 编著

责任编辑	申 佳
封面设计	赵 倩
责任印制	岳建宁

出版发行

出 版 人	薛文斌
地　　址	宁夏银川市北京东路 139 号出版大厦（750001）
网　　址	http://www.ygchbs.com
网上书店	http://shop129132959.taobao.com
电子信箱	yangguangchubanshe@163.com
邮购电话	0951-5014139
经　　销	全国新华书店
印刷装订	宁夏凤鸣彩印广告有限公司
印刷委托书号	（宁）0015159

开　　本	720 mm×980 mm　1/16
印　　张	22.5
字　　数	300 千字
版　　次	2020 年 8 月第 1 版
印　　次	2020 年 8 月第 1 次印刷
书　　号	ISBN 978-7-5525-5442-7
定　　价	98.00 元

版权所有　翻印必究

序

宁夏回族自治区彭阳县地理条件优越，土质、气温非常适宜辣椒生长。在彭阳生产的辣椒产品不仅色泽好，而且营养成分含量高、商品性好，深受市场青睐。辣椒生产地域主要分布于红河乡、城阳乡、白阳镇、古城镇、新集乡等12个乡镇。拱棚辣椒作为彭阳县农业优势主导型产业，起始于2006年宁夏设施农业大发展时期，至2013年，全县建有水泥桁架拱棚5万亩，均进行辣椒生产，亩均年收益6800元。目前，产品销往甘肃、陕西、内蒙古、四川、北京、深圳等省市，销售收入达3亿多元，彭阳县也成为西北地区辣椒主要产区之一。2015年，"彭阳辣椒"获得国家地理标志商标专有权，"彭阳辣椒"已成为宁夏蔬菜品牌产品之一，成为推动地方经济发展、带动农民脱贫致富的支柱产业。但由于水泥桁架型拱棚空间小、春提前秋延后温光不足、无法安装整装放风设备，加之种植年限均已达10年以上，土壤栽培障碍严重，产量、质量明显下降。另外，上市期集中于6月中旬至8月中旬，与全国辣椒集中上市期重叠，对特色优势产业销售及效益的削减问题日益突出，这些都严重制约宁夏辣椒产业发展。

2016年，宁夏农林科学院种质资源研究所承担实施了全产业链创新示范项目"宁夏特色瓜菜产业关键技术创新与示范"，其中的"拱棚辣椒标准化栽培技术试验示范"课题组与彭阳县蔬菜产业技术服务中心合作，在红河乡、新集乡、古城镇实施围绕"拱棚辣椒连年丰产栽培"为技术核心的标准化栽培技术试验示范工作。通过广泛引进国内外优良辣椒品种，筛选适合冷凉区拱棚栽培、抗病性强、丰产性好、适宜外运的牛角椒，以及适于干鲜两用、抗病性强、丰产性好的羊角椒优良品种，推广羊角椒、牛角椒、线椒、水果椒等系列品种及优质穴盘自根苗和嫁接苗；优化配置水泥拱架塑料大棚新材料、新设备，设计建造双层覆盖和大跨度全钢架拱棚2种；试验研究现有栽培条件下的不同定植时间、不同栽培方式、土壤培肥保育和连作障碍克服、滴灌水肥一体化以及病虫害绿色防控等关键技术，提出并建立拱棚辣椒标准化栽培技术模式；通过现场观摩、集中培训、农民科技示范户培养、订单种苗推广等形式，进行应用推广和辐射带动，目前，用"巨峰1号"、"朗月407"等品种更换了应用多年的"亨椒一号"，保障了品种优势；推广应用的9月底辣椒拉秧后套茬远缘作物燕麦、苏丹草、大蒜、三叶草、地豆及菠菜、小油菜栽培模式，结合应用全素有机底肥和土壤调节剂，较好地解决了土壤连作障碍问题；以涡流施肥灌为物化产品示范推广的水肥一体化技术，节省劳动力20%，节水30%，实现水肥量化管理；应用弥粉机喷施，改进白粉病、疫病等防控技术，较好地控制了病虫害发生，显著降低打药施药劳动强度。以上措施使整装技术得以完整落地和普及应用，基本解决了彭阳拱棚辣椒面临的栽培障碍问题，整体提高了栽培技术水平，为保障拱棚辣椒产业健康、持续、稳定发展以及提高冷凉区农民收入提供了坚实的科技支撑。

为总结课题组在辣椒栽培技术方面获得的经验，宁夏农林科学院种质资源研究所把项目实施过程中，因地制宜开展的一些科技攻关、技术难题、技术方法、应用范围、理论依据、各项管理措施及试验论文汇编成册，奉献给大家，供从事相关工作的同志们参考和借鉴。由于各种原因，书中难免有不足，敬请从事、关心、支持农业尤其是辣椒栽培的相关科研人员不吝赐教。

在本书的编写过程中，得到了宁夏农林科学院党政领导的高度重视，得到了种质资源研究所各位专家的悉心指导和同仁的大力协助。在此，向他们表示崇高的敬意与真挚的感谢。

目录

1. 不同栽培模式对连作辣椒生长及土壤养分的影响 / 001
2. 冷凉区拱棚辣椒牛角椒优新品种筛选试验 / 014
3. 宁夏地区优良牛角椒引进适应性试验 / 025
4. 宁夏地区优良羊角椒品种引进适应性试验 / 031
5. 优质羊角椒适应性栽培试验研究 / 042
6. 宁夏地区彩椒品种适应性筛选试验 / 050
7. 冷凉区拱棚辣椒不同新组合品比试验 / 060
8. 拱棚春提早、秋延后辣椒优新品种筛选试验 / 069
9. 不同年限连作对拱棚辣椒光合特性和果实品质的影响 / 083
10. 辣椒连作对拱棚土壤微生物多样性的影响 / 098
11. 微生物菌剂对拱棚连作辣椒生长、产量及品质的影响 / 145
12. 辣椒连作土壤微生物群落及土壤离子对微生物菌剂的响应 / 157
13. 有机肥与玉米秸秆配施对连作辣椒生长发育、产量及品质的影响 / 170
14. Study on Growth and Photosynthetic Characteristics of Pepper under Different Fertilization Modes / 182
15. Combined effects of organic fertilizer and straw on microbial community and functions of pepper continuous cropping soil / 192

16. 辣根素水乳剂对连作辣椒光合特性及土壤酶活性的影响 / 210

17. 辣根素水乳剂不同施药量处理对土壤微生物多样性的影响 / 222

18. 伴生燕麦对连作辣椒叶片保护酶活性及根际土壤微环境的影响 / 265

19. 嫁接砧木对拱棚连作辣椒生长发育、产量及品质的影响 / 276

20. 不同嫁接砧木对连作辣椒生长发育及 N、P、K 分配特性的影响 / 288

21. 不同穗砧组合对辣椒生理特性及对产量和品质的影响 / 308

22. 不同作物秸秆腐解对连作辣椒生长及根际环境的影响 / 329

23. 拱棚辣椒水肥一体化技术试验研究 / 343

不同栽培模式对连作辣椒生长及土壤养分的影响

高晶霞[1],吴雪梅[2],赵云霞[1],颜秀娟[1],谢 华[1]

(1.宁夏农林科学院种质资源研究所,宁夏银川 750002;
2.宁夏回族自治区彭阳县蔬菜产业发展服务中心,宁夏彭阳 756500)

摘 要:以不拉秧残茬越冬、辣椒复种燕麦、辣椒复种菠菜土壤为研究对象,探讨辣椒复种燕麦和辣椒复种菠菜对连作辣椒土壤养分、辣椒生长指标以及产量的影响,以期为缓解辣椒连作障碍提供理论依据。结果表明:与不拉秧残茬越冬处理相比,复种处理的pH和产量显著提高,复种处理的土壤全盐含量、全氮含量、速效氮含量、全磷、全钾含量、速效磷、速效钾养分和辣椒生长在整个生育期均显著低于对照。综合分析表明辣椒复种燕麦是冷凉区水泥拱架大棚连续丰产栽培的主要模式,对于改善辣椒土壤环境,降低土壤次生盐渍化,降低病虫害发病率,保障辣椒产业持续健康发展有着积极的意义。

关键词:不同栽培模式;辣椒;土壤养分;生长和产量

Effects of different cultivation modes on the growth and soil nutrients of continuous cropping pepper

Gao jing-xia[1], Wu xue-mei[2], Zhao yun-xia[1], Yan xiu-juan[1], Xie hua[1]

(1. *Institute of germplasm resources, ningxia academy of agriculture and forestry sciences, ningxia yinchuan 750002*; 2. *Ningxia hui autonomous region pengyang county vegetable industry development service center, ningxia pengyang 756500*)

Abstract: In this paper, the soil of overwintering with the stubble of rice seedling, oat and spinach were studied, the effects of compound oat and spinach on soil nutrient, growth index and yield of pepper in continuous cropping were studied,

In order to provide theoretical basis for easing the barrier of continuous cropping of pepper. The results show that: Compared with the overwintering treatment of the stubble of non-pull seedling,the pH and yield of multiple cropping treatment were significantly increased, The total salt content, total nitrogen content, available nitrogen content, total phosphorus, total potassium content, available phosphorus, available potassium nutrient and pepper growth in the whole growth period were significantly lower than the control.The comprehensive analysis shows that the double-cropping oat of pepper is the main mode of continuous high yield cultivation in cement arch greenhouse in cool area, It has positive significance for improving soil environment of pepper, reducing soil secondary salinization, reducing the incidence of diseases and insect pests, and ensuring the sustainable and healthy development of the industry.

Key words:Different cultivation modes; Chili; Soil nutrients; Growth and yield

辣椒(*Capsicum annuum L.*)属茄科辣椒属,为一年或有限多年生草本植物,原产于南美洲和中美洲等热带地区,是我国仅次于大白菜的第二大蔬菜作物。据农业部大宗蔬菜体系统计,近年来我国辣椒种植面积为150万~200万hm²,占全国蔬菜总种植面积的8%~10%[1]。然而,随着辣椒种植面积的扩大和连作年限的增加,连作障碍日益严重,轻者减产30%左右,严重的减产60%以上,甚至绝产[2]。连作障碍发生的原因很多,其主要原因是连续单一种植同一作物导致作物根际微生态平衡失调、微生物种群结构失衡、微生物总数和土壤酶活性下降[3]。因此,采用合理的耕作制度,如不同作物合理轮作、适当作物间作等措施,是提高土壤质量及作物产量的重要手段。

在辣椒生产中,复种有利于缓解连作障碍进而提高辣椒产量。因此,本研究主要从辣椒生长和土壤养分方面探讨复种对于辣椒连作障碍的影响,分析不同栽培模式下辣椒生长和土壤养分含量的差异,以期为缓解辣椒连作障碍提供理论依据,并且提供一定的实践指导。本试验在不改变冷凉地区拱棚辣椒产业模式的前提下,结合实际生产设计实施了拱棚辣椒不同栽培模式,增加了农户的生产效益,通过对不同栽培模式条件下对辣椒生长影响的研究,明确不同栽培模式对连作土壤、辣椒生长的影响,探索拱棚辣椒连续健康发展的途径。

1 材料和方法

1.1 试验地点

试验选在彭阳县新集乡沟口村新庄洼组3栋水泥拱架塑料大棚实施,每栋大棚面积均为540 m²(大棚长60 m,宽9 m,高2.7 m)。

1.2 供试材料

辣椒品种:亨椒新冠龙,宁夏嘉禾源种苗有限公司提供;菠菜、燕麦种子均购买自银川市西北农资城。

1.3 试验方法

试验设3个处理,处理1:辣椒复种燕麦;处理2:辣椒复种菠菜;对照(CK):不拉秧残茬越冬。辣椒品种于2013年2月2日在日光温室采用穴盘育苗,4月8日定植,结合整地施农家肥4 000 kg/667 m²,磷酸二铵30 kg/667 m²,尿素6 kg/667 m²。追肥2次,每次施磷酸二铵20 kg/667 m²,尿素5 kg/667 m²,硫酸钾10 kg/667 m²。采用膜下滴灌,垄高25 cm,垄面宽70 cm,垄沟宽60 cm,单株栽植,行距65 cm,株距35 cm,每667 m²定植2 773株。6月17日采收。各处理灌水,施肥时间及施肥量均一致,田间管理措施均一致。

1.4 调查及测定方法

1.4.1 土壤取样

跟踪监测辣椒整地前、定植前、坐果初期、始收期、盛果期各时间点的土壤情况,用土钻进行5点取样,每点取10 cm深土壤样品700 g。

测量方法:pH采用电位法、EC采用电导率测定仪、有机质采用重铬酸钾-硫酸氧化法、全氮采用半微量凯氏定氮法、碱解氮采用碱解扩散法、全磷采用钼锑抗比色法、速效磷采用钼锑抗比色法、速效钾采用火焰光度计法、全钾采用火焰光度法。

1.4.2 生长指标

每个大棚对角线定5个点,每点随机选取5株挂牌,在四母斗开花时每隔10 d测定1次株高和茎粗。株高是从辣椒茎基部到植株生长点的垂直距离;植株茎粗是茎秆中部茎粗。

1.4.3 产量

每个大棚每3畦为1个小区统计实产,每小区选5株挂牌统计单株结果数,称单果重,随机取20个果测量果长、果粗。

1.4.4 病害

分别在7月25日、8月5日、8月15日、8月25日调查辣椒白粉病的发病情况,分别对每个处理5点取样,每样点10株,记载病叶数,计算发病率。

1.5 数据处理

用Excel2013和DPS软件进行数据分析处理,不同处理间的多重比较采用Duncan's新复极差法($p \leqslant 0.05$)。

2 结果与分析

2.1 不同栽培模式对连作辣椒土壤全盐含量的影响

由图1可知,不同栽培模式下,土壤全盐含量随着辣椒的生长呈现先增长后降低的趋势且对照(CK)的土壤全盐含量始终高于复种处理,辣椒定植

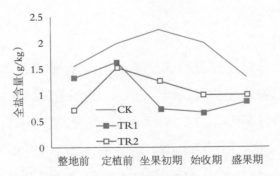

图1 不同栽培模式对连作辣椒土壤全盐含量的影响

Fig.1 Effects of different cultivation modes on soil total salt content of continuous cropping pepper

前,复种燕麦处理的土壤全盐含量高于复种菠菜处理的全盐含量,定植后,在辣椒的坐果初期到辣椒的盛果期,复种燕麦处理的全盐含量低于复种菠菜处理的全盐含量,相比较对照(CK),复种处理显著降低了土壤全盐含量,说明连续种植辣椒后,复种燕麦和菠菜,均能够减轻土壤盐分的积累,且辣椒定植后,复种菠菜处理对于降低土壤含盐量效果更好。

2.2 不同栽培模式对连作辣椒土壤pH的影响

由图2可知,不同栽培模式下,土壤的pH值随着辣椒的生长呈现先降低后增长的趋势,且CK的pH值始终低于复种处理,辣椒定植前,菠菜复种处理的pH值高于燕麦复种处理的pH值,定植后,在辣椒的坐果初期至辣椒的盛果期,燕麦复种处理的pH值高于菠菜复种处理的pH值。相比较对照,复种处理明显增加了土壤pH,说明复种燕麦和菠菜,均能够增加土壤pH。

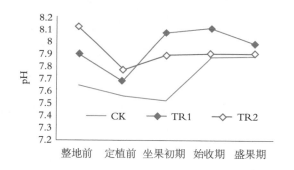

图2 不同栽培模式对连作辣椒土壤pH的影响

Fig.2 Effects of different cultivation modes on soil pH of continuous cropping pepper

2.3 不同栽培模式对连作辣椒土壤速效磷含量的影响

由图3可知,复种栽培模式下,连作辣椒土壤的速效磷含量整体呈增长的趋势,CK土壤的速效磷含量呈现先增长后降低再增长的趋势,但CK的土壤速效磷含量始终高于复种处理,定植前,复种菠菜处理的土壤速效磷含量高于复种燕麦处理的,定植后,在辣椒的坐果初期到始收期,复种燕麦处理土壤的速效磷含量高于复种菠菜处理土壤的速效磷含量,辣椒盛果期,复种处理土壤的速效磷含量相同。同对照相比,复种处理明显降低了土壤速效磷含量,说明连续种植辣椒后,复种处理能减轻土壤中速效磷的含量。

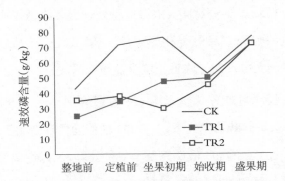

图3 不同栽培模式对连作辣椒土壤速效磷含量的影响

Fig.3 Effects of different cultivation modes on the content of available phosphorus in soil of continuous cropping pepper

2.4 不同栽培模式对连作辣椒土壤全磷含量的影响

由图4可知,不同栽培模式下,复种处理土壤的全磷含量在辣椒的整个生育期内无明显变化,CK的土壤全磷含量随着辣椒的生长呈现先增后降低的趋势,且CK的土壤全磷含量在辣椒的整个生育期内始终大于复种处理,说明在辣椒定植前复种处理可以明显降低土壤内的全磷含量,但不同的复种处理之间无明显差异。

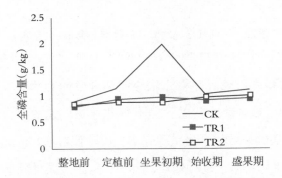

图4 不同栽培模式对连作辣椒土壤全磷含量的影响

Fig.4 Effects of different cultivation modes on soil total phosphorus content of continuous cropping pepper

2.5 不同栽培模式对连作辣椒土壤碱解氮含量的影响

由图 5 可以看出,在不同的栽培模式下,CK 的碱解氮含量先增加后降低,复种处理的碱解氮含量先增加后降低再增加,但在整个过程中,CK 的土壤碱解氮含量始终高于复种处理的土壤碱解氮含量。辣椒定植前,复种燕麦处理的土壤碱解氮含量大于复种菠菜处理的土壤碱解氮含量,定植后,从坐果初期到盛果期,复种菠菜处理的土壤碱解氮含量高于复种燕麦处理的土壤碱解氮含量。说明复种处理可以一定程度的降低土壤中碱解氮的含量,且辣椒定植后,复种菠菜处理对土壤中碱解氮的含量影响更大。

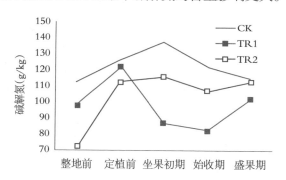

图 5 不同栽培模式对连作辣椒土壤碱解氮含量的影响

Fig. 5 Effects of different cultivation modes on soil alkali-hydrolyzed nitrogen content of continuous cropping pepper

2.6 不同栽培模式对连作辣椒土壤全氮含量的影响

由图 6 可知,在不同的栽培模式下,连作辣椒土壤全氮含量随着辣椒的生长呈现先增长后降低再增长的趋势,且 CK 的土壤全氮含量始终高于复种处理的土壤全氮含量,辣椒定植前,复种燕麦处理的土壤全氮含量明显高于复种菠菜处理,辣椒定植后,从辣椒坐果初期到盛果期,复种菠菜处理的土壤全氮含量高于复种燕麦处理的土壤全氮含量,说明复种处理可以降低土壤中全氮含量。

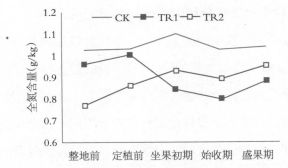

图6 不同栽培模式对连作辣椒土壤全氮含量的影响

Fig.6 Effects of different cultivation modes on soil total nitrogen content of continuous cropping pepper

2.7 不同栽培模式对连作辣椒土壤速效钾含量的影响

由图7可以看出,在不同的栽培模式下,连作辣椒土壤中速效钾的含量随着辣椒的生长呈现先升高后降低的趋势,整地前和辣椒坐果初期到盛果期,CK的土壤速效钾含量均高于复种处理,但是定植前复种处理的土壤速效钾含量高于复种处理。辣椒坐果初期前,复种燕麦处理的土壤速效钾含量高于复种菠菜处理的土壤速效钾含量,辣椒坐果初期后,复种燕麦处理土壤速效钾含量明显降低,低于复种菠菜处理土壤速效钾含量。说明在辣椒定植后,复种处理可以显著降低土壤中速效钾的含量。

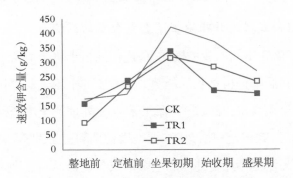

图7 不同栽培模式对连作辣椒土壤速效钾含量的影响

Fig.7 Effects of different cultivation modes on available potassium content in soil of continuous cropping pepper

2.8 不同栽培模式对连作辣椒土壤全钾含量的影响

由图8可以看出,不同栽培模式下,连作辣椒土壤中全钾含量随着辣椒的生长呈现先增长后降低的趋势,辣椒盛果期之前,CK的土壤全钾含量高于复种处理的土壤全钾含量,辣椒盛果期时,CK的土壤全钾含量低于复种处理的全钾含量,定植前到盛果期,复种菠菜处理的土壤全钾含量均高于复种燕麦处理的土壤全钾含量。说明在辣椒盛果期之前,复种处理可以降低土壤中全钾的含量,在辣椒盛果期时,复种处理可以增加土壤中全钾的含量。

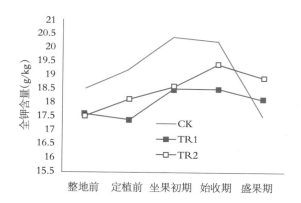

图8 不同栽培模式对连作辣椒土壤全钾含量的影响

Fig.8 Effects of different cultivation modes on soil total potassium content of continuous cropping pepper

2.9 不同栽培模式对连作辣椒生长指标的影响

复种燕麦和菠菜后,对于后茬辣椒的生长也产生了显著的影响。复种燕麦处理(TR1)植株株高比对照平均增加4.1 cm,提高4.3%;复种菠菜处理(TR2)比对照平均矮2.9 cm,降低3.1%。茎粗的变化趋势与株高相似,均随辣椒生育进程而增加,TR1的茎粗大于对照,8月29日,TR1茎粗最大,为2.56 cm,与对照相比增加0.2 cm,提高8.5%;TR2茎粗小于对照,降低2.5%。说明复种燕麦处理能够显著增加后茬辣椒的长势。

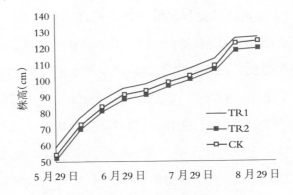

图 9 不同栽培模式对连作辣椒株高的影响

Fig.9 Influence of different cultivation modes on plant height of pepper in continuous cropping

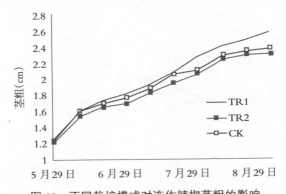

图 10 不同栽培模式对连作辣椒茎粗的影响

Fig.10 Effects of different cultivation modes on stem thickness of continuous cropping pepper

2.10 不同栽培模式对辣椒产量和构成因素的影响

由表1可以看出,辣椒单株结果数和单果质量是构成产量的2个主要因素,以复种燕麦处理的辣椒单株结果数较多、单果质量较大,因此其产量最高,单株产量为2.75 kg/株,比对照(CK)增加0.16 kg/株,提高6.2%。燕麦茬果长比重茬平均增加1.3 cm,提高5.2%;果粗增加0.09 cm,提高2.0%,单果重增加1.2 g,提高1.4%,畸形果率降低5.9%。菠菜茬与重茬差异不大。燕麦茬与菠菜茬果实比重茬颜色浓绿光泽度高,品质明显改变。

燕麦茬折合667 m² 产量最高,为7 363.7 kg,比对照增产448.7 kg,增幅6.5%;产值17 672.9 元/667 m²,较对照增收1 076.9 元/667 m²;菠菜茬产量较

低,比对照增产 135.8 kg/667 m²,增幅 2.0%。

表 1 不同栽培模式对辣椒单株结果数、单果质量和产量的影响

Table 1 Effects of different stubble on the number of pepper per plant, fruit quality and yield

处理 Treatment	果长 Fruit long /cm	果粗 Fruit coarse /cm	单株结果数 Number of fruit per plant /个	单果质量 Fruit quality /g	单株产量 Yield per plant（kg/株）	畸形果率 Deformity rate of fruit /%	折合 667 m² 产量 Equivalent to 667m² output（kg /667m²）
TR1	26.2	4.67	31.8	86.4	2.75	14.9	7 625.8aA
TR2	25.3	4.60	30.7	85.5	2.62	16.6	7 265.3bB
CK	24.9	4.58	30.5	85.2	2.59	20.8	7 182.1bB

2.11 不同栽培模式辣椒病害发生情况

2013 年进入 7 月份以来,连续降雨长达 15 d 左右,田间湿度大,光照不足,造成辣椒白粉病发生较早。据 7 月 25 日调查 3 种栽培模式辣椒白粉病都有不同程度发生,燕麦茬辣椒白粉病病叶率 2.3%,菠菜茬辣椒白粉病病叶率 3.7%;重茬辣椒白粉病病叶率 7.8%;至 8 月 5 日调查燕麦茬辣椒白粉病病叶率 3.5%,菠菜茬辣椒白粉病病叶率 5.8%;重茬辣椒白粉病病叶率 10.5%;至 8 月 15 日调查燕麦茬辣椒白粉病病叶率 4.6%,菠菜茬辣椒白粉病病叶率 7.5%;重茬辣椒白粉病病叶率 14.5%;至 8 月 25 日调查燕麦茬辣椒白粉病病叶率 5.7%,菠菜茬辣椒白粉病病叶率 8.8%;重茬辣椒白粉病病叶率 18.5%;试验表明通过复种燕麦,轮作倒茬种植可以有效降低辣椒病害的发生。

表 2 不同栽培模式辣椒病害发生情况

Table 2 Occurrence of pepper disease in different cultivation modes

处理 Treatment	不同时期辣椒白粉病 Powdery mildew of pepper in different periods/%			
	7 月 25 日	8 月 5 日	8 月 15 日	8 月 25 日
TR1	2.3	3.5	4.6	5.7
TR2	3.7	5.8	7.5	8.8
CK	7.8	10.5	14.5	18.5

3 结论与讨论

大量研究表明,许多作物连作会导致土壤理化性状变差、养分失衡等问题,而采用嫁接、施用微生物菌剂,以及合理的轮作、间作、填闲等栽培模式,能够提高土壤微生物种群多样性和稳定性,缓解连作障碍,促进作物生长,增加产量[7]。段曦等[8]采用嫁接的方式发现,嫁接辣椒根系分泌物组分变化是其减轻土传病害的重要机理之一,砧木和嫁接辣椒降低辣椒发病率,产量分别比对照高31.7%和38.3%。本试验采用复种燕麦和复种菠菜的栽培模式发现,复种燕麦处理和复种菠菜处理与对照相比土壤pH值增加,但显著降低了土壤全盐含量,说明在连续种植辣椒大棚,通过复种燕麦和菠菜,均能够减轻土壤盐分的积累;在定植前,复种燕麦处理的土壤全盐含量显著高于复种菠菜处理的土壤,但辣椒定植以后,复种燕麦处理的土壤全盐含量始终低于菠菜处理,说明复种燕麦处理对于降低土壤含盐量效果更好。

复种轮作能影响土壤速效养分的含量,起到平衡土壤养分、调节地力的作用[6]。试验结果表明,与对照相比,复种轮作降低了连作辣椒土壤中的全氮、全磷、全钾含量,速效磷含量先升高后降低再升高,速效钾含量和碱解氮含量先升高后降低,除速效磷含量外,复种菠菜处理辣椒后茬的全氮、全磷、全钾、速效钾和碱解氮含量均高于复种燕麦,说明复种菠菜对连作辣椒土壤养分的影响更大。

复种燕麦和菠菜后,对于后茬辣椒的生长燕麦处理优于菠菜处理,能够显著增加后茬辣椒的长势;复种燕麦后,比对照增产448.7 kg/667 m²,复种菠菜比对照增产135.8 kg/667 m²,显著提高了辣椒产量。综合分析不同栽培模式辣椒的生长、果实性状、产量、抗病性等相关指标,表明辣椒复种燕麦是冷凉区水泥拱架大棚连续丰产栽培的主要模式,虽然其当年比较效益低于一年三茬模式,但对于改善辣椒土壤环境,降低土壤次生盐渍化,降低病虫害发病率,保障产业持续健康发展有着积极的意义。

参考文献

[1] 王立浩,张正海,曹亚从,张宝玺."十二五"我国辣椒遗传育种研究进展及其展望.中国蔬菜,2016(1):1-7.

[2] 胡亮,陈宾.辣椒常见几种病害的发生及防治.现代园艺,2015(23):112-113.

[3] 袁龙刚,张军林.辣椒连作障碍的主要原因及其对策[J].中国农村小康科技,2006(2):32-33.

[4] 李威,程智慧,孟焕文,等.轮作不同蔬菜对大棚番茄连作基质中微生物与酶及后茬番茄的影响[J].园艺学报,2012,39(1):73-80.

[5] 王治林,乔俊卿,刘邮洲.苏北地区设施栽培茄果类蔬菜连作障碍成因分析及防治措施[J].江苏农业科学,2012,40(2):127-129.

[6] 田福发,张黎杰,周玲玲,刘金兵,姜若勇,吴绍军,王夏雯,余翔,孟佳丽.菇-菜轮作对连作根际土壤养分、微生物含量及辣椒死苗率的影响[J].江西农业学报,2016,28(06):46-49.

[7] 董宇飞,吕相漳,张自坤,贺洪军,喻景权,周艳虹.不同栽培模式对辣椒根际连作土壤微生物区系和酶活性的影响[J].浙江农业学报,2019,31(09):1485-1492.

[8] 段曦,孙晨晨,孙胜楠,等.嫁接辣椒根系分泌物对根腐病和青枯病的影响[J].园艺学报,2017,44(2):297-306.

冷凉区拱棚辣椒牛角椒优新品种筛选试验

高晶霞[1]，吴雪梅[2]，赵云霞[1]，王学梅[1]，谢　华[1]

（1. 宁夏农林科学院种质资源研究所，宁夏银川　750002；
2. 宁夏回族自治区彭阳县蔬菜产业发展服务中心，宁夏彭阳　756500）

摘　要：为筛选出适宜在冷凉区拱棚种植的辣椒新品种，以引进的 14 份辣椒优良牛角椒品种为试材，进行品种观察试验。结果表明：参试辣椒品种物候期、植物学性状、抗病性、果实性状、产量及产值均有差异。参试品种的开花期、坐果期早于对照的有"金惠 13-B"、"日本瑞崎"、"亨椒新冠龙"、"日本长金"、"红圣 202"、"硕源 3 号"6 个品种；晚于对照的有"中椒 106"、"长冠"、"国福 403"、"国福 308"、"红圣 101"、"京都椒王"6 个品种。采收期早于对照的有"红圣 202"、"硕源 3 号"、"金惠 13-B"、"亨椒新冠龙"、"日本长金"、"日本瑞崎"6 个品种，长约 1~5 d，其中"日本瑞崎"采收期最长，为 110 d。全生育期比对照"亨椒 1 号"长的有"亨椒新冠龙"、"金惠 13-B"、"日本瑞崎"3 个品种。参试品种"亨椒新冠龙"、"日本瑞崎"、"京都椒王"、"红圣 101"等品种植株生长势强，其余品种较强或中等；参试品种未发现枯萎病和疫病，但白粉病发病率普遍，14 个品种不同程度都有发病，但"亨椒新冠龙"、"长冠"、"日本瑞崎"几个品种发病率低于其他品种。"亨椒新冠龙"品种产量最高，为 4 643.6 kg/667 m^2，比对照"亨椒 1 号"增产 12.7%，"红圣 101"品种的产量最低为 2 156.8 kg/667 m^2。综上所述：参试品种"亨椒新冠龙"、"日本瑞崎"、"日本长金"、"长冠"辣椒性状典型，光泽度好，辣味适中，耐贮运，适宜拱棚辣椒栽培并推广应用。

关键词：冷凉区；拱棚；辣椒；优新品种；试验

Screening Test of New Superior Varieties of Gongpeng Pepper in Cool Area

Gao jing-xia[1], Wu xue-mei[2], Zhao yun-xia[1], Wang xue-mei[1], xie hua[1]

(1. *Institute of germplasm resources, ningxia academy of agriculture and forestry sciences, ningxia yinchuan 750002*; 2. *Ningxia hui autonomous region pengyang county vegetable industry development service center, ningxia pengyang 756500*)

Abstract: In order to select the new pepper varieties suitable for planting in the cool area, A variety observation experiment was carried out with 14 imported pepper varieties as test materials. the results show that: There were differences in phenological stage, botanical character, disease resistance, fruit character, yield and output value of pepper varieties, The flowering and fruiting stages of the tested varieties were earlier than those of the control, including 6 varieties of jinhui 13 -b, Japanese ruizaki, new crown dragon of heng jiao, Japanese changjin, hongsheng 202 and shuoyuan 3, Later than the control of pepper 106, long crown, guofu 403, guofu 308, red saint 101, Kyoto pepper king 6 varieties. The harvesting period was earlier than that of the control group, including 6 varieties: hongsheng 202, shuoyuan 3, jinhui 13 -b, hengjiao xinguanlong, Japanese changjin and Japanese ruizaki, which lasted for 1 to 5 days. The harvesting period of Japanese ruizaki was the longest, which was 110 days; In the whole growth period, there were 3 varieties of new crown dragon, jinhui 13 -b and Japanese ruizaki; The new crown dragon, Japanese ruizaki, Kyoto pepper king, red saint 101 and other varieties were strong, while the others were strong or medium; Blight and epidemic disease were not found in the tested varieties, but the incidence of powdery mildew was common, All the 14 varieties had different degrees of disease, but the incidence rate of new crown dragon, long crown and Japanese ruizaki were lower than other varieties, The highest yield was 4 643.6 kg/667 m^2 for the new guanlong variety of hengjiao, which was

12.7% higher than that of the no.1 variety of hengjiao, and the lowest yield was 2 156.8 kg/667 m² for the hongsheng101 variety.To sum up, the test varieties of new crown dragon, Japanese ruizaki, Japanese long gold, long crown pepper typical characteristics, good gloss, moderate spicy taste, storage and transportation, suitable for arch canopy pepper cultivation and application.

Key words: cool area; arch shed; pepper; excellent new variety; test

辣椒原产于秘鲁和墨西哥一带[1],因其色泽鲜艳、香辣味和富含维生素C而成为一种世界性蔬菜[2,3]。辣椒不仅是蔬菜,更是调味佳品和重要的天然色素,是制药原料和其他工业原料,种植效益高[4]。辣椒既适合露地栽培,也适合设施内栽培[5],目前,我国辣椒种植面积居蔬菜作物第二位,产值和效益居蔬菜作物之首[6]。

为筛选出适宜宁夏冷凉区拱棚春提早、秋延后栽培的优质、抗病、丰产、抗逆、色泽好的辣椒优良品种,2011年引进14份优良辣椒品种,在彭阳县红河乡上王村水泥拱架塑料大棚进行分类和田间品比观察,调查物候期、植株植物学性状、果实性状、抗病性和产量等主要指标,筛选出适合拱棚栽培的优良辣椒品种,丰富宁夏辣椒品种配置,提高优质品种的利用率。

1 材料与方法

1.1 供试品种

供试品种14个,"中椒106"由中国农业科学院蔬菜花卉研究所提供;"亨椒新冠龙"、"红圣101"、"红圣202"由北京中农绿亨种子科技有限公司提供;"国福308"、"国福403"由北京京研益农科技发展中心提供;"超大椒王"、"金惠13-B"由银川兴农良品农业科技有限公司提供;"硕源3号"由北京硕源种子有限公司提供;"长冠"、"日本瑞崎"由北京农瑞德农业科技有限公司提供;"日本长金"由寿光南澳绿亨农业有限公司提供;"京都椒王"由沈阳市九农高科科技有限公司提供;对照"亨椒1号"(CK)由北京中农绿亨种子科技有限公司提供。

1.2 试验方法

试验地点：选择在设施农业建设区的红河乡上王村水泥拱架塑料大棚中。大棚长 60 m，宽 9 m，高 2.7 m；土壤为中耕黑垆土，肥力中上等，前茬玉米，种植 3 棚。试验采用随机区组设计，每品种为 1 个小区，重复 3 次，小区面积 32.4 m²。采用起垄覆膜栽植，垄高 15 cm，垄宽 60 cm，垄沟 60 cm，每垄定植两行，单株栽植，株距 40 cm，每小区 114 株。

1.3 调查项目

观察记载物候期，植物学性状，病害发生情况，果实性状，并统计小区产量，折合 667 m² 产量。

1.4 数据分析

用 Excel2013 和 SPSS 软件进行数据分析处理，不同处理间的多重比较采用 Duncan's 新复极差法（$p \leq 0.05$）。

2 结果与分析

2.1 不同辣椒品种物候期比较

从表 1 可以看出，开花期、坐果期比对照"亨椒 1 号"（CK）早的有"金惠 13-B"、"日本瑞崎"、"亨椒新冠龙"、"日本长金"、"红圣 202"、"硕源 3 号"6 个品种；比对照"亨椒 1 号"（CK）迟的有"中椒 106"、"长冠"、"国福 403"、"国福 308"、"红圣 101"、"京都椒王"、"超大椒王"7 个品种。采收期比对照"亨椒 1 号"（CK）长的有"红圣 202"、"硕源 3 号"、"金惠 13-B"、"亨椒新冠龙"、"日本长金"、"日本瑞崎"6 个品种，长约 1~5 d，其中"日本瑞崎"采收期最长，为 110 d。全生育期比对照"亨椒 1 号"（CK）长的有"亨椒新冠龙"、"金惠 13-B"、"日本瑞崎"3 个品种。

2.2 不同辣椒品种植物学性状与抗病性比较

由表 2 可以看出，"红圣 101"品种的株高最高为 107.7 cm，"日本瑞崎"

表 1 不同辣椒品种物候期比较

Table 1 Comparison of phenological stages of different pepper varieties

品种 variety	出苗期 seeding stage (月/日)	定植期 planting date (月/日)	开花期 anthesis (月/日)	坐果期 fruit-set period (月/日)	采收始期 harvest the beginning (月/日)	采收末期 the end of the harvest (月/日)	定植至始收 planting to harvest (天)	采收期 picking time (天)	全生育期 growth duration (天)
中椒 106	3/11	4/20	6/4	6/9	6/28	10/3	68	97	206
亨椒新冠龙	3/8	4/20	5/27	6/3	6/21	10/8	61	109	214
国福 308	3/10	4/20	6/7	6/12	6/30	10/6	70	98	210
超大椒王	3/11	4/20	6/3	6/8	6/26	10/4	66	100	207
金惠 13-B	3/7	4/20	5/26	6/2	6/19	10/5	59	108	212
硕源 3 号	3/14	4/20	5/30	6/6	6/23	10/8	63	107	208
长冠	3/16	4/20	6/5	6/10	6/28	10/7	68	101	205
日本长金	3/12	4/20	5/28	6/4	6/21	10/8	61	109	210
日本瑞崎	3/10	4/20	5/26	6/3	6/20	10/8	60	110	212
京都椒王	3/16	4/20	6/7	6/13	7/2	10/6	72	96	204
红圣 101	3/13	4/20	6/4	6/9	6/30	10/8	70	100	209
红圣 202	3/13	4/20	5/31	6/5	6/25	10/8	65	105	209
国福 403	3/13	4/20	6/5	6/11	6/29	10/3	69	96	204
亨椒 1 号(CK)	3/11	4/20	6/1	6/8	6/26	10/8	66	104	211

表2 不同辣椒品种植物学性状和抗病性比较
Table 2 Comparison of botanical characters and disease resistance of different pepper varieties

品种 variety	株高 plant height /cm	茎粗 stem diameter /cm	门椒节位 the door bell pepper section /节	开展度 divergence /cm	分枝数 winding number /个	叶色 leaf colour	单株结果数 number of fruit per plant /个	植株长势 plants grow	发病率 powdery mildew /%
中椒106	91.3	2.22	10~12	101.6	2.40	深绿	24.2	较强	32.5
亨椒新冠龙	79.6	1.94	8~11	88	2.47	绿	25.3	强	21.2
国福308	87.7	2.24	8~12	93.6	3.00	绿	24.0	较强	29.8
超大椒王	89.5	2.08	8~11	103.1	2.53	绿	24.3	较强	42.6
金惠13-B	75.9	1.90	7~10	83.9	2.53	绿	23.6	中等	44.3
硕源3号	87.2	2.06	8~10	100.9	2.53	绿	23.2	较强	41.6
长冠	86.5	1.99	8~13	91.7	3.10	绿	29.3	较强	18.1
日本长金	75.3	1.90	8~11	92.0	2.62	绿	29.5	中等	18.7
日本瑞崎	74.8	1.99	8~11	83.8	2.67	绿	28.7	强	14.3
京都椒王	104.9	2.09	9~12	107.1	2.27	绿	21.6	强	32.3
红圣101	107.7	1.76	8~11	120.0	2.07	墨绿	50.9	强	45.5
红圣202	93.8	1.92	10~13	94.3	2.33	深绿	51.6	较强	43.1
国福403	95.9	1.88	9~11	93.1	2.13	深绿	72.9	较强	26.8
亨椒1号(CK)	95.5	1.86	8~11	99.8	2.47	绿	23.2	强	33.8

的株高最低为74.8 cm,其他品种株高范围在75.9~104.9 cm;"国福308"品种的茎粗最粗为2.24 cm,"红圣101"的茎粗最细为1.76 cm,其他品种的茎粗范围在1.86~2.22 cm。"红圣101"的开展度最大为120 cm,"日本瑞崎"和"金惠13-B"的开展度最小为83.8 cm和83.9 cm,其他品种的开展度范围在88~107.1 cm。"长冠"的分枝数最多为3.10个,"红圣101"的分枝数最少为2.07个,其他品种的分枝数范围在2.13~3.00个。"红圣101"的叶色为墨绿色,"中椒106"、"红圣202"和"国福403"的叶色为深绿色,其他品种的叶色均为绿色。单株结果数最多的是"国福403"为72.9个,单株结果数最少的是"京都椒王"为121.6个。"亨椒新冠龙"、"日本瑞崎"、"京都椒王"、"红圣101"等品种植株生长势强,其他品种较强或中等。参试的14个品种调查中未发现枯萎病和疫病,但白粉病发病率普遍;14个品种不同程度都有发病,但"亨椒新冠龙"、"长冠"、"日本瑞崎"几个品种发病率低于其他品种。

2.3 不同辣椒品种果实性状比较

从表3可以看出,"亨椒新冠龙"的果长最长为26.0 cm,"红圣101"和"中椒106"的果长最短为13.5 cm和13.9 cm,其他品种的果长范围在14.5~23.3 cm。"中椒106"品种的果实横径最大为5.22 cm,"国福403"的辣椒果实横径最小为137 cm,其他品种的辣椒果实横径范围在2.15~4.59 cm。"中椒106"和"京都椒王"的辣椒果肉最厚为0.39 cm,"国福403"的辣椒果肉最薄为0.17 cm,其他品种的辣椒果肉厚度范围在0.25~0.38 cm。"京都椒王"的单果质量最大为85 g,"国福403"的单果质量最小为16.1 g,其他品种的单果质量范围在21.1~83.6 g。"亨椒新冠龙"、"金惠13-B"、"国福308"、"超大椒王"、"硕源3号"、"长冠"果实与对照"亨椒1号"均为牛角形品种,耐贮运,尤其"亨椒新冠龙"品种果个大,果长为26.0 cm,顺直,光泽度好;"中椒106"为粗牛角形,耐贮运;"日本瑞崎"、"日本长金"、"京都椒王"为牛角形品种,"日本瑞崎"、"日本长金"辣味适中,"京都椒王"无辣味,3个品种均耐贮运。

表 3 不同辣椒品种果实性状比较

Table 3 Comparison of fruit characters of different pepper varieties

品种 variety	果实长度 fruit length /cm	果实横径 the fruit diameter /cm	心室数 number of ventricular /个	果肉厚度 the thickness of the pulp /cm	单果质量 fruit quality /g	果实外观、品质 fruit appearance and quality
中椒 106	13.9	5.22	3	0.39	74.4	粗牛角形,绿色,微辣,果面光滑,耐贮运
亨椒新冠龙	26.0	4.27	2~4	0.37	83.6	牛角形,绿色,微辣,果面微有皱褶,顺直,光泽度好,耐贮运
国福 308	22.1	4.34	2~4	0.37	82.8	牛角形,黄绿色,辣味适中,果基有皱,果形顺直,果面微有皱褶,有棱沟,耐贮运
超大椒王	22.5	4.29	3	0.36	79.4	牛角形,浅绿色,辣味适中,果面有皱褶,有棱沟
金惠 13-B	21.6	4.59	3~4	0.38	81.4	牛角形,绿色,微辣,果形顺直,果面光滑,光泽度好,耐贮运
硕源 3 号	21.7	4.44	2~3	0.37	79.7	牛角形,绿色,微辣,果面光滑,光泽度好,耐贮运
长冠	21.2	3.90	2~3	0.30	67.8	牛角形,黄绿色,辣味适中,果基微皱,果面有棱沟,有光泽,耐贮运
日本长金	21.6	3.94	2~3	0.34	68.5	牛角形,浅绿,辣味适中,果基微皱,果面光滑,皮厚,耐贮运
日都瑞崎	23.3	4.06	2~4	0.32	72.5	牛角形,浅绿色,辣味适中,果基微皱,果面光滑,光泽度好,耐运输
京都椒王	21.0	4.54	2~3	0.39	85.0	牛角形,淡绿色,无辣味,果面光亮,果形顺直,皮厚,耐运输
红圣 101	13.5	2.15	2~3	0.25	21.1	牛角形,鲜绿色,果面光滑,皮厚,光泽度好,耐运输
红圣 202	14.5	2.42	2~3	0.25	24.8	牛角形,深绿色,辣味香浓,皮厚,光泽度好,耐运输
国福 403	21.4	1.37	2~3	0.17	16.1	长指形,深绿色,辣味香浓,皮薄,果面光滑,顺直,果形美观,光泽度好,干椒品质优秀
亨椒 1 号(CK)	22.2	4.29	3~4	0.37	81.2	牛角形,黄绿色,微辣,果面微有皱褶,顺直,光亮,耐贮运

表 4 不同辣椒品种产量比较
Table 4 Yield comparison of different pepper varieties

品种 variety	小区产量 cell production/kg				前期折合产量 the early production /(kg/667 m²)	比对照增幅± increase over control /%	小区总产量 total producting of community/kg				折合 667 m² /(kg/667 m²)
	I	II	III	平均 average			I	II	III	平均 average	
中椒 106	68.4	63.1	67.9	66.5	1 368.3	-7.8	196.4	192.7	194.0	194.4	4 004.1
亨椒新冠龙	80.4	76.5	82.1	79.7	1 640.1	10.5	228.5	217.9	230.3	225.6	4 643.6
国福 308	80.5	76.8	79.4	78.9	1 624.3	9.4	215.2	198.7	204.6	206.2	4 244.2
超大椒王	75.7	61.1	71.7	69.5	1 430.8	-3.6	206.9	195.4	201.5	201.3	4 144.0
金惠 13-B	73.8	60.9	69.1	67.9	1 398.5	-5.8	207.2	190.7	204.8	200.9	4 135.8
硕源 3 号	74.1	66.5	64.3	68.3	1 406.1	-5.3	206.8	194.7	190.4	197.3	4 061.7
长冠	73.7	63.4	75.5	70.9	1 458.9	1.7	217.6	206.2	224.5	216.1	4 448.7
日本长金	68.4	62.1	66.7	65.7	1 353.2	-8.8	224.7	205.3	218.6	216.2	4 450.8
日本瑞崎	74.7	77.0	78.5	76.7	1 579.7	6.4	208.5	216.9	233.2	219.5	4 518.7
京都椒王	70.7	69.5	65.0	68.4	1 408.1	-5.1	201.6	194.7	190.4	195.6	4 026.7
红圣 101	20.5	21.1	17.2	19.6	403.5	-72.8	110.6	112.3	91.4	104.8	2 156.8
红圣 202	26.8	23.4	24.5	24.9	512.7	-65.5	134.5	120.1	127.6	127.4	2 622.7
国福 403	14.3	13.8	13.1	13.7	282.0	-81	117.3	118.9	109.5	115.2	2 371.6
亨椒 1 号(CK)	76.3	67.5	72.5	72.1	1 484.3		205.7	195.6	198.9	200.1	4 119.3

2.4 不同辣椒品种产量构成

从表4可知,"亨椒新冠龙"品种产量最高,为4 643.6 kg/667 m²,比对照"亨椒1号"增产12.7%;"日本瑞崎"、"日本长金"、"长冠"、"国福308"、"超大椒王"、"金惠13-B"等6个品种产量在4 135.8~4 518.7 kg/667 m²,均高于对照产量,增产幅度为0.4%~9.7%;其余品种"硕源3号"、"京都椒王"、"中椒106"、"红圣101"、"红圣202"、"国福403"等产量2 156.8~4 061.7 kg/667 m²,均低于对照产量,降低幅度为1.4%~47.6%。

从数据及方差分析结果看,前期产量"亨椒新冠龙"最高,"国福308"居第二位,但两者差异不显著,"日本瑞崎"居第三位,对照"亨椒1号"居第四位,"亨椒新冠龙"与对照"亨椒1号"产量差异达到显著水平,"亨椒新冠龙"前期亩产量达到1 640.1 kg,比对照"亨椒1号"提高10.5%;"国福403"、"红圣101"、"红圣202"前期产量最低。

从综合产量数据及方差分析结果看,综合产量"亨椒新冠龙"最高,"日本瑞崎"居第二位,"日本长金"居第三位,"长冠"居第四位,前四位之间产量差异不显著,但前四位产量均与对照"亨椒1号"达到显著水平;"亨椒新冠龙"、"日本瑞崎"、"日本长金"、"长冠"亩产量分别比对照"亨椒1号"提高12.7%、9.7%、8.1%、8.0%。

3 结论与讨论

试验结果表明:引进的14个辣椒品种,其中"亨椒新冠龙"果实长、光泽度好、耐贮运;"日本瑞崎"、"日本长金"辣味适中,耐贮运;"长冠"果实较长、光泽度好、耐贮运。4个品种坐果能力强,产量及效益显著高于对照"亨椒1号",应在进一步试验示范的基础上进行大面积推广种植。

在田间筛选试验的基础上,2011—2014年在彭阳县开展了广泛的示范推广工作,在古城镇、新集乡、红河乡等拱棚辣椒主产区以点带面开展示范,2011年推广面积5 000亩,2012年推广面积1万亩,2013年推广面积达到了

1.3万亩,2014年推广面积达到了2万亩,丰富了辣椒栽培品种,降低了目前拱棚辣椒生产中单一品种栽培的风险,逐步成为彭阳县拱棚辣椒栽培的主栽品种。

参考文献

[1] 黄智文,徐晓美,王恒明,等.辣椒品种的耐热性研究试验[J].上海蔬菜,2014(6):11-12.

[2] 戴雄泽,刘志敏.初论我国辣椒产业的现状及发展趋势[J].辣椒杂志,2005(2):1-6.

[3] 高怀春.辣椒果实维生素C含量变化的研究[M].泰安:山东农业大学,2004.

[4] 贺洪军.加工型辣椒绿色高产高效生产技术[M].北京:中国农业科学技术出版社,2015.

[5] 徐珊珊.辣椒对盐碱胁迫的生理反应及适应性机理研究[D].长春:吉林农业大学,2007.

[6] 王继榜.我国辣椒产业现状及发展趋势综述[J].安徽农学通报,2013,19(19):64.

宁夏地区优良牛角椒引进适应性试验

高晶霞[1],吴雪梅[2],赵云霞[1],谢 华[1],颜秀娟[1],王学梅[1]

(1. 宁夏农林科学院种质资源研究所,宁夏银川 750002;
2. 宁夏回族自治区彭阳县蔬菜产业发展服务中心,宁夏彭阳 756500)

摘 要:以引进的15个牛角椒优良品种为试材,在水泥拱架塑料大棚中进行品种观察试验,为宁夏地区牛角椒的优良品种更新换代提供数据基础。结果表明:参试品种"长冠"、"日本瑞崎"、"日本长金"牛角椒性状典型,抗病性强,红果色泽鲜艳,连续坐果力强,单果质量、产量较高,适宜在宁夏保护地栽培中推广应用。

关键词:牛角椒;适应性;宁夏

Experiment of Adaptability of Introduction Good Cayenne Pepper in Ningxia Region

GAO Jingxia[1], WU Xuemei[2], ZHAO Yunxia[1], XIE Hua[1], YAN Xiujuan[1], WANG Xuemei[1]

(1. *Institute of Germplasm Resources, Ningxia Academy of Agriculture and Forestry Sciences, Yinchuan, Ningxia 750002; 2. The Ningxia Hui Autonomous Rwgion, Pengyang County Agriculture Technology Extension and Service Center, Pingyang Ningxia 756500*)

Abstract: 15 varieties of introducing cayenne peppers were used as test materials, varieties characteristics of cayenne peppers in cement arch plastic greenhouse was studied by varieties observing experiment to provide the data basis for the excellent varieties of cayenne pepper in Ningxia region. The results showed that the varieties characteristics of 'Long Crown''Japan Ricky''Japan Long Golden' cayenne pepper, all were stronger resistance to disease, red color, stronger fruit setting ability, higher fruit weight, higher yield. The three varieties of cayenne peppers were suitable for protected cultivation in Ningxia.

Key words: cayenne pepper; adaptability; Ningxia

牛角椒是宁夏辣椒栽培的主要品种之一,设施栽培品种主要为荷兰、日本、法国等国外进口品种,具有抗逆性强,耐低温弱光,适宜长季节栽培和外销的特性,露地栽培主要以宁夏地方品种宁夏牛角椒为主。为加速品种更新换代,确保牛角辣椒产业的可持续发展和菜农收入的稳步增长,从2014—2015年陆续收集国内外及当地牛角椒品种资源35份,并对部分优良品种资源进行田间适应性筛选试验,以期获得适宜在宁夏保护地栽培的牛角椒新品种。

1 材料与方法

1.1 试验材料

供试品种共15个,"天宝F1"、"天禄日富"、"寿龙1号"(由山东瑞丰公司提供);"大阪金秀"(从山东引进);"欧特莱"(由山东瑞丰公司提供);"丰盛巨椒"(由安徽萧县盛丰种业公司提供);"日本吉川"(由盛隆种业提供);"世纪椒王"(由山东凯特种业提供);"长冠"、"日本瑞崎"(由北京农瑞德农业科技有限公司提供);"日本长金"(由寿光南澳绿亨农业有限公司提供);"京都椒王"(由沈阳市九农高科科技有限公司提供);"超大椒王"(由银川兴农良品农业科技有限公司提供);"国福308"(由北京京研益农科技发展中心提供);"亨椒1号"(CK)(由北京中农绿亨种子科技有限公司提供)。

1.2 试验方法

试验于2015年4月在彭阳县新集乡水泥拱架塑料大棚中进行。塑料大棚为水泥拱架结构,长60 m、宽9 m、高2.7 m;土壤为中耕黑垆土,肥力中上等,前茬为玉米,共种植 3 个棚室。采用随机区组设计,每品种为 1 个小区,小区面积 32.4 m²,重复 3 次。采用起垄覆膜栽植,垄高 15 cm、宽 60 cm,每垄定植 2 行,单株栽植,株距 40 cm。

1.3 项目测定

测定株高（用卷尺从根茎到茎生长点之间的距离）、茎粗（用游标卡尺测量）、始花节位、植株开展度、叶色、白粉病发病率、单果质量、果长、果宽、果柄长、果肉厚、中心柱长、果形、果色、小区产量及折合 667 m² 产量。

1.4 数据分析

用 Excel2013 和 DPS 软件进行数据分析处理,不同处理间的多重比较采用 Duncan's 新复极差法（$p \leqslant 0.05$）。

2 结果分析

2.1 不同牛角椒品种植物学性状及产量

从表 1 可以看出,参试的 15 个辣椒品种植物学性状有差异,"京都椒王"辣椒株高最高为 104.9 cm,"日本瑞琦"株高最矮为 74.8 cm;"国福 308"辣椒茎粗最粗为 2.24 cm,"亨椒 1 号"（CK）、"丰盛巨椒"茎粗最细,分别为 1.86 cm、1.89 cm;"京都椒王"辣椒的始花节位最高为 10.0 节,"天宝 F1"、"大阪金秀"的始花节位最低为 8.4 节,其他品种辣椒的始花节位范围在 8.6~9.6 节;"超大椒王"、"京都椒王"辣椒开展度最大为 68.3、67.1 cm;"天宝 F1"、"世纪椒王"、"日本瑞琦"辣椒的叶色均为浅绿色,其他辣椒品种的叶色均为绿色;"日本瑞琦"辣椒的小区产量、折合 667 m² 产量最高,分别为 219.5 kg、4 518.7 kg,"京都椒王"小区产量、折合 667 m² 产量最低,分别为 195.6 kg、4 026.7 kg;"超大椒王"、对照"亨椒 1 号"辣椒白粉病发病率最高,分别为 42.6%、45.5%,"日本瑞琦"辣椒白粉病发病率最低为 14.3%。

2.2 不同牛角椒品种果实性状比较

从表 2 可以看出,参试的 15 个辣椒品种果实性状有差异,"天宝 F1"辣

表 1 牛角椒品种植物学性状及产量
Table 1 Botanical characters and yield of horn pepper varieties

品种 Varieties	株高 Plant height /cm	茎粗 Stem diameter /cm	始花节位 Flower node	开展度 Development degree /cm	叶色 Leaf color	小区产量 Plot yield per 32.4 m² /kg	折合 667 m² 产量 Yield per 667 m² /kg	白粉病发病率 Morbidity of powdery mildew /%
天宝 F1	86.3	2.14	8.4	44.3	浅绿	202.3	4 165.3	32.5
欧特莱	86.5	2.01	8.6	44.5	绿色	204.1	4 201.4	21.2
天禄日富	87.3	2.06	9.0	46.3	绿色	209.4	4 311.5	29.8
丰盛巨椒	84.3	1.89	8.6	45.7	绿色	213.1	4 387.6	22.6
大阪金秀	92.0	2.06	8.4	48.3	绿色	213.8	4 401.2	18.7
寿龙 1 号	89.7	1.98	8.9	46.3	绿色	213.5	4 396.5	22.3
日本吉川	79.6	2.10	9.2	42.3	绿色	204.9	4 218.6	22.6
世纪椒王	82.3	2.13	9.1	45.0	浅绿	211.5	4 355.8	33.1
长冠	86.5	1.99	9.0	51.7	绿色	216.1	4 448.7	18.1
日本长金	75.3	1.90	9.5	52.0	绿色	216.2	4 450.8	18.7
日本端崎	74.8	1.99	9.6	53.8	浅绿	219.5	4 518.7	14.3
京都椒王	104.9	2.09	10.0	67.1	绿色	195.6	4 026.7	32.3
国福 308	87.7	2.24	9.6	46.8	绿色	206.2	4 244.2	29.8
超大椒王	89.5	2.08	8.9	68.3	绿色	201.3	4 144.0	42.6
亨椒 1 号(CK)	95.5	1.86	9.2	48.2	绿色	200.1	4 119.3	45.5

表 2 牛角椒品种果实性状比较

Table 2 Fruit character comparison of horn pepper varieties

品种 Varieties	单果质量 Fruit weight /g	单果长 Fruit length /cm	果柄长 Stalk length /cm	果宽 Fruit width /cm	果肉厚 Thick pulp /cm	中心柱长 Central column length	果形 Fruit shape	果色 Fruit color
天宝 F1	141.8	26.4	4.4	3.98	0.28	5.1	粗牛角	浅绿色
欧特莱	85.6	24.8	4.8	3.65	0.31	5.3	粗牛角	绿色
天禄日富	101.4	23.3	4.3	3.87	0.33	6.1	粗牛角	绿色
丰盛巨椒	94.1	19.8	5.0	3.54	0.37	5.8	粗牛角	绿色
大阪金秀	84.8	24.9	4.4	3.75	0.32	5.7	粗牛角	绿色
寿龙 1 号	98.4	29.7	5.1	4.12	0.34	6.2	粗牛角	绿色
日本吉川	61.4	21.3	4.2	4.31	0.27	5.5	粗牛角	绿色
世纪椒王	58.7	21.7	4.3	4.02	0.31	6.1	粗牛角	浅绿色
长冠	67.8	23.2	4.9	3.90	0.30	5.4	粗牛角	浅绿色
日本长金	68.5	24.0	5.1	3.94	0.34	5.8	粗牛角	绿色
日本瑞崎	72.5	23.3	4.5	4.06	0.32	5.2	粗牛角	浅绿色
京都椒王	85.0	21.0	4.1	4.54	0.39	5.5	粗牛角	绿色
国福 308	82.8	22.1	3.9	4.50	0.37	5.3	牛角形	黄绿色
超大椒王	79.4	22.5	3.6	4.45	0.36	5.6	牛角形	浅绿色
亨椒 1 号 (CK)	81.2	22.2	4.7	4.29	0.37	6.4	粗牛角	浅绿色

椒单果质量最大为141.8 g,"世纪椒王"单果质量最小为58.7 g,其他品种单果质量范围在61.4~101.4 g;"寿龙1号"单果长最长为29.7 cm,"丰盛巨椒"单果长最短为19.8 cm,其他品种单果长无明显差异,范围在21.0~26.4 cm;"寿龙1号"、"日本长金"辣椒的果柄长最长均为5.1 cm,"超大椒王"、"国富308"果柄长最短,分别为3.6、3.9 cm;"京都椒王"辣椒的果宽最宽为4.54 cm,"丰盛巨椒"辣椒的果宽最窄为3.54 cm,其他品种果宽差异不明显;"丰盛巨椒"、"京都椒王"、"国福308"、"超大椒王"、"亨椒1号"(CK)辣椒果肉厚最厚,范围为0.36~0.39 cm,"日本吉川"辣椒果肉厚最薄为0.27 cm,其他品种辣椒果肉厚范围在0.28~0.34 cm;对照"亨椒1号"辣椒中心柱最长为6.4 cm,"天禄日富"、"寿龙1号"、"世纪椒王"次之,其他品种中心柱差异不明显;除"国富308"、"超大椒王"果形为牛角形外,其他品种果形均为粗牛角形;"国富308"果色为黄绿色,"天宝F1"、"世纪椒王"、"长冠"、"超大椒王"、"亨椒1号"辣椒果色均为浅绿色,其他品种果色均为绿色。

3　结论

从田间观察结果看,"长冠"、"日本瑞崎"、"日本长金"牛角椒性状典型,抗病性强,红果色泽鲜艳,连续坐果力强,单果质量、产量较高,适宜在宁夏保护地栽培中推广应用。

参考文献

[1] 马瑞,惠浩剑,马守才,等.大棚辣椒品种引进观察试验[J].现代农业科技,2012(7):149-150.
[2] 周刚,刘琪,徐国友,等.大棚早熟栽培辣椒品比试验[J].长江蔬菜,2008(3):48-49.

宁夏地区优良羊角椒品种引进适应性试验

高晶霞[1]，吴雪梅[2]，牛勇琴[2]，高　昱[2]，颜秀娟[1]，裴红霞[1]，谢　华[1]

（1.宁夏农林科学院种质资源研究所，宁夏银川　750002；
2.宁夏回族自治区彭阳县蔬菜产业发展服务中心，宁夏彭阳756500）

摘　要：为筛选出适宜在宁夏拱棚种植的羊角椒新品种，以引进的16份羊角椒优良品种为试材进行品种筛选试验。结果表明：参试的16个羊角椒品种植物学性状、抗病性、果实性状、产量及产值均有差异，参试品种中"航椒5号"和"新抗七号"品种的生长势较好，"绿农辣子"和"航椒7号"品种的株高高于对照"平罗羊角"，植株开展度除"美特"和"陇椒5号"品种小于对照"平罗羊角椒"外，其他羊角椒品种的植株开展度均大于对照，"甘科4号"和"韩椒二号"品种的始花节位低于对照"平罗羊角椒"，其他羊角椒品种的始花节位均高于对照，坐果率高；参试品种中未发现病毒病和疫病，但白粉病发病率普遍，16个品种均有不同程度发病；参试品种中"陇研长椒501"和"新抗七号"品种的辣椒单果重较大，果实性状较好，"天椒2号"、"新抗7号"和"陇椒5号"品种产量最高，分别为3 533.1 kg/667 m²、3 462.3 kg/667 m²和3 577.3 kg/667 m²，比对照"平罗羊角椒"增产19.06%、16.67%和20.55%。综合分析结果：与对照相比，"陇椒5号"、"新抗七号"、"天椒2号"羊角椒性状典型，皱褶适中，绿色果色泽鲜绿，连续坐果力强，产量较高，适宜在拱棚栽培中推广应用。

关键词：羊角椒；品种筛选；宁夏地区；适应性

Adaptability Test of Introduction of Good Horn Pepper Varieties in Ningxia Region

Gao jingxia[1], Wu xuemei[2], Niu yongqin[2], Gao yu[2], Yan xiujuan[1], Pei hongxia[1], Xie hua[1]

(1. *Institute of Germplasm Resources, Ningxia Academy of Agriculture and Forestry Sciences, Yinchuan, Ningxia* 750002; 2. *Ningxia Hui autonomous Region Pengyang County Vegetable Industry Development Service Center, Pengyang Ningxia* 756500)

Abstract: In order to screen out new varieties of horn pepper suitable for planting in ningxia tunnel, 16 imported excellent varieties of Horn pepper were used as test material for selection. The results show that: There were differences in botanical characters, disease resistance, fruit characters, yield and output value of the 16 varieties tested. The growth potential of the tested cultivars navigation pepper. 5 and The new no. 7 were better, and the plant height of the cultivars green agriculture chiliand navigation pepper. 7 were higher than that of the control cultivars Pingluo lamb horn pepper, Plant development was higher than that of the control group except that the variety mete and long pepper no.5 was lower than that of the control group, The node position of the first flower of ganke no. 4 and Korean jiao no. 2 varieties was lower than that of the control group, while the node position of the first flower of other varieties was higher than that of the control group, and the fruit setting rate was higher; No virus disease and epidemic disease were found in the tested varieties, but the incidence of powdery mildew was common, and all 16 varieties had different degrees of disease; The single fruit weight of the test varieties oflong yan chang pepper 501 and new kang7 was larger and the fruit characters were better, Tianjiao no.2, xinkang no.7 and longjiao no.5 had the highest yields, which were 3 533.1 kg/667 m^2, 3 462.3 kg/667 m^2 and 3 577.3 kg/667 m^2, respectively, which were 19.06%, 16.67% and 20.55% higher than that of pingluo horn pepper. Comprehensive analysis results:

compared with the control group, longjiao no. 5, xinkangno. 7 and tianjiao no. 2 have typical characteristics, moderate fold, bright green fruit color, strong continuous fruiting force, high yield, suitable for application in arch cultivation.

Key words：Hhorn pepper；Variety screening；Ningxia region；Adaptive

羊角椒又名"鸡泽椒"，是一年生草本植物，以其尖上带钩、形如羊角而得名[1]，是一种重要的茄科蔬菜作物和调味品，富含胡萝卜素和维生素，具有很高的营养价值，深受消费者喜爱。在我国，利用设施优势进行早春茬辣椒生产，以解决北方地区冬季早春的新青椒供应[2]。羊角椒是西北地区辣椒栽培的主要种类，栽培地区主要集中在甘肃、新疆、宁夏、青海及陕西部分地区，由于其果肉厚、果皮薄、辣味适中，品质优良，市场价格一直高于牛角椒，具有区域品种优势。羊角椒也是宁夏栽培的主要蔬菜品种，栽培面积占整个蔬菜播种面积的20%左右。菜农在品种选择上具有盲目性，造成产量和品质的差距较大，影响生产[3]。因此，本试验通过引进最新最优的羊角椒品种，评价其植物学性状、果实商品性状、产量及产值，以期筛选出适宜宁夏种植的丰产、优质、高抗的羊角椒品种，为宁夏冷凉地区羊角椒优新品种的示范与推广提供科学依据。

1 材料与方法

1.1 供试材料

"天椒2号"(甘肃省天水市农科所)，"航椒4号"、"航椒5号"(天水绿鹏农业科技有限公司)，"亨椒龙行"、"亨椒龙爪"(北京中农绿亨种子科技有限公司)，"美特"(安徽砀山鸿丰种苗研究所)，"绿农辣子"(辽宁绿野公司)，"陇研长椒501"(兰州盛世农公司)，"新抗七号"(新疆天地禾种业)，"甘科4号"、"甘科5号"、"甘科6号"、"甘科10号"、"甘科16号"、"陇椒5号"(甘肃农科院)，"韩椒二号"(山西博大种苗)，共计16个品种，以平罗特大羊角椒为对照。

1.2 试验方法

本试验采用随机区组设计,3次重复。4月20日定植,株行距150 cm×35 cm,双行定植,每小区面积9.8 m²,栽培管理技术同常规。

1.3 调查项目

对不同辣椒品种的形态学特征观察与测定在田间进行,生理学特性及果实各性状于收获后在室内进行,同时对果实产量进行测定。测定项目如下:株高、始花节位、植株开展度、坐果率、白粉病发病率、单果质量、果长、果宽、果柄长、萼宽、果肉厚、中心柱长、果形、小区产量、折合667 m² 产量[4-7]。

1.4 数据处理

用Excel 2013和DPS软件进行数据分析处理,不同处理间的多重比较采用Duncan's新复极差法($p \leqslant 0.05$)。

2 结果与分析

2.1 不同羊角椒品种植物学性状及病害比较

从表1可以看出,不同羊角椒品种的株高、始花节位、植株开展度、坐果率、白粉病发病率均差异显著,"绿农辣子"和"航椒7号"品种的株高高于对照"平罗羊角椒",分别为78.2 cm和78.6 cm,较对照增加3.2~3.6 cm,其余品种的株高均小于对照"平罗羊角椒",范围在45~73.6 cm,较对照减少1.4~30 cm;植株开展度除"美特"和"陇椒5号"品种小于对照"平罗羊角椒"外,其余品种均大于对照,范围在60.8~77.6 cm,较对照增加1.8~18.6 cm;"甘科4号"和"韩椒二号"品种的始花节位低于对照"平罗羊角椒",分别为7.2节和6.4节,较对照减少0.8节和1节,其余品种的始花节位均高于对照,范围在7.8~9.2节,较对照增加0.4~1.8节;"亨椒龙行"、"亨椒龙爪"、"美特"和"航椒4号"品种的果形为细羊角,其余各品种的果形均为羊角形;"美特"品种的坐果率低于对照"平罗羊角椒",为26.4%,较对照降低7.2%,其余各品种的坐果率均高于对照,范围在33.9%~56.3%,较对照增加0.3%~22.7%;"美特"、"绿农辣子"和"陇研长椒501"品种的白粉病发病率分别为46.9%、52.8%和50.3%,较对照增加6.4%~12.3%,其余各品种的白粉病发病率均低于对照,范围在28.9%~38.4%,较对照减少2.1%~11.6%。各品种病毒病、疫病均没有发生。

表1 羊角椒品种植物学性状及病害调查
Table 1 Investigation of botanical characters and disease of amniocelis chinensis cultivar

处理 treatment	株高 plant height /cm	开展度 divergence /cm	始花节位 flower festival in the beginning /节	果形 fruit shape	坐果率 fruit setting rate /%	白粉病 powdery mildew /%
天椒2号	73.2c	65.8c	7.8b	羊角形	45.3c	30.1a
航椒5号	68.3b	64.2c	9.2c	羊角形	41.8b	29.8a
亨椒龙行	66.8b	70.8d	8.6bc	细羊角	36.6b	33.5b
亨椒龙爪	65.1b	61.6b	8.4bc	细羊角	34.0b	30.3a
美特	45.0a	46.6a	8.3bc	细羊角	26.4a	46.9c
绿农辣子	78.2c	62.7c	8.8c	羊角形	34.4b	52.8d
陇研长椒501	65.5b	62.8c	8.8c	羊角形	38.2b	50.3d
新抗七号	78.6c	77.6d	8.2b	羊角形	43.4c	28.9a
甘科4号	67.5b	64.6c	7.2b	羊角形	39.6b	38.4b
甘科5号	71.6c	66.2c	8.4bc	羊角形	41.1b	38.4b
甘科6号	73.6c	69.4c	8.8c	羊角形	39.6b	36.8b
甘科10号	70.7b	64.8c	7.8b	羊角形	34.5b	31.9a
甘科16号	66.4b	63.8c	8.2b	羊角形	33.9b	33.2a
陇椒5号	68.2b	58.8b	8.2b	羊角形	56.3c	34.3a
航椒4号	68.7b	67.8c	8.2b	细羊角	41.3b	33.6a
韩椒二号	65.6b	61.2b	6.4a	羊角形	41.2b	38.2b
平罗羊角CK	75.0c	59.0b	7.4b	羊角形	33.6b	40.5c

2.2 不同羊角椒品种果实性状比较

从表2可以看出,羊角椒品种单果重、果柄长、果宽、果肉厚、中心柱长差异均显著,"陇研长椒501"、"新抗七号"、"甘科6号"和"韩椒二号"品种的单果重均高于对照，分别为 85.1 g、66.7 g、64.8 g 和 63 g，较对照增加 5.2%~29.9%；其余品种的单果重小于对照，范围在 27.2~53.9 g，较对照减少 5.8~32.5 g；"天椒 2 号"、"亨椒龙爪"、"绿农辣子"、"新抗七号"、"甘科 5 号"、"甘科 10 号"、"甘科 16 号"和"航椒 4 号"品种的单果长大于对照"平罗羊角椒"，范围在 26.4~31.1 cm，较对照增加 2.1~6.8 cm，其余品种的单果长小于对照，范围在 19.8~23.6 cm，较对照减少 0.7~4.5 cm，"美特"品种的单果长与对照无显著性差异，均为 24.3 cm；"亨椒龙爪"、"美特"、"新抗七号"、"甘科 4 号"和"航椒 4 号"品种的果柄长大于对照"平罗羊角椒"，分别为 5.5 cm、5.9 cm、5.1 cm、6.2 cm 和 5.6 cm，较对照增加 0.7~1.4 cm，"天椒 2 号"、"陇研长椒 501"、"甘科 10 号"、"甘科 16 号"、"陇椒 5 号"和"航椒 5 号"品种的果柄长显著低于对照，范围在 3.8~4.3 cm，较对照减少 0.5~1.0 cm，剩余品种的果柄长与对照无显著性差异；"甘科 6 号"品种的果宽最大，为 4.8 cm，较对照增加 1.0 cm，其余品种的果宽均小于对照"平罗羊角椒"，其中"天椒 2 号"、"绿农辣子"、"陇研长椒 501"、"新抗七号"、"甘科 16 号"、"航椒 5 号"和"韩椒二号"品种的果宽与对照"平罗羊角椒"无显著性差异，对照"平罗羊角椒"的果宽显著大于"亨椒龙爪"、"美特"、"甘科 4 号"、"甘科 5 号"、"甘科 10 号"、"陇椒 5 号"和"航椒 4 号"品种，范围在 1.6~3.0 cm；"亨椒龙行"和"航椒 4 号"品种的辣椒果实萼宽显著小于对照"平罗羊角椒"，分别为 1.7 cm 和 1.9 cm，较对照减少 0.6 cm 和 0.8 cm，其余品种的辣椒果实萼宽与对照无显著性差异；"亨椒龙行"、"陇研长椒 501"、"甘科 4 号"和"陇椒 5 号"品种的果肉厚显著大于对照"平罗羊角椒"，较对照增加 0.04~0.12 cm，"甘科 10 号"和"航椒 4 号"品种的果肉厚显著小于对照"平罗羊角椒"，较对照减少 0.08 cm 和 0.1 cm，剩余品种的果肉厚与对照无显著性差异；"甘科 16 号"和"陇椒 5 号"的中心柱长显著小于对照"平罗羊角椒"，较对照减少 1.7 cm 和 2 cm，"甘科 4 号"、"甘科 5 号"、"甘科 6 号"、"甘科 10 号"和"韩椒二号"品种的中心柱长与对照无显著性差异，其余品种的中心柱长显著大于对照"平罗羊角椒"，范围在

5.2~9.8 cm,较对照增加了 0.3~4.9 cm。

表2 不同羊角椒品种果实性状调查

Table 2 Fruit character investigation of different amniocentric pepper varieties

品种 variety	单果重 weight of single fruit /g	单果长 fruit long /cm	果柄长 peduncle long /cm	果宽 fruit width /cm	萼宽 calyx wide /cm	果肉厚 pulp thick /cm	中心柱长 center column length /cm
天椒2号	48.2b	26.4b	3.9a	3.7c	2.4b	0.27b	7.7e
亨椒龙形	27.2a	22.4a	4.7b	1.6a	1.7a	0.31c	6.8d
亨椒龙爪	43.1b	30.7b	5.5c	2.7b	2.3b	0.26b	9.8f
美特	42.7b	24.3a	5.9c	2.7b	2.4b	0.26b	5.5c
绿农辣子	53.9c	27.3b	4.8b	3.7c	2.1b	0.22b	5.2c
陇研长椒501	85.1e	19.8a	3.9a	3.9c	2.9b	0.39d	6.2d
新抗七号	66.7d	30.9b	5.1b	3.4c	2.8b	0.27b	5.4c
甘科4号	45.6b	22.9a	6.2c	2.9b	2.3b	0.31c	4.7b
甘科5号	44.1b	27.9b	4.6b	2.9b	2.2b	0.24b	3.9b
甘科6号	64.8d	23.4a	4.8b	4.8d	2.7b	0.27b	3.8b
甘科10号	51.2b	27.4b	4.1a	3.0b	2.4b	0.19a	4.1b
甘科16号	50.4b	31.1b	4.1a	3.5c	2.5b	0.24b	3.2a
陇椒5号	43.9b	23.5a	3.8a	2.8b	2.3b	0.31c	2.9a
航椒4号	26.6a	27.7b	5.6c	1.9a	1.9a	0.17a	7.9e
航椒5号	40.7b	20.5a	4.3b	3.5c	2.3b	0.26b	5.8c
韩椒二号	63.0d	23.6a	4.7b	3.3c	2.4b	0.26b	4.1b
平罗羊角CK	59.7c	24.3a	4.8b	3.8c	2.5b	0.27b	4.9b

2.3 不同羊角椒品种产量比较

从图1可以看出,"天椒2号"、"新抗7号"和"陇椒5号"品种的小区产量最高,分别为51.91 kg、50.87 kg 和52.56 kg,比对照"平罗羊角椒"增产19.06%、16.67%和20.55%,"航椒5号"、"亨椒龙行"、"亨椒龙爪"、"陇研长椒501"、"甘科5号"、"甘科6号"、"甘科16号"和"韩椒二号"8个品种的小区产量在 43.78~49.85 kg，均高于对照"平罗羊角椒"的小区产量，增产幅度在 0.04%~14.33%；"甘科 10 号"和"航椒 4 号"品种的小区产量与对照无差异，分别为 43.63 kg 和 43.61 kg；其余的"美特"、"绿农辣子"和"甘科 4 号"3 个品

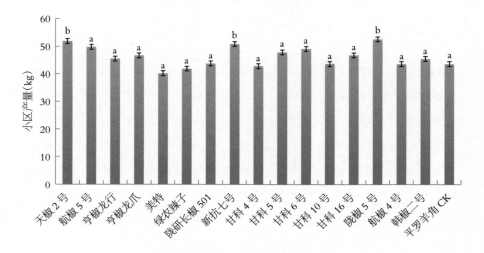

图 1 不同羊角椒品种小区产量比较

Fig. 1 Plot yield comparison of different kinds of yangjiao pepper

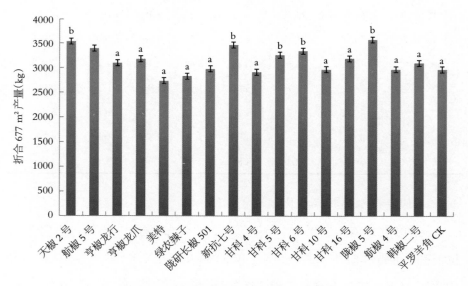

图 2 不同羊角椒品种折合 677 m² 产量比较

Fig.2 Yield comparison of 667 m² for different kinds of yangjiao pepper

种的小区产量分别为40.18 kg、41.85 kg和42.82 kg,均低于对照小区产量,降低幅度为7.8%、4.01%和1.79%。

从图2可以看出,"天椒2号"、"新抗7号"和"陇椒5号"品种的折合677 m²产量最高,分别为3 533.1 kg、3 462.3 kg和3 577.3 kg,比对照"平罗羊角椒"增产19.06%、16.67%和20.55%,"航椒5号"、"亨椒龙行"、"亨椒龙爪"、"陇研长椒501"、"甘科5号"、"甘科6号"、"甘科16号"和"韩椒二号"8个品种的折合677 m²产量在2 979.7 g~3 392.9 kg,均高于对照"平罗羊角椒"的折合677 m²产量,增产幅度在0.04%~14.33%;"甘科10号"和"航椒4号"品种的折合677 m²产量与对照无差异,分别为2 969.5 kg和2 968.2 kg;"美特"、"绿农辣子"和"甘科4号"3个品种的折合677 m²产量分别为2 734.7 kg、2 828.4 kg和2 914.4 kg,均低于对照的折合677 m²产量,降低幅度为7.8%、4.01%和1.79%。

3 结论

综合分析参试16个品种的植株性状、果实商品性状、产量及产值,发现"绿农辣子"和"航椒7号"品种的株高高于对照"平罗羊角椒",较对照增加3.2~3.6 cm,其余品种的株高均低于对照"平罗羊角椒";植株开展度除"美特"和"陇椒5号"品种小于对照"平罗羊角椒"外,其他品种均大于对照;"甘科4号"和"韩椒二号"品种的始花节位低于对照"平罗羊角椒",其余品种的始花节位均高于对照,较对照增加0.4~1.8节;"亨椒龙行"、"亨椒龙爪"、"美特"和"航椒4号"品种的果形为细羊角形,其他品种的果形均为羊角形;"美特"品种的坐果率低于对照"平罗羊角椒",其余各品种的坐果率均高于对照,较对照增加0.3%~22.7%;"美特"、"绿农辣子"和"陇研长椒501"品种的白粉病发病率高于对照,其他品种的白粉病发病率均低于对照,较对照减少2.1%~11.6%,各品种病毒病、疫病均没有发生;"陇研长椒501"、"新抗七号"、"甘科6号"和"韩椒二号"品种的单果重均高于对照,其他品种的单果重小于对照;"天椒 2 号"、"亨椒龙爪"、"绿农辣子"、"新抗七号"、"甘科 5 号"、"甘科 10 号"、"甘科 16 号"和"航椒 4 号"品种的果长大于对照"平罗羊角椒",其他品种的果长小于对照;"亨椒龙爪"、"美特"、"新抗七号"、"甘科 4 号"和"航椒 4 号"品种的果柄长大于对照"平罗羊角椒","天椒 2 号"、"陇研长椒 501"、"甘科 10 号"、"甘科 16 号"、"陇椒 5 号"和"航椒 5 号"品种的果柄长显著低于对照;"甘科 6 号"品种的果宽最宽,其他品种的果宽均小于对照"平罗羊角椒";"亨椒龙行"和"航椒 4 号"品种的辣椒果实萼宽显著小于对照"平罗羊角椒",其他品种的辣椒果实萼宽与对照无显著性差异;"亨椒龙行"、"陇研长椒 501"、"甘科 4 号"和"陇椒 5 号"品种的果肉厚显著大于对照"平罗羊角椒","甘科 10 号"和"航椒 4 号"品种的果肉厚显著小于对照"平罗羊角椒",其余品种的果肉厚与对照无显著性差异;"甘科 16 号"和"陇椒 5 号"的中心柱长显著小于对照"平罗羊角椒"。参试品种中"天椒 2 号"、"新抗 7 号"和"陇椒 5 号"品种产量最高,分别为 3 533.1 kg/667 m²、3 462.3 kg/667 m² 和 3 577.3 kg/667 m²,比对照"平罗羊角椒"增产 19.06%、16.67%和 20.55%。

综合分析结果:"陇椒 5 号"、"新抗七号"、"天椒 2 号"3 个羊角椒品种性状典型,皱褶适中,绿色果色泽鲜绿,连续坐果力强,产量较高,适宜在拱棚栽培中推广应用。

参考文献

[1] 杨小兰.五个羊角椒品种的耐盐性研究[J].北方园艺 2010(23):1-4.

[2] 刘连妹,屈海泳,王雪梅,徐虹,陈沁滨.大棚栽培早春辣椒品种比较试验[J].江苏农业科学,2008(1):147.

[3] 颜秀娟,何鑫,王学梅,高晶霞.宁夏地区优质羊角椒适应性栽培研究[J].安徽农业科学,2019,47(19):36-37+42.

[4] 梁合荣,田浩,吕少元,等.不同辣椒品种在遵义市的适应性研究[J].耕作与栽培,2014(2):30-31.

［5］赵宏,张智柱,舒建钢.不同辣椒新品种在贵阳地区的适应性研究[J].农技服务,2012,29(11):1208-1210.

［6］赵贞祥,张二喜,杨永岗,等.旱地辣椒新品种筛选和适应性试验[J].中国园艺文摘,2012(9):1-4.

［7］郭英,丁自立,姚明华,等.湖北省辣椒种质资源的鉴定与评价[J].辣椒杂志,2011(4):31-34.

优质羊角椒适应性栽培试验研究

高晶霞[1],谢 华[1]

(宁夏农林科学院种质资源研究所,宁夏银川 750002)

摘 要: 为筛选出适宜宁夏栽培的优质羊角椒品种,以12个羊角椒品种为供试材料,在宁夏拱棚进行栽培试验,对不同羊角椒品种株高、始花节位、叶色、白粉病发病率、单果质量、果实性状、小区产量(折合667 m² 产量)进行测定比较分析。结果表明:参试的12个羊角椒品种植物学性状、抗病性、果实性状及产量均有明显差异,"航椒8号"、"美特"、"娇龙"单果重、小区产量和折合667 m² 产量最大,分别为46.7 g、44.6 g、45.2 g、146.5 kg、145.2 kg和144.4 kg,3 015.91 kg、2 989.15 kg、2 972.68 kg。综合分析结果:"航椒8号"、"美特"、"娇龙"3个羊角椒品种性状典型、抗病性强、果实性状优、连续坐果力强、产量高,适宜宁夏保护地及露地栽培推广。

关键词: 羊角椒;适应性;栽培

Experimental Study on the Adaptive Cultivation of High Quality Capsicum

GAO Jingxia[1], XIE Hua[1]

(Institute of Germplasm Resources, Ningxia Academy of Agriculture and Forestry Sciences, Yinchuan Ningxia 750002, China)

Abstract: To select suitable for cultivation of high-quality in ningxia horn pepper varieties, with 12 horn pepper varieties were used as material, tunnel in ningxia, cultivation experiment of different horn pepper varieties of plant height, flower festival in the beginning, leaf color, incidence of powdery mildew, fruit quality, fruit characters, plot yield production (667 m²) for determination of the comparative analysis. Results show that the volunteers´ 12 horn pepper varieties of botany character, disease resistance, fruit properties and yield were significantly different, "Hang Jiao No.8", "Meite", "Jiaolong" fruit weight, yield and over 667 m² production is the largest, 46.7 g, 44.6 g and 45.2 g, 146.5 kg, 145.2 kg and 144.4 kg, 3 015.91 kg, 2 989.15 kg, 2 972.68 kg. The comprehensive analysis results showed that the three cultivars, Namely, "Hang Jiao No.8", " Meite" and " Jiaolong", had typical characters, strong disease resistance, excellent fruit characters, strong continuous fruiting ability and high yield, which were suitable for cultivation and popularization in protected areas and open fields in Ningxia.

Key words: Horn pepper; Adaptability; Cultivation

羊角椒又名"鸡泽椒"，以其尖上带钩、形如羊角而得名，具有很高的食用价值[1]，是西北地区辣椒栽培的主要种类，栽培地区主要集中在甘肃、新疆、宁夏、青海及陕西部分地区，由于其果肉厚、果皮薄、辣味适中，品质优良，市场价格与牛角椒相比一直处于领先地位，具有区域品种优势。羊角椒是宁夏栽培的主要蔬菜品种，栽培面积占整个蔬菜播种面积的20%左右。市场要求羊角椒品种熟性早，果实较长、较大，果形顺直、光滑，嫩果果色为黄色，商品性好，抗病性强，产量高，适合早春保护地栽培[2]。宁夏南部山区设施蔬菜经过多年发展，目前种植总面积已达到2万hm²，种植种类涉及辣椒、番茄、黄瓜、茄子、甜瓜等，种植效益十分显著，已成为蔬菜主产区农民增收致富的主导产业之一。但由于多年连作，病虫害种类增加，发病程度上升，品种退化十分严重[3-4]，因此，开展设施辣椒新品种引进筛选试验研究，及早选出早熟、丰产、优质、适应性、抗逆性强的辣椒新品种，为宁夏南部山区设施蔬菜可持续发展保驾护航，具有非常重要的意义。

1 材料与方法

1.1 试验材料

"陇椒16号"(甘肃省农科院)、"航椒8号"(甘肃航天育种工程技术中心)、"新选22号尖椒"(太谷县蔬菜良种繁育场)、"金元十号"(甘肃高台大华种业有限公司)、"美特"(安徽砀山鸿丰种苗研究所)、"22号尖椒"(山西黎来种业公司)、"陇椒3号"、"甘科6号"、"甘科8号"(甘肃蔬菜研究所)、"绿农辣子"、"娇龙"、"娇艳"(宁夏巨丰种苗有限责任公司),共计12个品种。

1.2 试验方法

1.2.1 试验设计

本试验采用随机区组设计,3次重复,日光温室穴盘育苗,于2016年3月10日播种,4月25日定植于拱棚,株行距70 cm×35 cm。每小区面积32.4 m²,左右双行定植,栽培管理条件同常规。对不同辣椒品种的形态学特征观察与测定在田间进行,果实各性状于收获后在室内进行。

1.2.2 测定项目

辣椒生长盛期,调查株高、始花节位、植株开展度、叶色、白粉病发病率、单果质量、果长、果宽、果柄长、果肉厚、中心柱长、果色、果形、小区产量、折合667 m²产量[5-7]。

1.3 数据分析

数据用Excel和SPSS软件进行单因素方差分析,多重比较采用LSR法(Duncan's法),显著水平P<0.05。

2 结果与分析

2.1 不同羊角椒品种植物学性状比较

从表1可以看出,参试的12个羊椒角品种植物学性状有明显差异,"陇椒16号"、"甘科8号"和"甘科6号"株高最高,分别为77.0 cm、76.2 cm和

71.8 cm,"美特"和"太空螺丝椒"株高最低,分别为58.3 cm和59.8 cm;"娇艳"和"甘科8号"辣椒开展度最大,分别为57.2 cm、59.1 cm,"金元十号"、"陇椒16号"和"新选22号尖椒"开展度最小,分别为44.3 cm、46.7 cm和47.8 cm;"金元10号"和"娇艳"始花节位最大,分别为9.4节和9.0节,"航椒8号"、"陇椒3号"和"绿农辣子"始花节位最小,分别为7.4节、6.9节和7.6节;"太空螺丝椒"叶色为深绿色,其他品种辣椒叶色均为绿色;"陇椒16号"辣椒白粉病的发病率最高为42.6%,"航椒8号"、"新选22号尖椒"、"太空螺丝椒"和"甘科8号"白粉病发病率最低,分别为19.1%、19.0%和20.8%。

表1 参试羊角椒品种植物学性状比较

Table 1 Comparison of botanical characters of the cultivars of capsicum chinense

品种 Varieties	株高 Plant height /cm	开展度 Development degree/cm	始花节位 Flower node/节	叶色 Leaf colo	白粉病发病率 Powdery mildew Morbidity/%
陇椒16号	77.0	46.7	8.9	绿色	42.6
航椒8号	69.3	54.7	7.4	绿色	19.1
新选22号尖椒	62.3	47.8	8.0	绿色	19.0
美特	58.3	56.1	8.3	绿色	26.9
金元十号	65.7	44.3	9.4	绿色	38.4
太空螺丝椒	59.8	51.2	8.2	深绿色	21.3
陇椒3号	69.0	56.3	6.9	绿色	39.2
甘科6号	71.8	51.0	8.7	绿色	39.6
甘科8号	76.2	59.1	8.9	绿色	20.8
绿农辣子	64.7	55.4	7.6	深绿	34.5
娇龙	67.1	54.3	8.7	绿色	19.6
娇艳	69.8	57.2	9.0	绿色	38.7

2.2 不同羊角椒品种果实性状比较

从表2可以看出,参试的12个羊角椒品种资源果实性状有明显差异,其中"娇龙"辣椒果长最长为28.0 cm,"陇椒16号"和"新选22号尖椒"果长最短，分别为 22.5 cm、22.0 cm，其他品种辣椒果长在 23.6~27.5 cm；"金元十号"辣椒果柄长最长为 6.9 cm，"甘科 6 号"和"甘科 8 号"辣椒果柄长最短，分别为 4.5 cm 和 4.0 cm，其他品种辣椒果柄长在 4.5~6.0 cm；"娇龙"、"陇椒 16 号"和"太空螺丝椒"果宽最宽，分别为 3.9 cm、3.6 cm 和 3.6 cm，"甘科 6 号"果宽最窄为 2.3 cm，其他羊角椒品种果宽在 2.5~3.4 cm；"太空螺丝椒"和"娇龙"果肉厚最厚，分别为 0.38 cm 和 0.35 cm，"甘科 6 号"和"绿农辣子"果肉厚最薄，为 0.21 cm 和 0.22 cm，其他羊角椒品种果肉厚在 0.24~0.31 cm；"陇椒 16 号"和"金元十号"中心柱最长，为 7.5 cm 和 7.6 cm，"太空螺丝椒"和"娇艳"中心柱最短为 4.9 cm，其他羊角椒品种中心柱范围在 5.2~7.0 cm；"太空螺丝椒"和"绿农辣子"的辣椒果色为深绿色，其他羊角椒品种的果色均为绿色；"太空螺丝椒"果形为似猪大肠，"新选 22 号尖椒"和"美特"果形似细羊角，其他羊角椒品种果形均为羊角形。

表 2 参试羊角椒品种果实性状调查

Table 2 Investigation on the fruit characters of variety resources of capsicum chinense

品种 Varieties	果长 Fruit length /cm	果柄长 Stalk length /cm	果宽 Fruit width /cm	果肉厚 Thick pulp /cm	中心柱长 Central column length /cm	果色 Fruit color	果形 Fruit shape
陇椒 16 号	22.5	5.5	3.6	0.27	7.5	绿色	羊角形
航椒 8 号	27.5	4.8	3.4	0.32	7.0	绿色	羊角形
新选 22 号尖椒	22.0	4.5	2.8	0.29	5.2	绿色	细羊角
美特	24.3	6.0	2.5	0.30	6.2	绿色	细羊角
金元十号	24.5	6.9	2.9	0.26	7.6	绿色	羊角形
太空螺丝椒	23.6	4.9	3.6	0.38	4.9	深绿色	猪大肠
陇椒 3 号	27.8	4.7	2.8	0.24	5.2	绿色	羊角形
甘科 6 号	26.8	4.5	2.3	0.26	6.1	绿色	羊角形
甘科 8 号	26.7	4.0	2.5	0.21	7.0	绿色	羊角形
绿农辣子	25.8	5.5	3.1	0.22	5.6	深绿色	羊角形
娇龙	28.0	5.8	3.9	0.35	6.3	绿色	羊角形
娇艳	27.3	4.9	3.2	0.29	4.9	绿色	羊角形

2.3 不同羊角椒品种单果重和产量比较

由图1可知,"航椒8号"、"美特"、"娇龙"单果重最大,分别为46.7 g、44.6 g、45.2 g,"金元十号"、"太空螺丝椒"、"甘科6号"、"绿农辣子"和"娇艳"单果重最小,分别为36.5 g、39.4 g、34.1 g、39 g、38.6 g,其他品种单果重分别在40~41.8 g。

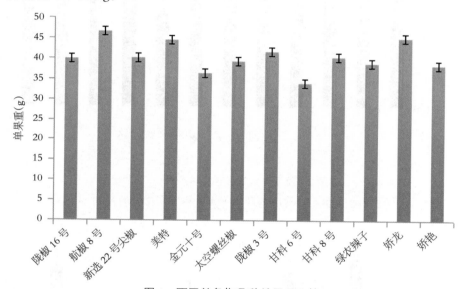

图1 不同羊角椒品种单果重比较

FIG. 1 Comparison of the weight of single fruit of different varieties of Capsicum chinense

由图2可知,"航椒8号"、"娇龙"和"美特"小区产量最高,分别为146.5 kg、145.2 kg和144.4 kg,"金元十号"、"太空螺丝椒"、"甘科6号"、"绿农辣子"和"娇艳"小区产量最低,分别为133.2 kg、134.5 kg、130.2 kg、134.8 kg、133.6 kg,其他品种小区产量分别在140~141.1 kg。

由图3可知,"航椒8号"、"娇龙"和"美特"折合667 m² 产量最高,分别为3 015.91 kg、2 989.15 kg、2 972.68 kg,"金元十号"、"太空螺丝椒"、"甘科6号"、"绿农辣子"和"娇艳"折合667 m² 产量最低,分别为2 742.11 kg、2 768.87 kg、2 680.35 kg、2 775.05 kg、2 750.35 kg,其他品种折合667 m² 产量分别在2 882.10~2 904.74 kg。

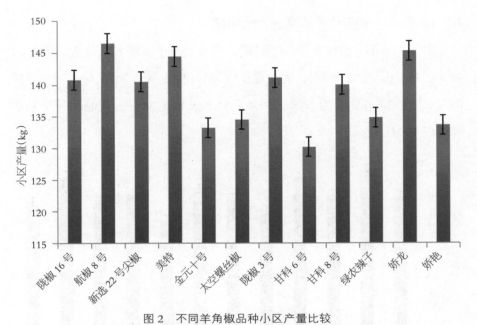

图 2 不同羊角椒品种小区产量比较

FIG. 2 Comparison of yields of different cultivars of Capsulosa in small areas

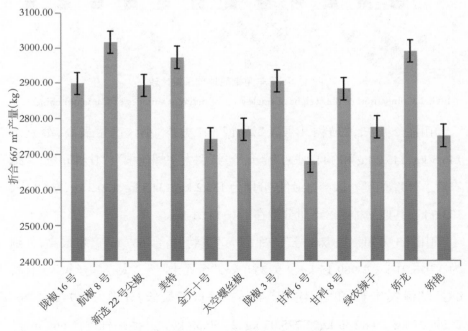

图 3 不同羊角椒品种折合 667 m² 产量比较

FIG. 3 Yield comparison of different varieties of capsicum annuum(667 m²)

3　结论

通过栽培调查结果可以得出:"航椒8号"、"美特"、"娇龙羊角椒"性状典型,抗病性强,皱褶适中,红果色泽鲜艳,连续坐果力强,果肉厚度适中,单果重、产量较高,适宜在宁夏保护地及露地栽培中推广应用。

参考文献

[1] 杨小兰.五个羊角椒品种的耐盐性研究[J].北方园艺,2010(23):1-4.

[2] 郑积荣,王慧俐,王佳明.设施羊角椒品种的引进筛选试验[J].浙江农业科学,2010(05):946-947.

[3] 王泽杰,吴华尧,范远丰,等.辣椒品种比较试验[J].现代农业科技,2011(16):12.

[4] 侯全刚,李江,马本元.日光温室辣椒品种比较试验[J].青海农林科技,2005(4):1-2.

[5] 梁合荣,田浩,吕少元,等.不同辣椒品种在遵义市的适应性研究[J].耕作与栽培,2014(2):30-31.

[6] 赵宏,张智柱,舒建钢.不同辣椒新品种在贵阳地区的适应性研究[J].农技服务,2012,29(11):1208-1210.

[7] 赵贞祥,张二喜,杨永岗,等.旱地辣椒新品种筛选和适应性试验[J].中国园艺文摘,2012(9):1-4.

[8] 詹永发,杨红,涂祥敏,等.辣椒品种资源的遗传多样性和聚类分析[J].贵州农业科学,2010,38(11):12-15.

[9] 冯刚刚,孙万斌,马晖玲,等.22个苜蓿品种在甘肃永登地区的生产适应性评价[J].草原与草坪,2016,36(4):64-66.

宁夏地区彩椒品种适应性筛选试验

高晶霞[1]，吴雪梅[2]，牛勇琴[2]，高　昱[2]，颜秀娟[1]，裴红霞[1]，谢　华[1]

（1. 宁夏农林科学院种质资源研究所，宁夏银川　750002；
2. 宁夏回族自治区彭阳县蔬菜产业发展服务中心，宁夏彭阳　756500）

摘　要：为筛选适宜宁夏种植的彩椒品种，引进10个常规彩椒品种，调查其物候期、植物学性状、果实性状及产量产值。结果表明："月黑"品种的生育期和对照相同，其他品种的生育期较对照延长2~8天；9个品种与对照相比均表现出强或较强的抗病性；"绿圆"、"绿舟"、"绿星"、"月黄"、"红曼"、"黄川"等6个品种的植株生长势强或较强，"月黑"、"月紫"与对照中等；"绿圆"、"绿星"、"绿舟"、"红曼"、"黄川"等5个品种单果质量均高于对照，果肉厚度以"绿圆"品种最厚。从果实外观、品质看，参试各品种果面光滑、味甜、肉厚，均耐贮运；"绿圆"折合667 m² 产量最高，"绿星"、"绿舟"、"月黄"、"红曼"、"黄川"等5个品种均比对照增产，"月黑"、"月白"、"月紫"等3个品种均比对照减产。综上所述："绿圆"、"绿星"、"月黄"3个品种综合性状好，增产潜力大，生长势较强，辣椒果实商品性好，折合667 m² 产量及效益显著高于对照茄门，应在进一步试验示范的基础上进行大面积推广种植。

关键词：宁夏；彩椒；品种筛选

Adaptability Screening Test of Color Pepper Varieties in Ningxia Region

Gao jingxia[1], Wu xuemei[2], Niu yongqin[2], Gao yu[2], Yan xiujuan[1], Pei hongxia[1], Xie hua[1]

（1. *Institute of Germplasm Resources, Ningxia Academy of Agriculture and Forestry Sciences, Yinchuan, Ningxia 750002*; 2. *Ningxia Hui autonomous Region Pengyang County Vegetable Industry Development Service Center, Pengyang Ningxia 756500*）

Abstract: In order to select the color pepper varieties suitable for planting in ningxia, 10 conventional color pepper varieties were introduced to investigate their phenological stage, botanical characters, fruit characters and output value. The results show that: The growth period was the same as that of the control, and the growth period of other varieties was extended by 2~8 days.Compared with the control, 9 varieties all showed strong or stronger disease resistance; The plant growth potential of 6 cultivars such as green circle, green boat, green star, moon yellow, red man, and Yellow River was strong or strong, and the black and purple of the moon were medium; The fruit quality of each of the five varieties was higher than that of the control group. The thickness of the pulp was the thickest in the green circle variety; Green circle equivalent to 667 m² yields the highest, green star, green boat, moon yellow, red man, huangchuan and other 5 varieties are all increased than the control, black, white, purple and other 3 varieties are reduced than the control.To sum up, the three varieties of green circle, green star and moon yellow have good comprehensive characteristics, great potential for increasing yield, strong growth potential, good commercial capsicum fruit, the yield and benefit of 667 m² is significantly higher than that of the control nightshade door, so it should be popularized and planted on a large area on the basis of further experiments and demonstration.

Key words: Ningxia; Color pepper; Variety screeing

彩椒(Capsicum frutescens var.grossum)又称为彩色甜椒,因其果实皮色呈现红、黄、橙、绿、紫等艳丽多彩的色泽而得名[1,2]。彩椒是一年生或多年生草本植物,由中南美热带原产的辣椒在北美经长期栽培和自然、人工选择后演化而成,具有果肉变厚、辣味消失、心室腔增多、果型变大[3],果实成熟后皮色呈不同颜色等特征。近年来因其味美、保健、观赏等功能而备受国内外客商的青睐,其产量和经济效益也十分可观[4]。其果实含有丰富的维生素A、B、C,糖类、纤维质、钙、磷、铁等营养物质,尤其是在成熟期,果实中的营养成分除维生素C含量增加外,其他营养成分均会增加5倍,具有预防心血管疾病、促进新陈代谢、抗老化等作用。彩椒以鲜食为主,是制作快餐食品的理想原料,是近年风靡国际市场的特色蔬菜品种之一,其经济价值和营养价值都高于普通椒[5-7]。随着人们生活水平的不断提高、消费习惯的变化,品质优良的甜辣椒品种,特别是外观好、维生素 C 含量高、营养价值高的彩色甜椒系列越来越引起人们的关注[8]。但是,目前我国市场上以国外彩椒品种占主导地位,特别是欧美地区品种,其价格较高,而我国自研的彩椒品种较少,不能满足市场需求[9]。

彩椒在宁夏的种植品种以国外进口品种为主,颜色有红、黄、橙、绿、紫、白等,在日光温室有少量种植。彭阳县位于宁夏回族自治区南部边缘,六盘山东麓,介于东经 106°32′~106°58′,北纬 35°41′~36°17′。年平均气温 7.4~8.5℃,无霜期 140~170 d,降水量 350~550 mm。属典型的温带半干旱大陆性季风气候,特别适合辣椒栽培。因此,本试验通过引进最新最优的彩椒品种,在彭阳县辣椒核心示范基地进行品比试验,评价其物候期、植物学性状、果实商品性状以及产量及产值,以期筛选出适宜宁夏种植的丰产、优质、高抗的彩椒品种,为宁夏彩椒优新品种的示范与推广提供参考依据。

1 材料与方法

1.1 供试材料

供试材料:品种为"月黄"、"月黑"、"红曼"、"月白"、"黄川"、"月紫"、"绿圆"、"绿舟"、"绿星"等 9 个彩椒品种,均由北京中农绿亨种子科技有限公司提供;对照品种为茄门甜椒,由中国农科院蔬菜花卉研究所提供。

1.2 试验地点

试验地点选在设施农业建设区的新集乡下马洼村水泥拱架塑料大棚。土壤为黄绵土,肥力中等水平,前茬作物为玉米。

1.3 试验设计

试验设 10 个处理,重复 3 次,随机区组排列,小区面积 35.1 m²。采用膜下滴灌,垄宽 70 cm,垄沟宽 60 cm,垄高 25 cm。单株栽植,行距 65 cm,株距

35 cm,每小区定植138株,田间管理按常规方法进行。

1.4 调查项目

观察记载物候期,农艺性状,并测定统计产量,折合667 m² 产量。

1.5 数据分析

用Excel 2013和SPSS软件进行数据分析处理,不同处理间的多重比较采用Duncan's新复极差法($p \leqslant 0.05$)。

2 结果与分析

2.1 参试彩椒品种的物候期及生育期

由表1可以看出,所有品种的播种期和出苗期相同,均为2月10日和3月1日;与对照茄门相比,"月黑"品种的开花期与坐果期与对照相同,均为6月15日与6月22日,其他8个品种的开花期与坐果期较对照品种延后2~

表1 参试彩椒品种的物候期及生育期

Table 1 Phenological period and growth period of colored pepper varieties tested

品种 variety	播种期 seeding time (月-日)	出苗期 period of emergence (月-日)	开花期 bloom (月-日)	坐果期 fruit-set period (月-日)	始收期 beginning to receive period (月-日)	末收期 closed at the end of the period (月-日)	生育期 period of duration (d)
月黄	2-10	3-1	6-17	6-24	7-15	9-24	226
月黑	2-10	3-1	6-15	6-22	7-13	9-22	224
红曼	2-10	3-1	6-18	6-25	7-16	9-26	228
月白	2-10	3-1	6-18	6-25	7-16	9-26	228
黄川	2-10	3-1	6-18	6-25	7-16	9-26	228
月紫	2-10	3-1	6-16	6-23	7-13	9-24	226
绿圆	2-10	3-1	6-20	6-27	7-17	9-30	232
绿舟	2-10	3-1	6-20	6-27	7-18	9-30	232
绿星	2-10	3-1	6-17	6-24	7-18	9-28	230
茄门(CK)	2-10	3-1	6-15	6-22	7-13	9-22	224

5 d;"月黑"与"月紫"品种的始收期与对照相同,其余品种的始收期较对照品种延后 2~5 d;"月黑"品种的生育期与对照茄门相同,为 224 d,其余 8 个品种均较对照长,较对照茄门延长 2~8 d。

2.2 参试品种植株主要性状与抗病性

由表 2 可以看出,参试的 9 个品种植株叶色均为绿色,与对照相比均表现出强或较强的抗病性。"绿圆"、"绿舟"、"绿星"、"月黄"、"红曼"、"黄川"等 6 个品种的植株生长势强或较强,"月黑"、"月紫"品种与对照的植株生长势中等。"月黑"、"月紫"2 个品种的株高矮于对照,分别为 52.3 cm 和 53.1 cm,较对照减少 1.1 cm 和 0.3 cm,其余 7 个品种的株高均比对照高,范围在 53.8~85.8,较对照增加 0.4~32.4 cm,提高 0.8%~60.7%。参试 9 个品种的茎粗均大于对照,其中"绿星"品种的茎粗最粗为 1.79 cm,较对照增加 0.55 cm,

表 2 参试彩椒品种的植株主要性状与抗病性
Table 2 Main characters and disease resistance of colored pepper varieties

品种 varieties	株高 plant height /cm	茎粗 thick stems /cm	第一花序节位 first inflorescence node /节	开展度 /cm	分枝数 branch number /个	叶色 leaf color	单株结果数 number of fruit per plant /个	植株长势 plants grow	抗白粉病 resistance to powdery mildew
月黄	63.4	1.43	7.7	58.0	2.4	绿	11.1	较强	较强
月黑	52.3	1.30	8.3	43.1	2.0	绿	8.7	中	较强
红曼	60.1	1.36	9.2	45.1	2.6	绿	10.3	较强	较强
月白	53.8	1.29	9.1	48.7	2.4	绿	10.5	中	强
黄川	60.1	1.48	7.6	42.5	2.4	绿	7.3	较强	较强
月紫	53.1	1.33	9.1	48.3	2.0	绿	9.6	中	较强
绿圆	79.8	1.75	9.6	69.6	2.8	绿	8.8	强	强
绿舟	82.0	1.71	9.6	71.9	2.7	绿	8.6	强	强
绿星	85.8	1.79	9.6	70.3	2.4	绿	8.3	强	强
茄门(CK)	53.4	1.24	9.8	42.3	2.0	绿	8.5	中	中

其余品种的茎粗范围在 1.30~1.75 cm,较对照增加 0.05~0.51 cm,提高 4.0%~41.1%。参试 9 个品种的第一花序节位均低于对照,对照品种的第一花序节位最大为 9.8 节,"黄川"和"月黄"品种的第一花序节位最小为 7.6 节和 7.7 节,较对照减少 2.2 节和 2.1 节,其余品种的第一花序节位较对照减少 0.2~1.5 节,降低 2.0%~15.3%。参试 9 个品种的植株开展度均大于对照,其中"绿舟"和"绿星"品种的植株开展度最大,分别为 71.9 cm 和 70.3 cm,较对照增加 29.6 cm 和 28 cm,其余品种的植株开展度范围在 43.1~69.6 cm,较对照增加 0.2~27.6 cm,提高 0.4%~65.24%。"月黑"、"月紫"2 个品种的分枝数为 2 个,与对照相同,其余 7 个品种的分枝数均高于对照,范围在 2.4~2.8 个,较对照增加 0.4~0.8 个。"黄川"和"绿星"品种的单株结果数低于对照,分别为 7.3 个和 8.3 个,其余 7 个品种的单株结果数均高于对照,范围在 8.6~11.1 个,较对照增加 0.1~2.6 个,提高 1.2%~30.1%。

2.3 参试品种果实性状比较

从表 3 可以看出,"月黄"和"月黑"2 个品种的果形为长灯笼,其余各品种的果形均为方灯笼形;"绿圆"品种的果实纵径最大为 10.5 cm,对照茄门的果实纵径最小为 6.9 cm,其余品种的果实纵径范围在 7.2~9.6 cm;"绿圆"品种的果实横径最大为 9.6 cm,"月黄"品种的果实横径最小为 5.1 cm,其他品种的果实横径范围在 7.2~9.3 cm;10 个品种的心室数均为 3~4 个;"绿圆"品种的果肉厚度最大为 0.73 cm,"月白"和"月紫"品种的果肉厚度最小为 0.52 cm,其余品种的果肉厚度范围在 0.54~0.70 cm;"绿圆"品种的单果质量最大为 248.9 g,"月白"品种的单果质量最小为 118.9 g,其他品种的单果质量范围在 124.6~245.4 g;"月黄"品种的青果果色为浅绿,"月黑"品种的青果果色为黑色,"月白"品种的青果果色为白色,"月紫"的青果果色为紫色,"黄川"品种的青果果色为绿色,其余品种的青果果色均为深绿色;"月黄"品种的老熟果果色为黄色,"月白"品种的老熟果果色为淡黄色,"黄川"品种的老熟果果色为橘黄色,"月黑"、"红曼"和对照茄门的老熟果果色为红色,"绿

表3 参试彩椒品种和果实性状比较

Table 3 Comparison of fruit characters of color pepper varieties

品种 varieties	果形 shape	果纵径 fruit longitudinal diameter /cm	果横径 fruit diameter /cm	心室数 number of ventricular /个	果肉厚度 the thickness of the pulp /cm	单果质量 fruit quality /g	果色 fruit color		果味 the fruit taste	果面 fruit surface
							青果 holly	老熟果 the ripe fruit		
月黄	长灯笼	8.0	5.1	3~4	0.54	151.4	浅绿	黄色	甜	光滑
月黑	长灯笼	8.6	7.4	3~4	0.56	150.0	黑色	红色	甜	光滑
红曼	方灯笼	8.3	7.8	3~4	0.55	155.8	深绿	红色	甜	光滑
月白	方灯笼	7.7	7.9	3~4	0.52	118.9	白色	淡黄	甜	光滑
黄川	方灯笼	7.2	8.9	3~4	0.62	205.6	绿色	橘黄色	甜	光滑
月紫	方灯笼	8.5	8.2	3~4	0.52	124.6	紫色	橘红	甜	光滑
绿圆	方灯笼	10.5	9.6	3~4	0.73	248.9	深绿	鲜红	甜	光滑
绿舟	方灯笼	9.6	9.3	3~4	0.66	233.9	深绿	鲜红	甜	光滑
绿星	方灯笼	9.6	8.5	3~4	0.70	245.4	深绿	鲜红	甜	光滑
茄门(CK)	方灯笼	6.9	7.2	3~4	0.60	155.4	深绿	红色	甜	光滑

圆"、"绿舟"和"绿星"的老熟果果色为鲜红色；参试各品种的果面光滑,味甜、肉厚,均耐贮运。

2.4 参试品种产量与产值比较

从表4可以看出,"绿圆"品种的折合产量最高,为5 168.8 kg/667 m²,比对照茄门增产65.5%,总产值为15 506.4 元/667 m²,较对照增收6 139.8 元/667 m²;"绿星"、"绿舟"、"月黄"、"红曼"、"黄川"等5个品种产量和产值均高于对照,产量范围在3 545.9~4 805.8 kg/667 m²,增产幅度为13.6%~53.9%;产值的范围在10 637.7~14 417.4 元/667 m²,增值幅度为13.6%~53.9%。"月黑"、"月白"、"月紫"等3个品种的产量和产值均低于对照,产量的范围在2 829.5~3 084.2 kg/667 m²,减产幅度为1.2%~9.4%;产值的范围在8 488.5~9 252.6 元/667 m²,减值幅度为1.2%~9.4%。

综合分析参试9个彩椒品种的生育期、农艺性状、果实性状及产量,以"绿圆"、"绿星"、"月黄"3个品种综合性状好,增产潜力大,生长势较强,椒果商品性好,折合667 m²产量及效益显著高于对照茄门,应在进一步试验示范的基础上进行大面积推广种植。

3 讨论与结论

综合分析参试9个品种的生育期、农艺性状、果实性状和产量及产值,发现"月黑"品种的开花、坐果期与对照茄门相同,其余8个品种均晚于对照茄门,较对照延后2~5 d;"月黑"品种的生育期与对照茄门相同,为224 d,其余8个品种均较对照长,较对照茄门延长2~8 d。参试的9个品种与对照相比均表现出强或较强的抗病性;"绿圆"、"绿舟"、"绿星"、"月黄"、"红曼"、"黄川"等6个品种的植株生长势强或较强,"月黑"、"月紫"与对照中等。"绿圆"、"绿星"、"绿舟"、"红曼"、"黄川"等5个品种单果质量均高于对照,果肉厚度以绿圆最厚,为0.73 cm。从果实外观、品质看,"月黄"和"月白"3心室多,其余7个品种均4心室多,参试各品种果面光滑,味甜、肉厚,均耐贮运。"绿

表 4 参试彩椒品种的产量与产值比较
Table 4 Comparison of yield and output value of color pepper varieties

品种 varieties	小区产量 cell production/kg			平均小区产量 average plot yield/kg	折合 667 m² 产量 equivalent to 667 m² output /kg	比对照土 compared with /%	667 m² 折合产值 equivalent to the output value /元	新复极差比较 comparison of new complex range 5% 1%	位次 seating arrangement
	Ⅰ	Ⅱ	Ⅲ						
月黄	201.7	214.5	209.6	208.6	3 964.0	27.0	11 892.0	c C	4
月黑	158.2	161.9	166.8	162.3	3 084.2	-1.2	9 252.6	fg E	9
红曼	206.7	194.5	197.4	199.5	3 791.1	21.4	11 373.3	d C	5
月白	150.2	160.8	155.9	155.6	2 956.8	-5.3	8 870.4	gh EF	9
黄川	183.5	189.6	186.8	186.6	3 545.9	13.6	10 637.7	e D	6
月紫	144.6	147.3	154.7	148.9	2 829.5	-9.4	8 488.5	h F	10
绿圆	276.0	272.5	267.4	272.0	5 168.8	65.5	15 506.4	a A	1
绿竹	241.5	247.8	249.7	246.3	4 680.4	49.9	14 041.2	b B	3
绿星	249.4	256.9	252.5	252.9	4 805.8	53.9	14 417.4	b B	2
茄门(CK)	159.5	167.6	165.7	164.3	3 122.2		9 366.6	f E	7

备注：单价 3.0 元/kg

圆"折合产量最高,为5 168.8 kg/667 m²,比对照茄门增产65.5%;总产值15 506.4元/667 m²,较对照增收6 139.8元/667 m²。"绿星"、"绿舟"、"月黄"、"红曼"、"黄川"等5个品种产量在3 545.9~4 805.8 kg/667 m²,均比对照增产,增产幅度为13.6%~53.9%;"月黑"、"月白"、"月紫"等3个品种产量在2 829.5~3 084.2 kg/667 m²,均比对照减产,减产幅度为1.2%~9.4%。

综合分析结果:"绿圆"、"绿星"、"月黄"3个品种综合性状好,增产潜力大,生长势较强,辣椒果实商品性好,折合667 m²产量及效益显著高于对照茄门,应在进一步试验示范的基础上进行大面积推广种植。

参考文献

[1] 钟小明.温室彩椒长季节栽培技术[J].技术与市场,2007(12):7-8.

[2] 张焕丽,朱永,霍红.春节观赏彩椒的引种及栽培[J].中国瓜菜,2007(2):14-15.

[3] 牛玉,史小美,李育军,等.彩椒品种的无土栽培试验筛选与评价[J].广东农业科学,2011(19):52-53.

[4] 李春艳.彩椒品种无土栽培筛选试验[J].湖北农业科学,2009,48(02):360-362.

[5] 何勇明,何金勇,陆文科.彩色甜椒品种比较试验[J].北方园艺,2012(6):30-32.

[6] 严泽生,陈先知.彩色甜椒品种比较试验[J].长江蔬菜,2004(10):50-51.

[7] 于恩晶,司力珊,范艳艳,等.日光温室不同彩色甜椒品种比较试验[J].上海交通大学学报(农业科学版),2008,26(5):471-473.

[8] 耿三省,陈斌,张晓芬.我国甜辣椒品种市场需求的变化趋势分析及建议[J].辣椒杂志,2006,18(3):40-41.

[9] 孟雅宁,严立斌,范妍芹.日光温室不同彩色甜椒品质比较[J].安徽农业科学,2016,44(14):25-26.

冷凉区拱棚辣椒不同新组合品比试验

高晶霞[1],马守才[2],吴雪梅[2],赵云霞[1],颜秀娟[1],王学梅[1]

(1. 宁夏农林科学院种质资源研究所,宁夏银川 750002;
2. 宁夏回族自治区彭阳县农业技术推广服务中心,宁夏彭阳 756500)

摘 要:研究了20个新组合羊角椒材料在拱棚的生长情况。结果表明:参试羊角椒品种物候期、植物学性状、抗病性、果实性状、产量及产值均有差异。参试羊角椒品种的开花期、坐果期、始收期以0020、0023辣椒最早,较其余19个品种分别提前1~9 d、1~11 d、1~12 d;0019辣椒株高最高、茎粗最粗,分别为130.4 cm、2.12 cm,较其余品种分别增加8.4~58.4 cm、0.15~0.84 cm,提高6.9%~81.1%、7.6%~65.6%;单株结果数以0016辣椒最多,为25.4个,较其余品种增加1.2~9.4个,提高5.0%~58.8%;0063辣椒结果数最少,仅为16.0个;0011、0012、0013、0016、0019、0020、0021辣椒品种抗病性强;参试的20个辣椒材料均为长羊角椒,味辣,皮薄,口感好;0016辣椒折合667 m^{2}产量及产值最高,为3 890.8 kg和8 559.8元,较其余品种增产11.4~2 072.2 kg、增收25.1~4 558.9元,增幅0.29%~113.9%。

关键词:冷凉区;拱棚;辣椒;新组合;试验

Cold Aegion Shed Of Different Pepper Varieties Test Of New Combination

GAO Jing-xia[1], Ma Shou-cai[2], Wu Xue-mei2, ZHAO Yun-xia[2],
YAN Xiu-juan[2], WANG Xue-mei[2]

(1. *NingxiaAcademyofagricultureandforestryplantresources*, *Ningxia*, *Yinchuan*, 750002;
2. *The Ningxia Hui Autonomous Region Pengyang County AgriculturalTechnology Extension and service center*)

Abstract: Study 20 new combinations of Different Pepper Cultivars in the growth of archshed,The results show that: The phenological period of hot pepper varieties,botanical characters,disease resistance,fruit yield and output value are different characters,Compared with the other 19 varieties to advance 1 to 9 days、11 days、12 days;the height of 0019 pepper is the highest,For the 130.4 cm, stem diameter is the crude, For the 2.12 mm, Compared with the other cultivars were increased 8.4~58.4 cm、0.15~0.84 mm, To improve the 6.9%~81.1%，7.6%~65.6%; Fruit number per plant, The 0016 most chilies, For the 25.4, Compared with the other cultivars increased 1.2~9.4, To improve the 5.0%~58.8%;the least number of 0063 pepper;0011、0012、0013、0016、0019、0020、0021 peppers varieties of disease resistance;20 pepper cultivars were long hot pepper,Spicy,thin skin,good taste;0016 pepper reduced 667 m^2 yield and output value of the highest, For the 3 890.8 kg and 8 559.8 yuan, Compared with the other cultivars increased by 11.4~2 072.2 kg, The income of 25.1~4 558.9 yuan,Increase in 0.29%~ 113.9%.

Key word：Cold region；Shed；Pepper；New combinations；Test

彭阳县位于宁夏回族自治区南部边缘,六盘山东麓,介于东经106°32′~106°58′,北纬35°41′~36°17′。年平均气温7.4~8.5℃,无霜期140~170 d,降水量350~550 mm。属典型的温带半干旱大陆性季风气候,特别适合辣椒栽培[1,2]。羊角辣椒适口性好、辣度适宜受到很多人的喜爱,也占有较大的市场份额[4,5],同时随着经济社会迅速发展、城市化加快和各地区人员的广泛交流促进了人们对农副产品多样性的需求[3],但是随着大棚辣椒栽培面积迅速扩大,辣椒栽培小气候条件发生了重大变化,加之品种单一,选择适合拱架大棚可控小气候条件下的辣椒新品种成为一项新课题。本试验结合当地实际情况,选择羊角椒新品种进行试验,鉴定新育成的螺丝辣椒品系及杂交组合的遗传稳定性、形态特征、生物学特性等,测试新品种的生产潜力,以期为新品种的审定、推广提供依据。

1 材料与方法

1.1 供试材料

供试 20 个辣椒新组合为 0011、0012、0013、0014、0015、0016、0017、0018、0019、0020、0021、0022、0023、0024、0025、0026、0027、0048、0062、0063 等辣椒新组合均由宁夏农林科学院种质资源研究所辣椒课题组提供。

1.2 试验方法

试验地点选在宁夏彭阳县设施农业建设区的新集乡沟口村水泥拱架塑料大棚实施。土壤为黄绵土,肥力中等水平,前作为辣椒。

试验采用大区观察栽培,共设 20 个处理,不设重复,小区面积 23.4 m²。采用膜下滴灌,垄宽 70 cm,垄沟宽 60 cm,垄高 25 cm。单株栽植,行距 65 cm,株距 35 cm,每小区定植 92 株。田间管理按常规方法进行。观察记载物候期、农艺性状,并测定产量。

2 结果与分析

2.1 不同羊角椒新组合材料物候期调查

从表 1 看出,参试羊角椒新组合材料开花期、坐果期、始收期以 0020 辣椒、0023 辣椒最早,较其余 17 份材料分别提前 1~9 d、1~11 d、1~12 d;其次以 0022 辣椒、0024 辣椒较早,较其余材料分别提前 1~8 d、1~10 d、1~10 d;以 0027 辣椒开花、坐果、始收最迟。采收结束以 0048 辣椒最早,较其余材料提前 1~8 d 采收结束;以 0016 采收结束最迟,较其余材料延后 1~8 d 采收结束。采收期以 0024 辣椒最长,为 124 d,较其余 18 份材料延后 1~14 d 采收结束。其次是 0023 辣椒、0016 辣椒,分别为 123 d、122 d,较其余 16 份材料分别延后 3~13 d、2~12 d 采收结束。0014 采收期为 116 d。生育期以 0016 最长,为 263 d,较其余 18 份材料生育期延后 1~8 d。0014 生育期为 259 d。

表1 不同新组合品种辣椒物候期比较

Table 1 Different varieties of pepper phenological period

品种 Varieties	播种期 Sowing period (月-日)	出苗期 Seedling stage (月-日)	定植期 Planting date (月-日)	开花期 Blossom period (月-日)	坐果期 The fruit setting stage (月-日)	始收期 The first harvest time (月-日)	末收期 At the last harvest period (月-日)	采收期 Harvest time (d)	生育期 Growth period (d)
0011	2-3	2-24	4-22	6-1	6-8	6-28	10-23	117	262
0012	2-3	2-24	4-22	5-31	6-7	6-27	10-23	118	262
0013	2-3	2-28	4-22	5-30	6-6	6-26	10-23	119	262
0014	2-3	2-24	4-22	5-30	6-6	6-26	10-20	116	259
0015	2-3	2-24	4-22	5-26	6-3	6-24	10-17	115	256
0016	2-3	2-24	4-22	5-30	6-6	6-26	10-24	122	263
0017	2-3	2-24	4-22	5-30	6-6	6-26	10-21	117	260
0018	2-3	2-28	4-22	5-31	6-7	6-27	10-20	115	259
0019	2-3	2-26	4-22	5-30	6-6	6-26	10-20	116	259
0020	2-3	2-24	4-22	5-24	5-31	6-21	10-21	122	260
0021	2-3	2-24	4-22	5-29	6-5	6-25	10-21	118	260
0022	2-3	2-24	4-22	5-25	6-1	6-22	10-20	120	259
0023	2-3	2-24	4-22	5-24	5-31	6-21	10-23	124	262
0024	2-3	2-24	4-22	5-25	6-1	6-22	10-23	123	262
0025	2-3	2-24	4-22	5-30	6-6	6-26	10-19	115	258
0026	2-3	2-24	4-22	5-29	6-5	6-25	10-21	118	260
0027	2-3	2-24	4-22	6-3	6-11	7-2	10-20	110	259
0048	2-3	2-24	4-22	5-29	6-5	6-25	10-16	113	255
0062	2-3	2-26	4-22	6-2	6-10	7-1	10-19	110	258
0063	2-3	2-26	4-22	5-30	6-6	6-26	10-19	115	258

2.2 不同羊角椒新组合材料辣椒植物学性状与抗病性

从表2可知,羊角椒株高以0019辣椒植株最高,为130.4 cm,较其余材料增加8.4~58.4 cm,提高6.9%~81.1%。其余18份新组合株高均在72.0~122.0 cm之间。茎粗以0019辣椒最粗,为2.12 cm,较其余材料增加0.15~0.84 cm,提高7.6%~65.6%;0015羊角椒最细,为1.28 cm。第一花序节位以0027羊角椒节位最高,为7.2节,较其余材料增加1.0~1.9节;0011羊角椒节位最低,为5.3节。分枝数以0017、0025和0026羊角椒最多,为3.0个,较其余品种增加0.2~1.0个,提高7.1%~50%,其余18份材料均2.0~2.8个。开展度以0019辣椒最大,为74.6 cm,较其余材料增加3~20.8 cm,提高4.2%~38.7%;0023最小,为53.8 cm。单株结果数以0016羊角椒最多,为25.4个,较其余材料增加1.2~9.4个,提高5.0%~58.8%;0063羊角椒结果数最少,仅为16.0个。0011、0012、0013、0016、0019、0020、0021等7个羊角椒材料抗病性强,0025、0026、0048、0063等4个羊角椒材料抗病性弱,其余9份材料抗病性中等。从植株长势上看,0011、0012、0013、0016、0018、0019、0020、0021、0022、0024、0027等11个新组合羊角椒材料植株长势强,其余8个新组合材料植株长势中等。0025、0026、0048、0063植株长势弱。

2.3 不同新组合羊角椒材料辣椒果实性状

从表3可知,参试的20份材料均为长羊角椒,味辣,皮薄,口感好;0011、0015、0022、0023、0024、0025、0026、0027、0048、0062共10份材料果色为浓绿色,0017、0018、0019、0063共4份材料果色为浅绿,0012、0013、0016、0020、0021共5份材料果色为黄绿色;0014牛角椒果色为浅绿色,味辣,肉厚,耐贮运。羊角椒果长以0025羊角椒最长,为32.8 cm,较其余材料增加0.1~11.8 cm,提高0.3%~56.2%;果粗以0063羊角椒最粗,为4.5 cm,较其余材料增加0.55~1.92 cm,提高13.9%~74.4%;果肉厚度以0019羊角椒最厚,为0.32 cm,较其余材料增加0.04~0.17 cm,提高14.3%~113.3%;单果质量以0023羊角椒最重,为69.2 g,较其余材料增加0.6~23.6 g,提高0.87%~51.8%,其余18份羊

表 2 不同组合羊角椒品种辣椒植物学性状与抗病性

Table 2 Different varieties of pepper botany characters and disease resistance

品种 Varieties	株高 Plant height /cm	茎粗 Stem diameter /cm	第一花序节位 The first flowers order node/节	开展度 Plant expansion /cm	分枝数 Branch number /个	叶色 Leaf color	单株结果数 Single node the number of fruit/个	植株长势 Plant long potential	抗白粉病 Anti white flydisease
0011	95.0	1.76	5.3	71.6	2.0	绿	21.8	强	强
0012	113.2	1.97	5.6	68.6	2.0	绿	24.2	强	强
0013	108.8	1.68	5.8	66.8	2.0	绿	22.0	强	强
0014	82.4	1.74	5.6	67.4	2.0	绿	18.4	中等	中等
0015	87.0	1.28	6.0	60.0	2.0	绿	24.2	中等	中等
0016	110.4	1.67	5.4	64.4	2.0	绿	25.4	强	强
0017	111.6	1.54	5.8	69.0	3.0	绿	21.2	中等	中等
0018	116.2	1.55	5.6	69.0	2.6	绿	24.8	强	中等
0019	130.4	2.12	6.2	74.6	2.6	绿	21.6	强	强
0020	107.8	1.65	5.4	70.8	2.0	绿	20.2	强	强
0021	113.0	1.55	6.0	68.0	2.0	绿	21.2	强	强
0022	104.8	1.49	5.8	71.6	2.0	绿	21.5	中等	中等
0023	83.0	1.44	6.2	53.8	2.0	绿	22.0	中等	中等
0024	103.2	1.82	5.8	71.0	2.0	绿	23.0	强	弱
0025	74.6	1.29	5.8	59.2	3.0	绿	20.8	中等	弱
0026	99.2	1.30	5.6	66.4	3.0	绿	20.6	中等	中等
0027	122.0	1.55	7.2	70.6	2.4	绿	23.4	强	弱
0048	72.0	1.49	5.8	65.2	2.8	绿	14.8	中等	中等
0062	84.0	1.41	6.0	62.6	2.6	绿	21.0	中等	中等
0063	73.8	1.47	5.6	66.4	2.6	绿	16.0	中等	弱

表 3 不同新组合羊角椒品种辣椒果实性状
Table 3 Different varieties of pepper fruit characters

品种 Varieties	果长 Fruit length /cm	果粗 The fruit thick/cm	心室数 The number of ventricle/个	果肉厚度 Pulp thickness /cm	单果质量 Simple fruit quality/g	果实外观、品质 The appearance, Quality of fruit
0011	30.6	3.50	2~3	0.20	68.6	长羊角椒,浓绿,辣味浓,果肩皱褶多,果面有棱沟
0012	28.5	3.63	2~3	0.18	59.2	长羊角椒,黄绿,辣味强,果肩有皱褶,果面有棱沟
0013	30.0	3.10	2~3	0.23	60.9	长羊角椒,黄绿,味辣,果肩有皱褶,羊角形弯曲
0014	25.8	4.62	3~4	0.33	88.7	大羊角椒,浅绿,味辣,果面光滑,果肩微有皱褶
0015	27.5	3.63	2~3	0.18	50.8	长羊角椒,浓绿,辣味浓,果肩有皱褶,呈螺丝形变曲
0016	31.7	3.23	2~3	0.23	61.5	长羊角椒,黄绿,辣味强,果肩皱褶多且深
0017	27.9	3.69	2~3	0.20	62.0	长羊角椒,浅绿,辣味浓,果肩微有皱褶,果面粗糙且有皱褶
0018	25.1	4.11	2~3	0.25	60.0	长羊角椒,浅绿,味辣,果肩有皱褶深,多弯曲
0019	27.8	3.76	2~3	0.32	68.7	长羊角椒,浅绿,味辣,果肩皱褶深
0020	32.7	3.34	3~4	0.20	66.0	长羊角椒,黄绿,极辣,果肩皱褶深且多
0021	29.8	3.41	2~3	0.19	56.9	长羊角椒,黄绿,味辣,果肩有皱褶深皱褶,果肩皱褶多
0022	29.9	3.95	2~3	0.23	51.4	长羊角椒,浓绿,辣味浓,果肩皱褶多,呈丝螺形弯曲
0023	32.6	3.54	2~3	0.28	69.2	长羊角椒,浓绿,辣味浓,果肩皱褶多,果面有棱沟
0024	31.2	3.60	3~4	0.20	67.7	长羊角椒,浓绿,辣味浓,果肩深皱褶,变曲
0025	32.8	3.69	2~3	0.21	61.7	长羊角椒,浓绿,辣味适中,果肩有皱褶,果面有棱沟
0026	31.4	3.62	2~3	0.20	56.0	长羊角椒,浓绿,辣味适中,果肩有皱褶,变曲
0027	31.1	2.58	2~3	0.24	49.9	长羊角椒,浓绿,辣味浓,果肩皱褶多,果面有棱沟
0048	32.1	3.56	2~3	0.15	65.9	长羊角椒,浓绿,辣味适中,果肩皱褶多,果面有棱沟
0062	31.4	2.77	2~3	0.15	54.3	长羊角椒,浓绿,辣味浓,果面皱褶多且有棱沟
0063	21.0	4.50	2~3	0.20	45.6	长羊角椒,浅绿,辣味适中,果肩皱褶且有棱沟

角椒材料均 45.6~68.6 g,0063 单果重最小,为 45.6 g。0014 果长 25.8 cm,果粗 4.62 cm,肉厚 0.33 cm,单果重 88.7 g。

2.4 不同羊角椒材料产量与产值比较

从表 4 可知,参试 20 份新组合羊角椒材料中,以 0016 羊角椒折合 667 m² 产量及产值最高,为 3 890.8 kg 和 8 559.8 元,较其余材料增产 11.4~2 072.2 kg

表 4 不同新组合羊角椒辣椒材料产量与产值
Table 4 Varieties of hot pepper yield and output value

品种 Varieties	23.4 m² 小区产量 Cell production /kg	折合 667 m² 产量 Equivalent to 667 m² output/kg	667 m² 产值 Production value /元	排序 Rank
0011	130.7	3 725.5	8 196.1	4
0012	125.2	3 568.7	7 851.1	7
0013	117.1	3 337.9	7 343.4	8
0014	142.7	4 067.6	3 660.8	
0015	107.4	3 061.4	6 735.1	12
0016	136.5	3 890.8	8 559.8	1
0017	114.9	3 275.1	7 205.2	10
0018	130.1	3 708.4	8 158.5	5
0019	129.7	3 697.0	8 133.4	6
0020	116.6	3 323.6	7 311.9	9
0021	105.5	3 007.2	6 615.8	13
0022	96.6	2 753.5	6 057.7	16
0023	133.1	3 793.9	8 346.6	3
0024	136.1	3 879.4	8 534.7	2
0025	112.3	3 201.0	7 042.2	11
0026	100.8	2 873.2	6 321.0	15
0027	102.0	2 907.4	6 396.3	13
0048	85.2	2 428.6	5 342.9	17
0062	99.7	2 841.9	6 252.2	14
0063	63.8	1 818.6	4 000.9	18

和增收 25.1~4 558.9 元,增幅 0.29%~113.9%。其次以 0024、0023 羊角椒折合 667 m² 产量及产值较高,分别为 3 879.4 kg、3 793.9 kg 和 8 534.7 元、8 346.6 元,分别较其余材料增产 153.9~2 060.8 kg、增幅 4.1%~113.3%;68.4~1 975.3 kg、增幅 0.18%~108.6%。以 0063 羊角椒折合 667 m² 产量及产值最低,分别为 1 818.6 kg,4 000.9 元。0014 牛角椒折合 667 m² 产量及产值分别为 4 067.6 kg,3 660.8 元。

3 结论

综合分析参试 20 份羊角辣椒材料的生育期、农艺性状、果实性状及产量,以 0016、0024、0023、0011 共 4 个新组合羊角椒材料综合性状表现好,增产潜力大,生长势强,椒果商品性好,适口性好,市场销售旺,抗病性强,折合 667 m² 总产量及效益显著高于其余 16 份材料,可在塑料大棚早春茬生产中进一步试验示范的基础上进行大面积推广种植,0063 羊角椒果实短,抗病性差,建议淘汰,其余 15 份材料有待进一步试验观察。

参考文献

[1] 马瑞,惠浩剑,马守才,等.大棚辣椒品种引进观察试验[J].现代农业科技,2012(07).

[2] 周刚,刘琪,徐国友,等.大棚早熟栽培辣椒品比试验[J].长江蔬菜,2008(3):48-49.

[3] 朱广凯.早春大棚辣椒的综合防治[J].北京农业实用技术,2011(3):15.

[4] 徐佩娟,何铁海,董阳辉.辣椒品种对比试验[J].农业科技通讯,2008(12).

[5] 向幼衡.辣椒品比试验研究[J].湖南农业科学,2006(04).

拱棚春提早、秋延后辣椒优新品种筛选试验

高晶霞[1]，吴雪梅[2]，牛勇琴[2]，高 昱[2]，裴红霞[1]，谢 华[1]

（1.宁夏农林科学院种质资源研究所，宁夏银川 750002；
2.宁夏回族自治区彭阳县蔬菜产业发展服务中心，宁夏彭阳 756500）

摘 要：为筛选出拱棚春提早、秋延后栽培辣椒品种，以引进的19个辣椒优良品种为试材，在拱棚中进行品种观察试验。结果表明：参试辣椒品种物候期、植物学性状、抗病性、果实性状、产量及产值均有差异。参试牛角椒品种的开花期、坐果期、始收期以"鑫牛角"品种最早，"朗月206"品种最晚；参试羊角椒品种的开花期、坐果期、始收期以"甘科10号"品种最早、"7062"、"7063"和"华美105"品种最晚。参试牛角椒品种中"朗月407"品种的果实性状较好，"盛龙"品种的果实性状较差；参试羊角椒品种中"0701"品种的果实性状较好，"7062"品种的果实性状较差。参试牛角椒品种中"宝盛6号"品种的单株产量、小区产量、折合亩产量、产值均最高，分别为3.61 kg/株、1 256.3 kg/23.4 m^2、10 474.4 kg/667 m^2、21 682元/667 m^2，比对照产量增幅16.3%，产值增加3 041元/667 m^2，"日本瑞琦"品种的单株产量、小区产量、折合亩产量、产值均最小，分别为2.72 kg/株、946.6 kg/23.4 m^2、7 892.3 kg/667 m^2、16 337.1元/667 m^2；参试羊角椒品种中，"0701"品种的单株产量、小区产量、折合亩产量、产值均最高，分别为3.07 kg/株、1 188.11 kg/23.4 m^2、8 904.1 kg/667 m^2、15 938.3元/667 m^2，比对照产量增幅59.9%，产值增加5 970.5元/667 m^2，"7062"品种的单株产量、小区产量、折合亩产量、产值最低，分别为1.72 kg/株、665.6/23.4 m^2、4 988.3 kg/667 m^2、8 929.1元/667 m^2。综合分析结果：参试品种"朗月407"、"鑫牛角30"、"0701"、"37-94"的辣椒性状典型，坐果力强，单果质量、产量较高，适宜拱棚春提前、秋延后栽培，进行示范推广应用。

关键词：辣椒；春提早、秋延后；适应性；试验

Screening Test of New Superior Varieties of Pepper in Early Spring and Late Autumn in Arch Shed

Gao jing-xia[1], Wu xue-mei[2], Niu yong-qin[2], Gao yu[2],

Pei hong-xia[1], Xie hua[1]

(1. *Ningxia Academy of Agriculture and Forestry Plant Resources, Ningxia, Yinchuan,* 750002; 2. *The Ningxia Hui Autonomous Region, Pengyang County AgriculturalTechnology Extension and Service Center, Pengyang, Ningxia* 756500)

Abstract: In order to screen out the early spring and late autumn cultivation of pepper varieties in the arch shed, 19 excellent varieties of pepper were introduced as test materials, and the variety observation experiment was carried out in the arch shed. The results showed that there were differences in phenological stage, botanical character, disease resistance, fruit character, yield and output value. In the flowering, fruiting and harvesting stages, the "xin niu jiao" variety was the earliest, the "lang yue 206" variety was the latest, and the "ganke 10" variety was the earliest, "7062", "7063" and "huamei 105" varieties were the latest.In the test, the fruit characters of " lang yue 407 "and shenglong were better, while those of "0701" and "7062 "were worse. Volunteers horn pepper varieties of Julius baer 6 varieties, yield of plot yield, reduced area yield and output value are the highest, 3.61 kg/strain, 1 256.3 kg/23.4 m^2 to 10 474.4 kg/m^2, 667 m^2, 21 682 yuan/667 m^2, compared with 16.3% growth in production, output value 3041 yuan/667 m^2, "Japan's march" varieties, yield of plot yield, reduced area yield and output value are minimal, They were 2.72 kg/plant, 946.6 kg/23.4 m^2, 7 892.3 kg/667 m^2 and 16 337.1yuan/667 m^2, respectively. Among the tested varieties, "0701" had the highest yield per plant, plot yield, equivalent per mu yield and output value, which were 3.07 kg/plant, 1 188.11 kg/23.4 m^2, 8 904.1 kg/667 m^2 and 15 938.3 yuan/667 m^2, which increased by 59.9% compared with the control yield, and the output value increased by 5 970.5 yuan/

667 m². "7062" had the lowest yield per plant, plot yield, equivalent per mu yield and output value. They were 1.72/plant, 665.6/23.4 m², 4 988.3 kg/667 m², 8 929.1 yuan/667 m², respectively. Comprehensive analysis results: the test varieties "lang yue 407", "xin niu jiao 30", "0701", "37-94" showed typical characteristics of pepper, with strong fruit setting force, high fruit quality and high yield.

Key words: pepper; spring earlier; fall later; adaptability; test

辣椒,又叫番椒,是茄科辣椒属植物,为一年或多年生草本植物[1],原产于秘鲁和墨西哥一带[2],因其色泽鲜艳、香辣味和富含维生素C而成为一种世界性蔬菜[3,4]。其中维生素C含量在蔬菜中居第一位,既是人们爱吃的蔬菜,又是主要的调味品之一,在我国南北方广泛种植,现已成为全国栽培面积最大的蔬菜作物之一[5]。辣椒既适合露地栽培,也适合设施内栽培[6]。辣椒营养丰富,且具有较高的药用价值。根据测定,辣椒果实中含有的维生素A、B高于黄瓜、番茄、茄子等果菜类,特别是维生素C的含量比以上菜类高4~7倍,每100 g鲜辣椒含维生素C 170~360 mg,最高可达460 mg,辣椒作为调味品,是因为辣椒中含辣椒素,少量食用,能健脾开胃,增进食欲,帮助消化,而且还有驱寒除湿、舒血活络等药用功能,对关节炎、冻疮、青蛇咬伤、腋臭等也有一定疗效,因此被誉为"健康食品"[7]。目前,我国辣椒种植面积居蔬菜作物第二位,产值和效益居蔬菜作物之首[8]。为筛选出适宜宁夏拱棚春提早、秋延后栽培的优质、抗病、丰产、抗逆、色泽好的辣椒优良品种,本试验引进19份优良辣椒品种,在彭阳县辣椒核心示范基地进行分类和田间品比观察,分别调查物候期、植株生态性状、果实性状指标、抗病性(辣椒病毒病田间鉴定)和产量等主要指标,丰富宁夏辣椒品种配置,提高优质辣椒品种的利用率。

1 材料与方法

1.1 供试品种

本试验共引进19个品种,其中牛角椒10个,分别为"鑫牛角"(北京鑫阳光农业科技有限公司)、"鑫牛角30"(北京鑫阳光农业科技有限公司)、"盛龙"(寿光万盛种业有限公司)、"宝盛6号"(山东省昌乐群信种苗有限公司)、"日本瑞崎"(北京农瑞德农业有限科技公司)、"巨峰1号"(宁夏巨丰种苗有限责任公司)、"超级椒王"(安徽省萧县振丰种业有限公司)、"坂田神椒"(寿光市绿帆农业科技有限公司)、"朗月407"和"朗月206";羊角椒品种9个,分别为"37-94"(辽宁依农农业科技有限责任公司)、"6139"(宁夏农林科学院种质资源研究所)、"7061"(宁夏农林科学院种质资源研究所)、"螺美7号"(宁夏农林科学院种质资源研究所)、"0701"(宁夏农林科学院种质资源研究所)、"7062"(宁夏农林科学院种质资源研究所)、"7063"(宁夏农林科学院种质资源研究所)、"甘科10号"(甘肃省农业科学院)和"华美105"(酒泉市华美种子有限责任公司)。

1.2 测定方法与项目

1.2.1 辣椒物候期及抗病性的观察

观察辣椒的播种期、定植期、始花期、盛花期、坐果期、始收期、末收期以及辣椒病毒病和白粉病的发病率。

1.2.2 辣椒形态指标的测定

定植后,分别在辣椒的始花期、盛果期和盛收期利用卷尺和游标卡尺对各品种辣椒植株的株高、茎粗等形态指标进行测量。株高指标以茎基部至生长点的高度为标准,茎粗以茎基部以上3~5 cm高度的茎粗为标准。

1.2.3 辣椒果实性状的测定

辣椒的盛果期,利用卷尺和游标卡尺测定辣椒果实的果长、果粗、果肉厚、果柄长、萼片宽以及观察辣椒果实的果色和果味。

1.2.4 辣椒产量的测定

在辣椒收获期间,使用电子天平在每次收获时称量生产区域中的成熟绿色果实,并统计单株结果数量,最终根据栽培密度折算亩产量。

1.3 数据分析

用 Excel 2019 和 DPS 软件进行数据分析处理,不同处理间的多重比较采用 Duncan's 新复极差法(p≤0.05)。

2 结果与分析

2.1 不同辣椒品种物候期比较

从表1可知,参试的19个品种的播种期、定植期和末收期相同,均在1月17日、3月23日和10月9日,但不同辣椒品种的始花期、盛花期、坐果期、始收期有差异,参试牛角品种中,"鑫牛角"品种均早于其他品种,分别为4月17日、5月1日、4月25日和5月22日,"朗月206"品种最晚,分别为4月30日、5月15日、5月7日和6月5日,其他品种的始花期范围在4月18日~4月28日,盛花期的范围在5月2日~5月13日,坐果期的范围在4月26日~5月5日,始收期的范围在5月23日~6月3日;羊角椒品种中,"甘科10号"品种均早于其他辣椒品种,分别为4月18日、5月2日、4月26日和5月23日,"7062"、"7063"和"华美105"辣椒品种最晚,分别为5月1日、5月16日、5月8日和6月5日,其他品种的始花期范围在4月19日~4月30日,盛花期的范围在5月3日~5月15日,坐果期的范围在4月27日~5月7日,始收期的范围在5月24日~5月28日。

2.2 不同辣椒品种植株主要性状与抗病性比较

从表2可以看出,不同辣椒品种在不同时期,植株生长指标有差异,始花期时,参试牛角椒品种中,"鑫牛角30"辣椒株高最高为36.8 cm,"宝盛6号"、"朗月407"辣椒株高次之,为34.5 cm、34.1 cm,"盛龙"株高最低为21.8 cm,其他辣椒品种株高范围在27.9~30.4 cm,"宝盛6号"、"朗月407"辣

表 1 不同辣椒品种物候期比较

Table 1 Comparison of phenological stages of different pepper varieties

	品种 Variety	播种期 Seeding time (月-日)	定植期 Planting date (月-日)	始花期 Early flowering season (月-日)	盛花期 Full-blossom period (月-日)	坐果期 Fruit-set period (月-日)	始收期 Beginning to receive period (月-日)	末收期 Closed at the end of the period (月-日)
牛角椒	鑫牛角	1月17日	3月23日	4月17日	5月1日	4月25日	5月22日	10月9日
	鑫牛角30	1月17日	3月23日	4月21日	5月5日	4月29日	5月26日	10月9日
	盛龙	1月17日	3月23日	4月23日	5月8日	5月3日	5月28日	10月9日
	宝盛6号	1月17日	3月23日	4月18日	5月2日	4月26日	5月23日	10月9日
	日本瑞崎	1月17日	3月23日	4月27日	5月12日	5月4日	6月2日	10月9日
	巨峰1号	1月17日	3月23日	4月27日	5月12日	5月4日	6月2日	10月9日
	超级椒王	1月17日	3月23日	4月28日	5月12日	5月4日	6月2日	10月9日
	坂田神椒	1月17日	3月23日	4月28日	5月13日	5月5日	6月3日	10月9日
	朗月407	1月17日	3月23日	4月28日	5月13日	5月5日	6月3日	10月9日
	朗月206	1月17日	3月23日	4月30日	5月15日	5月7日	6月5日	10月9日
羊角椒	37-94	1月17日	3月23日	4月19日	5月3日	4月27日	5月24日	10月9日
	6139	1月17日	3月23日	4月23日	5月8日	4月31日	5月28日	10月9日
	7061	1月17日	3月23日	4月30日	5月15日	5月7日	6月5日	10月9日
	螺美7号	1月17日	3月23日	4月23日	5月8日	4月31日	5月28日	10月9日
	0701	1月17日	3月23日	4月21日	5月5日	4月29日	5月26日	10月9日
	7062	1月17日	3月23日	5月1日	5月16日	5月8日	6月5日	10月9日
	7063	1月17日	3月23日	5月1日	5月16日	5月8日	6月5日	10月9日
	甘科10号	1月17日	3月23日	4月18日	5月2日	4月26日	5月23日	10月9日
	华美105	1月17日	3月23日	5月1日	5月16日	5月8日	6月5日	10月9日

椒茎粗最粗，分别为9.17 mm、8.15 mm，"鑫牛角"、"坂田神椒"辣椒茎粗最细，分别为6.45 mm、6.94 mm，其他辣椒品种的茎粗分别在7.33~7.90 mm；参试羊角椒品种中，"甘科10号"和"37-94"品种的辣椒株高最高，分别为40.6 cm、40.4 cm，"7061"、"0701"辣椒株高最矮，分别为27.0 cm、27.2 cm，其他羊角椒系列辣椒品种株高分别在28.1~36.0 cm，"7061"、"0701"辣椒茎粗最粗，分别为8.78 mm、8.27 mm，"华美105"、"7062"茎粗最细，为6.70 mm、6.71 mm，其他辣椒品种茎粗分别在7.02~7.50 mm。盛果期，牛角椒品种中，"鑫牛角30"、"朗月407"辣椒株高最高，分别为71.4 cm、72.6 cm，"鑫牛角"、"超级椒王"、"巨峰1号"辣椒株高最矮，分别为61.0 cm、61.8 cm、60.2 cm，其他品种株高分别在62.8~68.0 cm，"鑫牛角30"、"超级椒王"、"朗月407"辣椒茎粗最粗，分别为13.56 mm、13.35 mm、13.39 mm，其他品种茎粗分别在10.03~13.31 mm；羊角椒品种中，"37-94"、"甘科10号"辣椒株高最高，为78.4 cm、77.0 cm，"0701"、"7062"、"7063"辣椒株高最矮，分别为60.6 cm、59.8 cm、61.6 cm，其他辣椒品种株高分别在65.6~74.6 cm，"7061"茎粗最粗为13.96 mm，"甘科10"、"7062"茎粗最细，分别为12.13 mm、11.58 mm，其他辣椒品种茎粗分别在12.19~13.22 mm。盛收期，牛角椒品种中，"鑫牛角30"和"朗月407"品种的株高最高，分别为108 cm、103 cm，"坂田神椒"、"巨峰1号"品种的株高最低，分别为82.4 cm、84.6 cm，其他辣椒品种分别在87.0~101.2 cm，"鑫牛角30"、"朗月206"辣椒茎粗最粗，分别为16.97 mm、16.84 mm，"盛龙"辣椒茎粗最细，为13.53 mm，其他辣椒品种茎粗分别在14.29~16.58 mm；羊角椒品种中，"37-94"辣椒株高最高为133.8 cm，"0701"株高最低为85.4 cm，其他品种株高分别在97.8~114.4 cm，"华美105"辣椒茎粗最粗，为17.63 mm，"37-94"茎粗最细，为13.68 mm，其他辣椒品种茎粗分别在15.40~16.84 mm。参试牛角椒品种的分支数分别在2.0~3.0个，门椒节位分别在8.0~11.0节，辣椒叶色为深绿、绿色，"超级椒王"、"朗月206"病毒病发病率分别为4.8%、8.5%，"朗月407"、"鑫牛角30"白粉病病叶率较重，分别为

表 2 不同辣椒品种植株主要性状与抗病性

Table 2 Main characters and disease resistance of different pepper varieties

	品种 Variety	始花期 Early flowering season		盛果期 Full productive age		盛收期 Sheng closed period		分枝数 Branching number /个	门椒节位 The door bell pepper section /节	叶色 Leaf colour	病毒病病株率 Virus strain rate /%	白粉病病叶率 Powdery mildew leaf rate/%
		株高 Plant height/cm	茎粗 Stem diameter /mm	株高 Plant height /cm	茎粗 Stem diameter /mm	株高 Plant height /cm	茎粗 Stem diameter /mm					
牛角椒	鑫牛角	29.6	6.45	61.0	12.12	100.6	15.61	2.0	8.0	绿色	0.0	14.6
	鑫牛角 30	36.8	7.33	71.4	13.56	108.0	16.97	2.2	11.0	绿色	0.0	22.1
	盛龙	21.8	7.14	64.8	10.03	94.0	13.53	3.0	10.0	深绿	0.0	20.3
	宝盛 6 号	34.5	9.17	66.0	12.38	89.8	15.29	2.4	10.0	绿色	0.0	16.5
	日本瑞崎	29.1	7.88	66.2	12.48	87.0	16.17	2.4	10.0	深绿	0.0	13.2
	巨峰 1 号	28.0	7.90	60.2	12.17	84.6	14.29	2.6	10.0	深绿	4.8	15.4
	超级椒王	27.9	7.61	61.8	13.35	100.6	16.58	2.0	10.0	深绿	0.0	12.8
	坂田神椒	29.4	6.94	62.8	12.21	82.4	15.24	3.0	11.0	深绿	0.0	15.6
	朗月 407	34.1	8.15	72.6	13.39	103.0	14.71	3.0	10.0	绿色	0.0	14.9
	朗月 206	30.4	7.38	68.0	13.31	101.2	16.84	3.0	11.0	深绿	8.5	23.4
	37-94	40.4	7.15	78.4	12.52	133.8	13.68	2.0	9.0	深绿	0.0	11.2
	6139	32.4	7.34	73.2	12.97	114.4	16.45	2.2	10.0	绿色	4.5	12.7
	7061	27.0	8.78	65.6	13.96	110.8	15.74	2.6	11.0	深绿	6.4	25.6
羊角椒	螺美 7 号	36.0	7.50	74.6	12.19	105.4	16.00	2.6	10.0	深绿	6.9	24.3
	0701	27.2	8.27	60.6	13.22	85.4	16.28	2.4	10.0	深绿	0.0	23.7
	7062	28.1	6.71	59.8	12.13	97.8	15.40	2.6	10.0	绿色	0.0	23.8
	7063	29.0	7.08	61.6	13.05	100.8	16.84	2.0	9.0	深绿	0.0	30.5
	甘科 10 号	40.6	7.02	77.0	11.58	104.6	15.61	2.2	8.0	绿色	4.7	26.2
	华美 105	31.2	6.70	69.6	13.14	112.0	17.63	2.2	12.0	绿色	0.0	16.5

23.4%、22.1%；参试羊角椒品种的分枝数分别在 2.0~2.6 个，门椒节位分别在 8.0~12.0 节，辣椒叶色深绿、绿色，"6139"、"7061"、"螺美 7 号"和"甘科 10 号"品种的病毒病发病率分别为 4.5%、6.4%、6.9%、4.7%，"6139"、"螺美 7 号"、"7063"和"甘科 10 号"品种的白粉病发病率较重，分别为 25.6%、24.3%、30.5%、26.2%。

2.3 不同辣椒品种果实性状比较

从表 3 可知，不同辣椒品种果实性状有差异，牛角椒品种中，"朗月 407"品种果长最长为 28.3 cm，"朗月 206"品种果长次之为 26.2 cm，"坂田神椒"品种果长最短为 22.56 cm，其他辣椒品种果长分别在 23.0~25.3 cm；羊角椒品种中，"37-94"品种果长最长为 35.9 cm，"螺美 7 号"品种果长次之，为 34.66 cm，"7062"果长最短为 27.42 cm，其他辣椒品种果长范围在 30.0~33.8 cm。牛角椒品种中，"超级椒王"品种果粗最粗为 4.88 cm，"朗月 407"品种果粗次之，为 4.73 cm，"日本瑞琦"品种果粗最细为 3.56 cm，其他辣椒品种果粗分别在 3.66~4.55 cm；羊角椒品种中，"0701"品种果粗最粗为 3.43 cm，"37-94"和"7063"品种果粗最细，分别为 2.43 cm、2.31 cm，其他羊角椒品种果粗范围在 2.50~3.11 cm。牛角椒品种中，"超级椒王"品种果肉厚最厚为 4.61 mm，"朗月 407"品种果肉厚次之，为 4.51 mm，"盛龙"品种果肉厚最薄为 3.25 mm，其他辣椒品种果肉厚分别在 3.67~4.45 mm。羊角椒品种中，"甘科 10 号"品种果肉厚最厚为 2.86 mm，"7061"品种果肉厚最薄为 2.12 mm，其他辣椒品种果肉厚范围在 2.29~2.61 mm。牛角椒品种中，"朗月 407"品种单果质量最大为 106 g，"盛龙"品种单果质量最小为 67.79 g，其他牛角椒品种单果质量分别在 80.30~101.56 g；羊角椒品种中，"0701"羊角椒品种单果质量最大为 80.82 g，"螺美 7 号"品种单果质量最小为 40.26 g，其他辣椒品种单果质量分别在 46.94~60.90 g。牛角椒品种中，"盛龙"、"宝盛 6 号"品种单株结果数最多，分别为 41.2 个、45.0 个，"超级椒王"品种的单株结果数最少为 30.2 个，其他辣椒品种单株结果数分别在 30.8~33.6 个；羊角椒品种中，"37-94"品种的

表 3 不同辣椒品种果实性状比较

Table 3 Comparison of fruit characters of different pepper varieties

	品种 Variety	果长 Fruit long /cm	果粗 Fruit coarse /cm	果肉厚 Pulp thick /mm	单果质量 Fruit quality /g	单株结果数 Number of fruit per plant/个	果柄长 Pedunclelong /cm	萼片宽 Sepals wide /cm	果色 Fruit color	果味 Fruity
牛角椒	鑫牛角	24.98	4.55	4.35	92.55	30.8	5.24	2.44	翠绿色	微辣
	鑫牛角 30	24.78	4.51	4.45	101.56	31.0	6.00	2.42	翠绿色	微辣
	盛龙	23.68	3.66	3.25	67.79	41.2	5.56	2.27	黄绿色	微辣
	宝盛 6 号	23.0	4.13	4.00	80.30	45.0	5.58	2.43	绿色	微辣
	日本瑞崎	25.3	3.56	3.83	81.07	33.6	6.10	2.44	黄绿色	微辣
	巨峰 1 号	23.0	4.16	3.67	93.11	32.4	5.90	2.41	绿色	微辣
	超级椒王	23.0	4.88	4.61	101.46	30.2	6.18	2.50	绿色	微辣
	坂田神椒	22.56	4.48	3.67	94.41	31.6	5.70	2.23	绿色	微辣
	朗月 407	28.3	4.73	4.51	106.0	30.8	7.84	2.79	黄绿色	微辣
	朗月 206	26.2	4.54	4.20	98.53	31.5	6.46	2.48	绿色	微辣
	37–94	35.9	2.43	2.49	46.94	52.4	5.22	2.19	翠绿色	微辣
	6139	30.0	3.11	2.12	60.90	36.8	6.18	2.33	翠绿色	辣味浓
	7061	33.8	3.04	2.30	50.96	37.2	5.02	2.45	翠绿色	微辣
羊角椒	螺美 7 号	34.66	2.68	2.61	40.26	47.8	4.88	2.50	绿色	无味
	0701	31.74	3.43	2.45	80.82	38.0	5.82	2.82	绿色	辣味浓
	7062	27.42	2.6	2.31	51.80	33.2	4.66	2.23	翠绿色	辣味浓
	7063	31.46	2.84	2.86	47.22	48.6	4.90	2.62	翠绿色	辣味浓
	甘科 10 号	31.94	3.32	2.50	60.67	36.2	4.06	2.52	深绿色	辣味强
	华美 105	32.14	2.58		54.87	43.0	5.80	2.29	绿色	辣味浓

单株结果数最多为52.4个，"7063"品种单株结果数次之，为48.6个，"7062"品种单株结果数最少为33.2个，其他辣椒品种单株结果数范围在36.8~47.8个。牛角椒品种中，"朗月407"品种的果柄长最长为7.84 cm，"鑫牛角"品种的果柄长最短为5.24 cm，其他辣椒品种分别在5.56~6.46 cm；羊角椒品种中，"6139"品种的果柄长最长，为6.18 cm，"甘科10"品种的果柄长最短为4.06 cm，其他辣椒品种果柄长分别在4.66~5.80 cm。牛角椒品种中，"朗月407"品种萼片最宽为2.79 cm，"坂田神椒"品种萼片最窄，为2.23 cm，其他辣椒品种萼片宽分别在2.27~2.50 cm；羊角椒品种中，"0701"羊角椒品种萼片最窄为2.82 cm，"37-94"品种萼片宽最小为2.19 cm，其他辣椒品种萼片宽分别在2.29~2.62 cm。牛角椒品种中，"鑫牛角"和"鑫牛角30"品种的辣椒果色为翠绿色，"盛龙"、"日本瑞崎"和"朗月407"品种的辣椒果色为黄绿色，"宝盛6号"、"巨峰1号"、"超级椒王"、"坂田神椒"和"朗月206"品种的辣椒果色为绿色，果味均微辣；羊角椒品种中，"37-94"、"6139"、"7061"、"7062"和"7063"品种的辣椒果色为翠绿色，"螺美7号"、"0701"和"华美105"品种的辣椒果色为绿色，"甘科10号"品种的辣椒果色为深绿色，"37-94"、"6139"和"螺美7号"品种的果味为微辣，"7061"、"7062"、"7063"和"华美105"品种辣味浓，"甘科10号"品种辣味强，"0701"品种无辣味。

2.4 不同辣椒品种产量与产值比较

从表4可以看出，不同辣椒品种产量产值有差异，参试牛角椒品种中，"宝盛6号"单株产量、小区产量、折合亩产量、产值均最高，分别为3.61 kg/株、1 256.3 kg/23.4 m²、10 474.4 kg/667 m²、21 682元/667 m²，比对照产量增幅16.3%，产值增加3 041元/667 m²，"朗月407"单株产量、小区产量、折合亩产量、产值次之，分别为3.26 kg/株、1 134.5 kg/23.4 m²、9 458.9 kg/667 m²、19 579.9元/667 m²，比对照产量增幅5%，产值增加938.9元/667 m²，"日本瑞琦"单株产量、小区产量、折合亩产量、产值均最小，分别为2.72 kg/株、946.6 kg/23.4 m²、7 892.3 kg/667 m²、16 337.1元/667 m²；参试羊角椒品种中，

表 4 不同辣椒品种产量与产值比较

Table 4 Comparison of yield and output value of different pepper varieties

	品种 Variety	单株产量 Yield per plant /kg	小区产量 Cell production /kg	折合 667 m² 产量 Equivalent to 667 m² output/kg	比对照增产 Specific stimulation±/kg	比对照增幅 Specific increase±/%	折合 667 m² 产值 Output value equivalent to 667 m²/元	位次 Precedence
牛角椒	鑫牛角	2.85	991.8	8 269.1	-736.2	-8.2	17 117	8
	鑫牛角 30	3.15	1 096.2	9 139.6	134.3	1.5	18 919	3
	盛龙	2.79	970.9	8 095.0	-910.3	-10.1	16 756.7	9
	宝盛 6 号	3.61	1 256.3	10 474.4	1469.1	16.3	21 682	1
	日本瑞崎	2.72	946.6	7 892.3	-1113	-12.4	16 337.1	10
	巨峰 1 号	3.02	1 050.0	8 754.4	-250.9	-2.8	18 121.6	6
	超级椒王	3.06	1 064.9	8 878.6	-126.7	-1.4	18 378.7	5
	坂田神椒	2.98	1 037.0	8 646.0	-359.3	-4	17 897.2	7
	朗月 407	3.26	1 134.5	9 458.9	453.6	5	19 579.9	2
	朗月 206	3.10	1 080.1	9 005.3	—	—	18 641.0	4
	37-94	2.46	952.0	7 134.7	1566.1	28.1	12 771.1	2
	6139	2.24	866.9	6 496.9	928.3	16.7	11 629.5	4
	7061	1.90	735.3	5 510.6	-58	-1	9 864.0	11
	螺美 7 号	1.92	743.0	5 568.6	—	—	9 967.8	10
羊角椒	0701	3.07	1 188.1	8 904.1	3335.5	59.9	15 938.3	1
	7062	1.72	665.6	4 988.3	-580.3	-10.4	8 929.1	12
	7063	2.29	886.2	6 641.5	1072.9	19.3	11 888.3	3
	甘科 10 号	2.20	851.4	6 380.7	812.1	14.6	11 421.5	6
	华美 105	2.36	913.3	6 844.6	1276.0	22.9	12 251.8	5

"0701"辣椒单株产量、小区产量、折合亩产量、产值均最高,分别为 3.07 kg/株、1 188.1 kg/23.4 m²、8 904.1 kg/667 m²、15 938.3 元/667 m²,比对照产量增幅 59.9%,产值增加 5 970.5 元/667 m²,"37-94"辣椒单株产量、小区产量、折合亩产量、产值次之,分别为 2.46 kg/株、952.0 kg/23.4 m²、7 134.7 kg/667 m²、产值 12 771.1 元/667 m²,比对照产量增幅 28.1%、产值增加 2 803.3 元/667 m²,"7062"辣椒单株产量、小区产量、折合亩产量、产值最小,分别为 1.72 kg/株、665.6 kg/23.4 m²、4 988.3 kg/667 m²、8 929.1 元/667 m²。

3 结论与讨论

本试验为了筛选出适宜宁夏拱棚栽培的优质、抗病、丰产、抗逆、色泽好的辣椒优良品种,根据宁夏地区的气候特点、消费习惯等,我们搜集、引进国内外不同地区的 19 个辣椒品种进行田间适应性品比试验。赵维[9]等研究选出"川辣 2 号"品种适宜在毕节地区推广种植;陈久爱[10]等研究发现"苏椒 27 号"的前期产量、总产量、抗病性、商品性状均优于对照品种"苏椒 5 号",而且"苏椒 27 号"早熟、果大肉厚、果表面微皱、辣度中等、品质好,适应性强,综合性状表现突出,在市场上很受欢迎,值得在本地春季设施大棚推广种植;赵明国[11]等研究发现"松江红 8 号",适宜在莱芜地区种植;高昱[12]等研究发现,"泰斗"、"朗月 2 号" 2 个品种的综合性状较好,增产潜力大,生长势较强,椒果商品性好,折合亩产量及效益显著高于对照"亨椒 1 号",应在进一步试验示范的基础上进行大面积推广种植;张立青[13]等研究发现"苏椒 15 号"、"农大 24 号"、"中椒牛角 1 号"、"神剑 23"和"金樽"具有产量高、商品性好、抗病性强等特点,是北京地区春大棚春季栽培首选品种;蔡明清[14]研究发现"鄂红椒 108"产量与对照"中椒 6 号"存在显著差异,从果形上看果柄处无凹陷,炭疽病发生明显较轻,商品性状好,并在试验地周围几家农户小面积示范种植,得到了周边广大农户及蔬菜经销商认可,有望成为鄂西南山区主栽品种"中椒 6 号"的替代品种。

本试验在相同的栽培和管理条件下，配套水肥一体化技术，通过对19个不同辣椒品种的物候期、植株形态指标、果实性状指标等综合评定得出以下结论：参试牛角椒品种"朗月407"、"鑫牛角30"，羊角椒品种"0701"、"37-94"辣椒性状典型，青果色泽鲜亮，坐果力强，单果质量、产量较高，适宜宁夏拱棚春提早、秋延后栽培，并推广应用。

参考文献

[1] 毛海文.试析高海拔地区日光温室辣椒无公害栽培技术[J].农业开发与装备,2016(8):134-135.

[2] 黄智文,徐晓美,王恒明,等.辣椒品种的耐热性研究试验[J].上海蔬菜,2014(6):11-12.

[3] 戴雄泽,刘志敏.初论我国辣椒产业的现状及发展趋势[J].辣椒杂志,2005(2):1-6.

[4] 高怀春.辣椒果实维生素C含量变化的研究[M].泰安:山东农业大学,2004.

[5] 邹学校.辣椒高效栽培技术[M].长沙:湖南科技出版社,1997.

[6] 徐珊珊.辣椒对盐碱胁迫的生理反应及适应性机理研究[D].长春:吉林农业大学,2007.

[7] 张海丽,黄文美.辣椒新品种引种适应性研究[J].耕作与栽培,2015(S1):30+13.

[8] 王继榜.我国辣椒产业现状及发展趋势综述[J].安徽农学通报,2013.

[9] 赵维,程娜,付毅,罗丽琴.蔬菜大棚专用辣椒品种筛选试验探讨[J].现代园艺,2019(06):4-5.

[10] 陈久爱,古松,李丹家,袁向阳,万志芳.阳新县春季设施大棚辣椒新品种比较试验[J].长江蔬菜,2015(12):15-16.

[11] 赵明国,尹爱国,李立国.19个辣椒新品种筛选试验[J].蔬菜,2018(06):75-78.

[12] 高昱,吴雪梅,海生广,马德俊,孟苞.大棚牛角辣椒品种筛选试验[J].农业科技通讯,2014(05):131-134.

[13] 张立青,徐全明,孙志远,徐峥,陈宗玲,钟连全.春大棚12个牛角椒品种比较试验[J].种子世界,2018(07):90-92.

[14] 蔡明清.辣椒新品种引进筛选试验[J].中国果菜,2018,38(04):25-28.

不同年限连作对拱棚辣椒光合特性和果实品质的影响

高晶霞[1],牛勇琴[2],吴雪梅[2],王学梅[1],裴红霞[1],谢 华[1]

(1. 宁夏农林科学院种质资源研究所,宁夏银川 750002;
2. 宁夏回族自治区彭阳县蔬菜产业发展服务中心,宁夏彭阳 756500)

摘 要:选取连作1 a、3 a、5 a 和7 a 的土壤,以相邻玉米田土壤为对照,进行盆栽辣椒试验。研究不同年限连作对辣椒叶绿素含量、光合特性、果实品质以及生理生化特性的影响。结果表明:第五年的辣椒叶绿素含量、荧光参数 F0、Fm、Fv 和 F0/Fm 最高,分别为 65.17 SPAD、693.9、2813.0、2118.9 和 0.249,第七年荧光参数 Fv/Fm 和 CK 荧光参数 Fv/F0 最大,分别为 0.76 和 3.21;CK 的辣椒叶片净光合速率最高为 21.23 ummol·m^{-2}·s^{-1},第七年辣椒叶片蒸腾速率、气孔导度和细胞间 CO_2 浓度最高,分别为 5.98 mmol·m^{-2}·s^{-1}、1 467.50 mol·m^{-2}·s^{-1} 和 345.03 μmol/mol;第一年辣椒叶片可溶性糖含量、可溶性蛋白含量最高,分别为 0.71%、1.18%,第七年辣椒叶片的过氧化氢酶活性最强为 0.061 μg/g·min,果实维生素 C 含量最高为 125.03 g/100 g;第一年辣椒叶片的丙二醛含量最高为 0.26 mmol/g,CK 辣椒叶片的丙二醛含量最低为 0.11 mmol/g;第三年辣椒果实的可溶性糖含量最高为 0.86 g/100 g,CK 辣椒果实的可溶性蛋白含量最高为 0.41 g/100 g,第一年和第七年的最低为 0.27 g/100 g。综合分析结果:随着连作年限的增加,辣椒叶绿素含量、光合特性、果实品质以及生理生化特性表现出先增后降的趋势,为防止拱棚辣椒连作障碍提供理论依据。

关键词:不同年限;辣椒;连作;光合作用;果实品质

Effects of different years of continuous cropping on photosynthetic characteristics and fruit quality of gongpeng pepper

GAO Jing-xia[1], NIU Yong-qin[2], WU Xue-mei[2], WANG Xue-mei[1], PEI Hong-xia[1], XIE-hua[1]

(1. Institute of germplasm resources, ningxia academy of agriculture and forestry sciences, Ningxia yinchuan, 750002; 2. Ningxia hui autonomous region pengyang county vegetable industry development service center, Ningxia Peng Yang, 756500)

Abstract: Select the soil of continuous cropping 1, 3, 5 and 7 a, The adjacent corn field soil was used as the control, Pot pepper experiment. The effects of different years on chlorophyll content, photosynthetic characteristics, fruit quality and physiological and biochemical characteristics of pepper were studied. The results show that: The chlorophyll content and fluorescence parameters F0, Fm, Fv and F0/Fm were the highest in pepper in 5 a, 65.17 SPAD, 693.9, 2813, 2118.9 and 0.249, respectively, The fluorescence parameters Fv/Fm and CK fluorescence parameters Fv/F0 were the largest, They were 0.76 and 3.214, respectively; The highest net photosynthetic rate of CK pepper leaves was 21.23 ummol·m^{-2}·s^{-1}, The transpiration rate, stomatal conductance and intercellular CO2 concentration were the highest in the leaves of pepper 7 a, 5.98 mmol·m^{-2}·s^{-1}, 1467.50 mol·m^{-2}·s^{-1} and 345.03 mol/mol, respectively; The soluble sugar content and soluble protein content of pepper leaves in no. 1 a were the highest, which were 0.71% and 1.18%, respectively, The strongest catalase activity in the leaves of pepper 7 a was 0.061 g/g·min, and the highest content of vitamin C in the fruits was 125.03 g/100 g; The highest malondialdehyde content in the leaves of pepper no. 1 a was 0.26 mmol/g, The lowest malondialdehyde in CK pepper leaves was 0.11 mmol/g; The soluble sugar content of the third a pepper fruit was 0.86 g/100 g, The soluble protein content of CK pepper fruit was the highest at 0.41 g/100 g, the minimum for 1a and 7a is 0.27 g/100 g. The results showed that the chlorophyll content, photosynthetic characteristics, fruit quality and physiological and biochemical characteristics of pepper increased first and then decreased with the increase of continuous cropping years,It provides theoretical basis for taking effective measures to ease the obstacle of continuous cropping of capsicum arch.

Key words:different years; pepper; continuous cropping; photosynthesis; fruit quality

辣椒是茄科辣椒属一年生草本植物,富含维生素及氨基酸等营养物质,具有解热、镇痛、增加食欲、帮助消化、降脂减肥,预防肿瘤等功效[1],其果实营养丰富、味道鲜美而在世界各地广泛栽培,年种植面积130万 hm^2 以上[2]。设施大棚能克服低温季节对作物生长的限制且方便种植管理，在人们的生产生活中扮演着不可或缺的角色[3],目前辣椒已成为设施栽培的主要蔬菜种类之一,是我国许多省市县的重要经济支柱作物[4]。辣椒属于不耐连作的茄果类蔬菜之一[5]。但由于设施农业的集约化生产,许多地区出现了不同程度的连作障碍,严重制约着设施生产的可持续发展[6]。据上海、无锡、南京、淮阴、济南、北京等地的调查,玻璃温室如使用不当,2~3 a就出现连作障碍,塑料大棚约5 a出现程度不同的连作障碍,减产在20%~50%,甚至达到70%。因此,土壤连作障碍已成为设施辣椒可持续生产的主要限制因子之一。

拱棚辣椒是宁夏南部山区特色优势主导产业,但因为固定设施,辣椒栽培的全产业链模式已较为完善，产业对连续优质丰产栽培的技术需求成为必然,土壤连作障碍成为拱棚辣椒预期产量质量形成的瓶颈问题,尤其是因连作导致植株死株及发育不良,目前尚无针对性的系统研究和成熟技术。本研究选取连续种植辣椒 1 a、3 a、5 a、7 a 的大棚土壤进行盆栽试验,以相邻玉米田为对照,对辣椒生理指标、光合特性、品质指标进行测定和分析,旨在为研究辣椒连作障碍机理和克服辣椒连作障碍提供科学依据。

1 材料和方法

1.1 材料与处理

选取彭阳县新集乡白河村连作辣椒 1 a、3 a、5 a、7 a 的拱棚土壤为不同处理,以相邻玉米田土壤为对照(CK)。在塑料大棚内用所取土样进行辣椒盆栽试验,设 5 个处理,每个处理 10 盆,重复 3 次,供试品种为"巨峰 1 号"(宁夏巨丰种苗有限公司提供),于 2019 年 1 月 20 日育苗,4 月 15 日进行盆栽,定植后 30d 各处理分别取样测定。

1.2 测定方法

1.2.1 辣椒叶片渗透物质含量和抗氧化酶活性

可溶性糖含量采用蒽酮法测定;可溶性蛋白采用考马斯亮蓝 G-250 染色法测定;CAT 活性采用 Dhindsa 等的方法测定。MDA 含量采用改进的硫代巴比妥酸法测定。

1.2.2 辣椒叶片叶绿素含量测定

定植 30 d 后,标记辣椒生长点下第二片展开功能叶,用日本 SPAD-502Plus 叶绿素测定仪进行叶绿素含量的测定。

1.2.3 光合和荧光的测定

定植 30 d 后,采用便携式光合测定系统(Li-6400,美国)于 10:00~11:00 进行光合参数的测定,净光合速率(Pn)、气孔导度(Gs)、胞间 CO_2 浓度(Ci)和蒸腾速率(Tr)由光合测定系统直接读出,每叶重复测定 3 次,取平均值。荧光采用 PAM2100 调制荧光仪(德国 Walz 公司生产)进行测定。

1.2.4 辣椒果实品质测定

维生素 C 含量采用 2,6-二氯靛酚法;可溶性总糖含量采用苯酚法测定;可溶性蛋白含量采用考马斯高蓝 G-250 染色法;可滴定酸含量采用 NaOH 滴定法。每个样品重复测定 3 次。

1.3 数据处理

数据用 Excel 2019 和 DPS 软件进行单因素方差分析，多重比较采用 LSR 法(Duncan's 法)，显著水平P<0.05。

2 结果与分析

2.1 连作对辣椒叶绿素含量的影响

叶绿素是作物进行光合作用的必要物质,光饱和点以下,叶片叶绿素含量与光合速率呈正相关[7]。由图1可知,不同的处理辣椒叶片叶绿素含量有差异,处理3辣椒叶片的叶绿素含量最高为65.17 SPAD,处理2辣椒叶片叶绿素含量最低为56.61 SPAD，处理4和CK的辣椒叶片叶绿素含量差异不明显,分别为62.96 SPAD 和64.33 SPAD,与CK相比,叶绿素含量在连作1 a、3 a 年和7 a 时均下降,在连作5 a 有所上升。

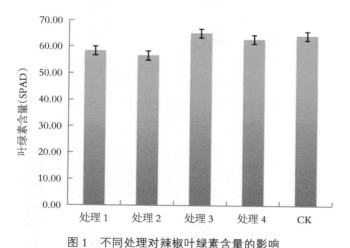

图1 不同处理对辣椒叶绿素含量的影响

Fig.1 Effects of different treatments on chlorophyll content of pepper

2.2 连作对辣椒叶片光合作用的影响

由图2~5可知,不同处理的辣椒叶片蒸腾速率(Tr)有差异。处理4辣椒叶片的蒸腾速率最高为5.98 mmol·m^{-2}·s^{-1},处理2的辣椒叶片蒸腾速率最低为4.64 mmol·m^{-2}·s^{-1}, 处理1和处理3的辣椒叶片蒸腾速率差异不明显;不同的处理辣椒叶片净光合速率(Pn)有差异,CK 辣椒叶片净光合速率最高为 21.23 μmol·m^{-2}·s^{-1},处理 1 辣椒叶片净光合速率最低为 19.41 μmol·m^{-2}·s^{-1},处理 3 和处理 4 的叶片净光合速率差异不明显,分别为 19.67 μmol·m^{-2}·s^{-1} 和 19.74 μmol·m^{-2}·s^{-1}。净光合速率的降低必然导致同化物积累量的下降,使植株的生长势减弱和干物质积累减小。不同处理的辣椒叶片气孔导度(Gs)有差异。处理 4 的辣椒叶片气孔导度最高为 1 467.50 mmol·m^{-2}·s^{-1},处理 1

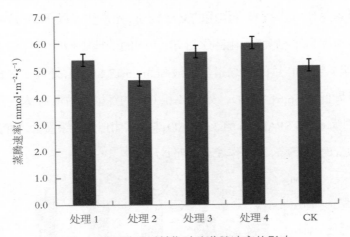

图 2　不同处理对辣椒叶片蒸腾速率的影响

Fig.2 Effects of different treatments on transpiration rate of pepper

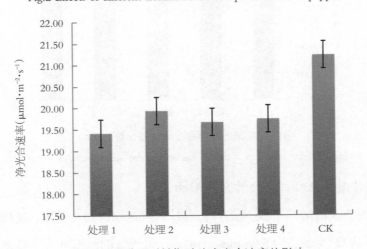

图 3　不同处理对辣椒叶片净光合速率的影响

Fig. 3 Effects of different treatments on net photosynthetic rate of pepper

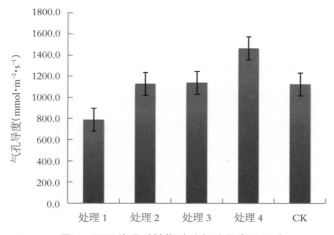

图 4 不同处理对辣椒叶片气孔导度的影响
Fig. 4 Effects of different treatments on stomatal conductance of pepper

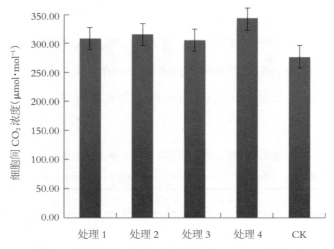

图 5 不同处理对辣椒叶片细胞间 CO_2 浓度的影响
Fig. 5 Effects of different treatments on intercellular CO_2 concentration in pepper

的辣椒叶片气孔导度最低为 788.20 mmol·m^{-2}·s^{-1}，处理 2、处理 3 和 CK 的气孔导度差异不明显。气孔导度（Gs）的降低会使光合底物传导能力降低，必然影响光合作用的正常进行。不同处理的辣椒叶片细胞间 CO_2 浓度（Ci）有差异，处理 4 辣椒叶片细胞间 CO_2 浓度最高为 345.03 μmol·mol^{-1}，CK 辣椒叶片胞间 CO_2 浓度最低为 276.27 μmol·mol^{-1}，处理 1、处理 2 和处理 3

的辣椒叶片细胞间 CO_2 浓度（Ci）差异不明显，分别为 308.70 $\mu mol \cdot mol^{-1}$、316.20 $\mu mol \cdot mol^{-1}$ 和 305.43 $\mu mol \cdot mol^{-1}$。

2.3 连作对辣椒叶片荧光的影响

由表1可知，不同处理辣椒叶片荧光参数有差异，处理3的荧光参数F0最大为693.93，处理1的荧光参数F0最小为648，其他处理的荧光参数F0分别在654.5~673.27；处理3的荧光参数Fm最大为2 812.83，处理1的荧光参数Fm最小为2 685.2，其他处理的荧光参数Fm分别为2 734.57~2 783.1；处理3的荧光参数Fv最大为2 118.9，处理1的荧光参数Fv最小为2 037.2，其他处理的荧光参数Fv分别在2 067.17~2 109.83；不同处理辣椒叶片荧光参数F0/Fm无差异，处理3的荧光参数F0/Fm最大为0.249，处理4和CK的荧光参数F0/Fm最小为0.24；不同处理辣椒叶片荧光参数Fv/Fm无差异，处理4和CK的荧光参数Fv/Fm最大为0.76，其他处理的荧光参数Fv/Fm分别在0.751~0.756；处理4的荧光参数Fv/F0最大的为3.214，处理3的荧光参数Fv/F0最小为3.092，其他处理的荧光参数Fv/F0无明显差异，分别在3.14~3.168。

表1 不同处理对辣椒叶片荧光参数的影响

Table 1 Effects of different treatments on fluorescence parameters of pepper leaves

处理/指标	F0	Fm	Fv	F0/Fm	Fv/Fm	Fv/F0
处理1	648	2685	2037.2	0.244	0.756	3.168
处理2	667.4	2735	2067.2	0.247	0.753	3.140
处理3	693.9	2813	2118.9	0.249	0.751	3.092
处理4	654.5	2750	2095.5	0.240	0.760	3.214
CK	673.3	2783	2109.8	0.240	0.760	3.160

2.4 连作对辣椒叶片渗透物质含量和抗氧化酶活性的影响

由图6~9可知，不同处理对辣椒叶片可溶性糖含量有较明显差异，处理1辣椒叶片可溶性糖含量最高为0.71%，CK可溶性糖含量最低为0.41%，其他处理辣椒可溶性糖含量在 0.45%~0.69%；不同处理对辣椒叶片可溶性蛋白含量有明显差异，处理 1 辣椒叶片可溶性蛋白含量最高为 1.18%，其他处理辣椒可溶性蛋白含量在 0.12%~0.3%；不同处理对辣椒叶片过氧化氢酶活性有明显差异，处理 4 辣椒叶片过氧化氢酶活性最高为 0.061 μg/g·min，处理 3 辣椒叶片过氧化氢酶活性最低为 0.018 μg/g·min，其他处理辣椒叶片过氧化氢酶活性无明显差异，在 0.036~0.039 μg/g·min；不同处理对辣椒叶片丙

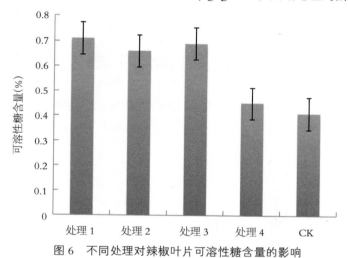

图 6 不同处理对辣椒叶片可溶性糖含量的影响

Fig. 6 Effects of different treatments on soluble sugar content of pepper leaves

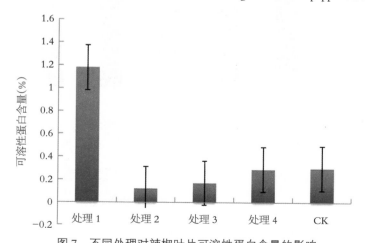

图 7 不同处理对辣椒叶片可溶性蛋白含量的影响

Fig. 7 Effects of different treatments on soluble protein content of pepper leaves

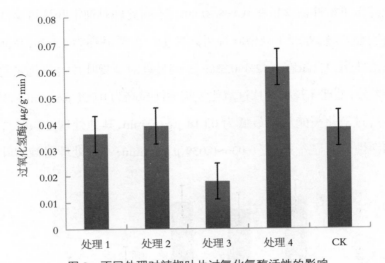

图 8　不同处理对辣椒叶片过氧化氢酶活性的影响

Fig. 8 Effects of different treatments on catalase activity in pepper leaves

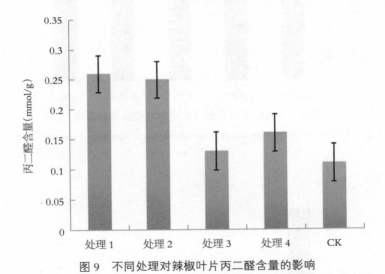

图 9　不同处理对辣椒叶片丙二醛含量的影响

Fig. 9 Effects of different treatments on malondialdehyde content in pepper leaves

二醛含量有明显差异,处理 1 和处理 2 辣椒叶片丙二醛含量最高,分别为 0.26 mmol/g 和 0.25 mmol/g,处理 3、处理 4 和 CK 辣椒叶片丙二醛含量无明显差异,分别为 0.13 mmol/g,0.16 mmol/g 和 0.11 mmol/g。

2.5 连作对辣椒果实品质的影响

由图 10~12 可知,在辣椒采收盛期,不同处理对辣椒果实品质有影响,处理 4 辣椒果实维生素 C 含量最高为 125.03 g/100 g,处理 2 辣椒维生素 C 含量次之,为 107.33 g/100 g,其他 2 个处理辣椒果实维生素 C 含量为处理 3>处理 1,处理 1 辣椒维生素 C 含量最低,为 83.33 g/100 g,其中处理 2 和

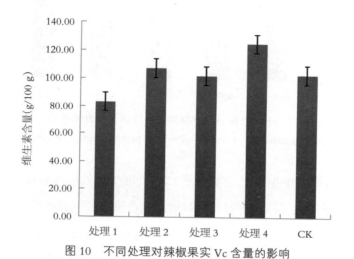

图 10　不同处理对辣椒果实 Vc 含量的影响

Fig. 10 Effects of different treatments on Vc content of pepper fruitssugar

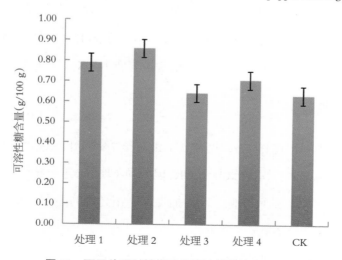

图 11　不同处理对辣椒果实可溶性糖含量的影响

Fig. 11 Effects of different treatments onsoluble content of pepper fruits

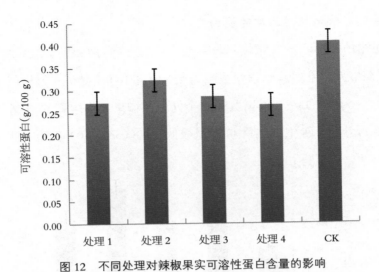

图 12　不同处理对辣椒果实可溶性蛋白含量的影响

Fig.12 Effects of different treatments on soluble protein content of pepper fruits

处理 4 的辣椒果实维生素 C 含量高于 CK,处理 1 和处理 3 的辣椒果实维生素 C 含量低于 CK;不同处理对辣椒可溶性糖含量有较明显差异,处理 2 辣椒果实可溶性糖含量最高为 0.86 g/100 g,CK 可溶性糖含量最低为 10.63 g/100 g,其他处理辣椒可溶性糖含量在 0.64~0.79 g/100 g,且各个处理的辣椒可溶性糖含量均高于 CK;不同处理对辣椒可溶性蛋白含量有明显差异,CK 辣椒可溶性蛋白含量最高,为 0.419 g/100 g,其他处理可溶性蛋白含量在 0.27~0.32 g/100 g,且各个处理辣椒可溶性糖含量均低于 CK。

3　讨论与结论

连作条件下,连作 1 a、连作 3 a 和连作 7 a 的叶绿素含量低于对照(CK),连作 5 a 的叶绿素含量高于 CK,但差异不明显,连作导致叶片光合色素质量分数的降低,从而降低了叶片捕捉和利用光能的能力,同时也影响了光能在叶绿体中的分配。这可能是由于连作造成土壤理化性状恶化,造成植株根系的吸收障碍,使其体内营养水平降低,从而导致叶片叶绿素质量分数降低[8]。

光合作用是植物形成生物产量的生理代谢基础[9]。连作条件下各个处理的蒸腾速率均高于对照,净光合速率均低于对照,细胞间 CO_2 浓度均高于对照,连作 1 a,气孔导度低于对照,连作 3 a 以上时,气孔导度升高。这可能是前茬辣椒根系分泌物对下茬辣椒的生长起到激发效应,但具体情况还有待于进一步研究[10]。通常认为,叶片净光合速率降低是由气孔因素和非气孔因素所致,依据 Farquhar 等的理论[11],可以认为长期连作造成辣椒的净光合速率降低是由非气孔因素引起的。

蛋白质是植物体生命过程中重要的结构物质和功能物质,其代谢受多种因素的影响和调控,越来越多的证据表明,变化了的环境因子或环境胁迫都会影响蛋白质代谢[12]。叶片可溶性蛋白的含量直接影响到酶的活性,并最终影响叶片的生理功能。细胞中的可溶性糖是大多数植物碳运输的主要形式,参与作物的库源调节,是光合作用生产的直接产物[13]。本试验研究发现,连作 1 a 的可溶性蛋白明显高于对照,其他连作年限的可溶性蛋白含量均低于对照,不同连作年限的辣椒叶片可溶性糖含量均高于对照,可能是在连作条件下辣椒植株通过植物体内渗透调节物质合成途径中一些酶的活性增强,促进了渗透调节物质的合成,从而增强了辣椒植株的渗透调节能力。

超氧化物歧化酶(SOD)、过氧化物酶(POD)和过氧化氢酶(CAT)都是细胞膜系统的保护酶,可在逆境胁迫时增强活性,加快对活性氧的清除,具有维持活性氧代谢平衡,保护膜结构的功能,是植物能忍耐外界不良环境的机理之一[14,15]。本试验研究发现,连作 5 a 辣椒叶片的过氧化氢酶活性明显低于对照,其他连作年限的过氧化氢酶活性均高于对照,这表明在连作 5 a 时,辣椒植株可以通过提高内保护酶(过氧化氢酶)活性来维持活性氧代谢平衡,保持膜系统的稳定。

Culter[16]认为,小麦叶片 MDA 的积累速率可代表组织中总的清除自由基能力的大小,MDA 积累越多,表明组织的保护能力越弱,衰老程度越深。本试验研究发现,不同连作年限的丙二醛含量均高于对照,这说明在连作条件下,辣椒植株体内积累了大量的 MDA,其膜脂过氧化作用加剧,衰老速度提高。

有学者认为,连作年限较长时,将导致其维生素 C 和可溶性固形物的质量分数均显著降低[17]。本试验连作条件下,不同处理辣椒果实可溶性蛋白均低于对照,可溶性糖含量均高于对照,连作 3 a 和连作 7 a 的维生素 C 含量高于对照,连作 1 a 和连作 5 a 的维生素 C 含量低于对照,这是由于连作显著降低了辣椒叶片叶绿素质量分数,影响其光合性能,从而影响辣椒果实的品质。

参考文献

[1] 唐胜球,董小英,邹晓庭. 辣椒素研究及其应用[J]. 江西饲料,2003(3):13-16.

[2] 邹学校. 中国辣椒[M]. 北京:中国农业出版社,2002:1-60.

[3] 刘蕾. 我国设施农业发展现状与对策分析[J]. 农业科技与装备,2013(4):57-58.

[4] 马艳青. 我国辣椒产业形势分析[J]. 辣椒杂志,2011(1):1-5.

[5] 张周让,上官金虎. 宝鸡市辣椒产业发展现状、问题与对策[J]. 陕西农业科学,2005(4):64-66.

[6] 杨琳,王克雄,王斐. 宁南山区设施辣椒连作障碍因子及防控措施[J]. 现代农业科技,2016(5):115-118.

[7] 贺洪军,造泽生,尚爱军,等. 西瓜光合特性研究[J]. 中国西瓜甜瓜.1994(1):18-20.

[8] 梁更生,赵春燕,赵国良,等. 连作对大棚辣椒生长发育及品质的影响[J]. 甘肃农业科技,2016(5):50-53.

[9] 陈有根,章敏,王冬良,等. 不同连作年限土壤对甜瓜幼苗叶绿素含量和光合特性的影响[J]. 2009,37(3):956-958.

[10] FARQUHARG D,SHARKEY T D.Stomatal conductance and photosynthesis [J]. Annual Review of Plant Physiology,1982,33:317-345.

[11] 李锐,李生泉,范月仙. 不同抗冷级别的棉苗低温诱导蛋白比较[J]. 棉花学报,2010,22(3):254-259.

[12] Longstreth D J. Nutrient influence on leaf photosynthesis effects on nitrogen phosphorus and potassium for Gossipy hirsutrum[J]. Plant Physiology,1980,65(3):541-543.

[13] 郑佳秋,顾闽峰,郭军,祖艳侠,吴永成,梅燚,季芳芳. 涝渍胁迫下辣椒的生理特性

[J].江苏农业学报,2012,28(03):617-621.

[14] 黄玉茜,韩晓日,杨劲峰,刘小虎,白洪志.连作胁迫对花生叶片防御酶活性及丙二醛含量的影响[J].吉林农业大学学报,2013,35(06):638-645.

[15] 李会云,郭修武.盐胁迫对葡萄砧木叶片保护酶活性和丙二醛含量的影响[J].果树学报,2008(02):240-243.

[16] Culter R G. Antioxidant, aging and longevity[J]. Free Radicals in Biology & Medicine, 1984(6):371-428.

[17] 吴凤芝,孟立君,文景芝.黄瓜根系分泌物对枯萎病菌菌丝生长的影响[J].中国蔬菜.

辣椒连作对拱棚土壤微生物多样性的影响

高晶霞[1],牛勇琴[2],吴雪梅[2],王学梅[1],裴红霞[1],谢 华[1]

(1. 宁夏农林科学院种质资源研究所,宁夏银川 750002;
2. 宁夏回族自治区彭阳县蔬菜产业发展服务中心,宁夏彭阳 756500)

摘 要:本研究选取连续种植辣椒 1 a、3 a、5 a、7 a 的大棚土壤进行盆栽试验,以相邻玉米田为对照,采用高通量测序技术,测定土壤 16S(细菌)和 ITS(真菌),分析土壤微生物群落变化。结果表明:16S 分析结果,对不同处理组样本的 OTUs 和 Tags 数量的统计,总 Tags 数量平均 89 245 条,有物种注释的 Tags 数量平均 70 870 条,没有物种注释的 Tags 为 0 条,unique tags 的数量平均为 67 027 条,总丰度为 1 的 OTUs 所对应的 Tags 会被过滤掉,平均为 4044 条;A 处理组 vsB 处理组 vsC 处理组 vsD 处理组 vs 对照 CK:A 处理组特有 1002 条 OTUs,B 处理组特有 1 244 条 OTUs,C 处理组最低,特有 563 条 OTUs,D 处理组和对照 CK 特有 831 条 OTUs 和 803 条 OTUs;不同处理组样本门水平的注释率在 95% 以上;Xanthomonadaceae 为各处理组样本在科水平占比最高的菌群,为第一优势菌群,B 处理组和 D 处理组较 A 处理组、C 处理组和对照(CK)组较低,同目水平一致。Tepidisphaeraceae 在对照 CK 组中最高,在各处理组中显著降低,其中 A 处理组最低,按照 A、B、C、D 处理组的顺序进行升高。Blastocatellaceae 在对照 CK 中较高,在各处理组中不同程度降低,有可能是该菌群的降低增加的连作障碍,可作为候选菌群进行研究。Comamonadaceae 在 A 处理组和 B 处理组中较 CK 组低,C 处理组和 D 处理组较对照(CK)组不变,该菌群没有随着连作年限的增加而增加,在对照(CK)组中含量偏高,但在 A 处理组和 B 处理组含量很低,未能解释该原因,可能存在互作关系;Alpha 多样性分析,A 处理组最大值 9.997 9,最小值 9.267 2,B 处理组最大值 10.024 8,最小值 6.107 2,C 处理组最大值 9.408 1,最小值 8.896 8,D 处理组最大值 9.746 5,最小值 9.084 8,对照(CK)组最大值 9.149 9,最小值 9.036 4。B 处理组波动最大,和 B-1 样本特异性有关;Beta 多样性分析,二维PCA图可以看出C处理组和对照(CK)组最为接近,D处理组最远,其次是B处理组和A处理组,各个样本除了B-1例外,其他样本聚类效果良好。跟进连作的特点,C处理组与对照(CK)最为接近,说明该组的菌群由于连作,先趋于平缓,后逐渐和对照(CK)接近,D处理组由于连作时间长,和对照(CK)距离远;物种差异分析结果,A处理组中显著差异的前3类菌群分别是 Erythrobacteraceae、Hyphomicrobiaceae 和 Devosia,B处理组中差异显著的物种有5种,但前3类菌群差异显著,分别是 Clostridiales、Bacteroidia 和 Bacteroidales。C 处理组前 3 类显著差异的菌群是 Caulobacterales、Caulobacteraceae 和 Mexicana,D 处理组前 3 类显著差异的菌群是 RB41、Thermoactinomyces 和 Chloroflexia,而对照CK组中前3类显著差异的菌群是 Luteimonas、Burkholderiales 和 Stenotrophobacter。ITS 分析结果,A 处理 vsB 处理 vsC 处理 vsD 处理 vs 对照CK:A处理组特有65条OTUs,B处理组最低,特有98条OTUs,C处理组特有84条OTUs,D处理组和对照CK特有87条OTUs和134条OTUs,5组相比CK组特有最多,也符合真菌和连作时间的增加菌种的变化特点;物种分类分析,每个处理样本在门水平的注释率在45%~75%;科水平上的物种分布堆叠图,Lasiosphaeriaceae为科水平丰度最高的菌群,在对照(CK)组样本中占比达到了25%,其余处理组样本中丰度平均在5%左右。Nectriaceae为主要关注菌群,在对照(CK)组、A处理组和B处理组中的丰富较高,随着连作年限的增加含量显著降低,Pichiaceae的丰度情况和Nectriaceae的正好相反,在C处理组和D处理组中的丰度最高,在A处理组、B处理组和对照(CK)中的丰度显著降低,该两组菌群可作为后期的主要研究对象,对连作的菌群问题进行详细的解释;Alpha 多样性分析,A 处理组最大值 5.157,最小值 4.484,B 处理组最大值 5.001,最小值 4.388,C 处理组最大值 5.357,最小值 4.291,D 处理组最大值 4.771,最小值 3.967,对照 CK 组最大值 5.530,最小值 5.255。对照(CK)组的shannon值最高,从盒形图明显可以看出随着处理时间的增加shannon值越来越低;Beta 多样性分析, 基于 Unweighted Unifrac 距离的目水平 UPGMA 聚类树,从图中可以看出各组样品的聚类效果显著,A 处理组和B处理组聚为一类,C处理组和D处理组聚为一类,对照CK组单独一类,说明A处理组和B处理组的处理时间接近且较短,C处理组和D处理组较远;物种差异分析,对照(CK)组的差异菌群最多,其次是A处理组的样本,D处理组没有显著差异的菌群,C处理组和D处理组各有一类菌群,有效的说明连作时间越长,真菌的菌群越趋于单一,甚至于多组间比较没有差异菌群,B 处理组和 C 处理组的 Chaetomium_aureum 和 Hrysosporium_carmichaelii 可作为有益菌群来关注,连作时间的增加,该两类菌群逐渐降低到无,同理 A 处理组中这五类真菌也可作为关注点,可优先关注得分较高的菌群。对照(CK)组的菌群种类是最多的,说明空白对照的菌群丰度最高,连作时间增加,菌群丰度降低,菌群趋于单一化,减少的菌群就可以作为我们改良土壤或者监控土壤变化的有力依据,可为土壤改良打下良好的数据基础。

关键词:不同年限;拱棚;微生物多样性

Effects of Pepper Continuous Cropping on Soil Microbial Diversity In Arch Canopy

Gao jingxia[1], Niu yongqin[2], Wu xuemei[2], Wang xuemei[1], Pei hongxia[1], Xie hua[1]

(1. *Institute of germplasm resources, ningxia academy of agricultural andforestrysciences, yinchuan, ningxia* 750002, *China*; 2. *Vegetableindustrydevelopment service center of pengyang county, ningxia hui autonomous region, pengyang* 756500, *China*)

Abstract: In this study, greenhouse soil with continuous planting of pepper 1 a, 3 a, 5 a and 7 a was selected for pot experiment, High-throughput sequencing technology was used to determine soil 16S (bacteria) and ITS (fungi) and analyze the changes of soil microbial community. The results show that: 16S analysis results show that: The number of OTUs and Tags in samples of different processing groups is 89 245 on average, the number of Tags with species annotation is 70 870, the number of Tags without species annotation is 0, the number of Tags with unique Tags is 67 027 on average, and the number of Tags corresponding to OTUs with a total abundance of 1 will be filtered out, with an average of 4 044; Treatment group A vsB treatment group vsC treatment group vsD treatment group vs control group CK: treatment group A unique 1 002 OTUs, treatment group B unique 1 244 OTUs, C treatment group the lowest, unique 563 OTUs, D treatment group and control group CK unique 831 OTUs and 803 OTUs; The annotation rate of sample gate level in different treatment groups was more than 95%. Xanthomonadaceae was the flora with the highest proportion of samples at the family level in each treatment group, and was the first dominant flora. The treatment groups B and D were lower than the treatment groups A, C and CK, with the same order level. Tepidisphaeraceae was the highest in the control CK group, and significantly decreased in each treatment group, of which the lowest was in the treatment group A, and increased in the order of A, B, C and D treatment groups. Blastocatellaceae was higher in control CK, and decreased to different degrees in each treatment group, which may be a continuous cropping obstacle to the decrease and increase of this flora, and can be studied as a candidate flora.

Comamonadaceae was lower in the treatment group A and the treatment group B than the CK group, and the treatment group C and the treatment group D than the control group CK group. The flora did not increase with the increase of continuous cropping years, and the content was higher in the control group CK, but low in the treatment group A and the treatment group B, which could not explain the reason. According to Alpha diversity analysis, the maximum value of 9.997 9 and the minimum value of 9.267 2 in group A, the maximum value of 10.024 8 and the minimum value of 6.107 2 in group B, the maximum value of 9.408 1 and the minimum value of 8.896 8 in group C, the maximum value of 9.746 5 and the minimum value of 9.084 8 in group D, and the maximum value of 9.149 9 and the minimum value of 9.036 4 in group CK. The fluctuation was the largest in the B treatment group, which was related to the specificity of b−1 samples; According to the analysis of Beta diversity, the two−dimensional PCA plot shows that the C treatment group is the closest to the CK control group, the D treatment group is the furthest, and the B treatment group and A treatment group are the next. All samples except b−1 have good clustering effect. Following up the characteristics of continuous cropping, the C treatment group was the closest to the control CK, indicating that the flora of this group tended to be flat first and then close to the control CK due to continuous cropping. The D treatment group was far away from the control CK due to the long continuous cropping time; According to the results of species difference analysis, the first three bacterial communities with significant differences in treatment group A were Erythrobacteraceae, Hyphomicrobiaceae and Devosia, and there were 5 species with significant differences in treatment group B, but the first three bacterial communities were Clostridiales, Bacteroidia and Bacteroidales. Caulobacterales, Caulobacteraceae and Mexicana were the first three significantly different flora in the C treatment group, RB41, Thermoactinomyces and Chloroflexia were the first three significantly different flora in the D treatment group, while Luteimonas, Burkholderiales and Stenotrophobacter were the first three significantly different flora in the CK control group. ITS analysis results, A treatment vsB vsC treatment vsD treatment vs control CK: A treatment group of 65 unique OTUs, minimum B treatment group, 98 special OTUs, C treatment group of 84 unique OTUs, D treatment group and the contrast specific article 87 OTUs and CK 134 OTUs, 5 compared with CK group the most unique, also accord with the additional strains of fungi and continuous cropping time change characteristic; For taxonomic analysis of species, the annotation rate of each treated sample at the gate level was between 45% and 75%; According to the stacking diagram of species distribution at the family level, Lasiosphaeriaceae is the flora with the highest family level abundance, accounting for 25% in the samples of the control CK group, and 5% on average in the samples of the other treatment groups. Nectriaceae as the main focus on flora, in contrast to CK group, A rich in treatment group and B group is higher, along with the increasing length of continuous cropping content decreased significantly, the abundance and the opposite of Nectriaceae Pichiaceae, in treatment group C and D the highest abundance in treatment group, in treatment group and B group and A control CK decreased significantly in abundance, the two groups of bacteria can be used as the main research object, in the late to flora of continuous cropping is A detailed explanation; In the analysis of Alpha diversity, the maximum value of group A was 5.157, the minimum value was 4.484, the maximum value was 5.001, the minimum value was 4.388, the maximum value was 5.357, the minimum value was 4.291, the maximum value was 4.771, the minimum value was 3.967, and the maximum value was 5.530, the minimum value was 5.255.Compared with CK group, shannon value was the highest, and it was obvious from the box diagram that shannon value decreased with the increase of treatment time;Beta diversity analysis, based on the Unweighted Unifrac eye level of the distance UPGMA clustering tree, can be seen from the diagram of each sample clustering effect is remarkable, A get together for A class of treatment group and B group, C gathered for A class of treatment group and D group, comparing CK group separate category, suggests A treatment group and B group close to the processing time and short, C treatment group and D group far away;The CK control group had the most bacterial species, indicating that the blank control group had the highest bacterial abundance, the continuous cropping time increased, the bacterial abundance decreased, and the bacterial population tended to be simplified. The reduced bacterial population could serve as a strong basis for us to improve the soil or monitor the soil changes, and lay a good data foundation for soil improvement.

辣椒是茄科辣椒属一年生草本植物,富含维生素及氨基酸等营养物质,具有解热、镇痛,增加食欲、帮助消化,降脂减肥,预防肿瘤等功效[1],其果实营养丰富、味道鲜美而在世界各地广泛栽培,年种植面积130万 hm^2 以上[2]。设施大棚能克服低温季节对作物生长的限制且方便种植管理,在人们的生产生活中扮演着不可或缺的角色[3],目前辣椒已成为设施栽培的主要蔬菜种类之一,是我国许多省市县的重要经济支柱作物[4]。辣椒属于不耐连作的茄果类蔬菜之一[5]。但由于设施农业的集约化生产,许多地区出现了不同程度的连作障碍,严重制约着设施生产的可持续发展[6]。据上海、无锡、南京、淮阴、济南、北京等地的调查,玻璃温室如使用不当,2~3 a就出现连作障碍,塑料大棚约5 a出现程度不同的连作障碍,减产在20%~50%,甚至达到70%。因此,土壤连作障碍已成为设施辣椒可持续生产的主要限制因子之一。

拱棚辣椒是宁夏南部山区特色优势主导产业,但因为固定设施,辣椒栽培的全产业链模式已较为完善,产业对连续优质丰产栽培的技术需求成必然,土壤连作障碍成为拱棚辣椒预期产量质量形成的瓶颈问题,尤其是因连作导致植株死株及发育不良,目前尚无针对性的系统研究和成熟技术。本研究选取连续种植辣椒 1 a、3 a、5 a、7 a 的大棚土壤进行盆栽试验,以相邻玉米田为对照,采用高通量测序技术,测定土壤 16S 和 ITS,分析土壤微生物群落变化,旨在为研究辣椒连作障碍机理和克服辣椒连作障碍提供科学依据。

1 材料和方法

1.1 材料与处理

选取彭阳县新集乡白河村连作辣椒 1 a、3 a、5 a、7 a 的拱棚土壤为不同处理,以相邻玉米田土壤为对照(CK)。在塑料大棚内用所取土样进行辣椒盆栽实验,设 5 个处理,每个处理 10 盆,重复 3 次,供试品种为"巨峰 1 号"(宁夏巨丰种苗有限公司的品种提供),于 2019 年 1 月 20 日育苗,4 月 15 日进行盆栽,定植后 30 d 各处理分别取样测定。

1.2 测定方法

土壤样品中细菌 16S rRNA 和真菌 ITS2 基因的高通量测序由广州 GeneDenovo 生物信息技术有限公司利用 Illumina HiSeq2500 平台进行。利用引物对 16S V3-V4 区域进行扩增:341F(CCTACGGGNGGCWGCAG)和 806R (GGACTACHVGGGTATCTAAT)。使用正向引物 KYO2F(GATGAAGAACGYAGYRAA) 和反向引物 ITS4RT(CCTCCGCTTATTGATATGC)扩增 ITS2 区。所有 PCR 反应进行的 15 μL Phusion 高保真 PCR 反应混合液(新英格兰生物学实验室);正向和反向引物 0.2 μM,大约 10 ng 模板 DNA。热循环包括 98℃初始变性 1 min,98℃变性 10 s,50℃退火 30 s,72℃延伸 30 s,共 30 个循环。72℃ 5 min,将等量的 SYB green 缓冲液(含 SYB green)与 PCR 产物混合,2%琼脂糖凝胶电泳检测。PCR 产物按等密度比混合。然后用 Qiagen 凝胶提取试剂盒(Qiagen,德国)对混合 PCR 产

物进行纯化。使用 truseq®DNA PCR-Free Sample Preparation Kit（Illumina，USA）按照制造商的建议添加索引代码，生成测序文库。在 Qubit 2.0 荧光仪（Thermo Scientific）和安捷伦 2100 生物分析仪（Agilent Bioanalyzer 2100）上进行库质量评估。

2 结果分析

2.1 土壤 16S 数据分析

2.1.1 不同处理组样本数据处理分析

图 1 为数据预处理的分布图，reads QC filter 为低质量 reads，右边为红色图标，占比小于 1%；Non-overlap 为没有 overlap 的未组装 reads，右边为紫色图标，占比小于 2%；tag QC filter 为未通过"tag 过滤"的 tags，低质量、未组装和未通过的 tags 总量小于总数据量的 5%；Chimera 为嵌合体；effective tags 为

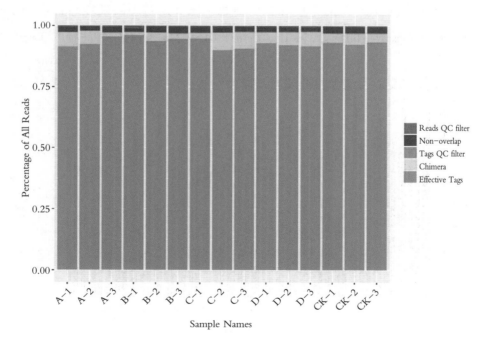

图 1　不同处理样本数据预处理分布图（百分比）

Fig.1 Distribution diagram of pretreatment of sample data with different processing（percentage）

有效用于后续分析的 tags,每个样品的占比都在 90%以上,有效 Tags 数量占据原始 PE Reads 的百分比从样本 A-1 到 CK-3 分别为 91.39%、92.31%、95.45%、96%、93.71%、94.52%、94.70%、90.12%、90.75%、92.96%、92.14%、91.73%、93.17%、92.48%和 93.34%。有效 Tags 数据量是足够进行后续分析,可靠数据。Tags 丰度统计显示经过 Tags 质控后得到的高质量 Tags 数量最低的为 88000 条,最高的为 99 458 条,Total length 最低的为 3700 万,Max length 为 474,Minlength 最低 314,最高 425,N50 为 460 左右,N90 为 441。从上述统计数据来看,下机数据足够,过滤后有效数据充足,为后续分析做好了基础。

2.1.2 不同处理组样本 OTU 分析

为了研究样本的物种的组成多样性信息,用 Uparse 软件对所有样本的全部 Effective Tags 序列聚类,默认提供以 97%的一致性(Identity)将序列聚类成为 OTUs(Operational Taxonomic Units)结果,并计算出每个 OTU 在各个样本中的 Tags 绝对丰度和相对信息。图 2 是对不同处理组样本的 OTUs 和 Tags 数量的统计,总 Tags 数量平均 89 245 条,有物种注释的 Tags

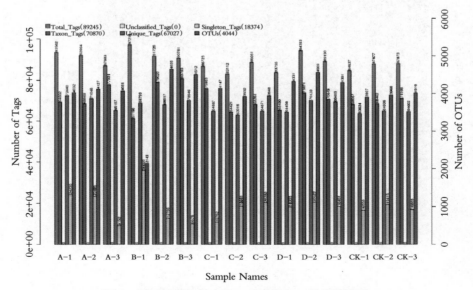

图 2　不同处理组样本的 OTUs 和 Tags 数量统计图

Figure.2 Quantity statistics of OTUs and Tags for different processing samples

数量平均70 870条,没有物种注释的Tags为0条,unique tags数量平均为67 027条,总丰度为1的OTUs所对应的Tags会被过滤掉,平均为4 044条。

不同生境下微生物群落,其物种分布存在一定程度的相似性和特异性。在多分组(或样本)情况下,为了解不同分组(或样本)之间OTU差异情况,我们可以根据OTU丰度信息开展韦恩图分析,从而了解不同样本或者分组之间OTU的共有或者特有信息。选取比较分组平均丰度大于1的所有OTU(即高丰度并集OTU)进行韦恩图分析。图3是各处理组之间共有和特有OTUs韦恩图和各个集合数量统计柱状图,空白对照CKvsA处理组:A处理组样本特有2 213条OTUs,空白对照CK样本特有1 796条OTUs,两处理组共有2 276条OTUs;空白对照CKvsB处理组:B处理组特有2 331条OTUs,空白对照CK特有1 585条OTUs,两处理组共有2487条OTUs;空白对照CKvsC处理组:C处理组特有1 554条OTUs,对照CK特有1 672条OTUs,两处理组共有2 400条OTUs;空白对照CKvsD处理组:D处理组特有1 931条OTUs空白对照CK特有1 561条OTUs,两处理组共有2 511条OTUs。A处理组vsB处理组vsC处理组vsD处理组vs对照CK:A处理组特有1 002条OTUs,B处理组特有1 244条OTUs,C处理组最低,特有563条OTUs,D处理组和对照CK特有831条OTUs和803条

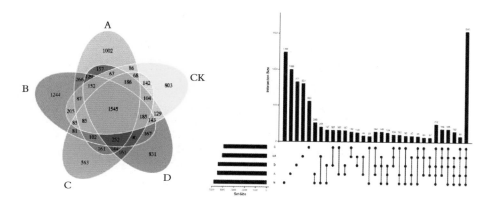

图3 不同处理组样本OTUs比较韦恩图
Fig.3 Comparison of sample OTUs of different treatment groups

OTUs。右图下方连线表示共有,单独一点表示特有。

2.1.3 不同处理组样本物种分类分析

微生物物种分类一般分为界、门、纲、目、科、属、种7个等级,而每个OTU代表某类型分类水平集合。因此根据OTU序列信息进行物种注释,能将分析结果与实际生物学意义进行关联,从而研究群落中物种的变化关系等内容。

物种注释方法:Uparse 在构建 OTUs 过程中会选取代表性序列(OTUs 中丰度最高的那条 Tag 序列),将这些代表性序列集合用 RDP Classifier 的 Naïve Bayesian assignment 算法,与 silva 数据库进行物种注释(设定置信度的阈值为 0.8~1)。表 1 为物种注释表,每个处理样本纲水平注释都在 60 000 条

表1 不同处理组样本物种注释 Tags 数量统计表
Table 1 Statistical table of quantity of Tags of sample species in different processing groups

处理样本	Domain	Phylum	Class	Order	Family	Genus	Species
A-1	69 332	68 886	65 507	60 705	53 636	33 723	3 445
A-2	68 569	67 981	63 967	58 253	52 082	28 098	2 480
A-3	77 931	77 003	72 396	65 888	59 382	35 598	3 759
B-1	61 196	60 949	60 528	44 313	43 340	29 784	4 147
B-2	79 020	77 766	73 619	66 932	58 211	36 125	2 376
B-3	80 783	79 203	74 243	67 316	59 370	31 343	3 782
C-1	75 963	75 597	70 625	65 894	58 591	34 333	2 485
C-2	64 421	63 983	60 255	56 300	50 985	31 739	4 131
C-3	68 281	67 952	64 585	61 216	55 639	34 848	3 289
D-1	65 400	64 873	60 417	55 525	48 112	29 711	3 356
D-2	73 871	73 363	68 905	64 112	56 428	38 083	2 297
D-3	70 649	69 909	65 783	60 957	53 839	35 218	2 928
CK-1	68 087	67 858	63 493	58 954	50 633	28 767	1 038
CK-2	68 462	68 073	64 352	60 326	51 989	29 976	1 300
CK-3	71 088	70 724	66 998	62 550	54 376	31 339	1 326

OTUs以上,目水平,除了处理组中B-1样本44 313条,其余都在55 000条以上,属水平平均30 000条,种水平1 038条到4 147条不等。

图4是不同处理组样本在各分类水平上序列构成柱形图,从图中可以看出门水平注释率在95%以上,纲水平注释率除了B-1样本,注释率都在90%以上,目水平注释率都在85%以上,科水平注释率都在75%以上,注释率较高,属水平注释率都在50%左右。

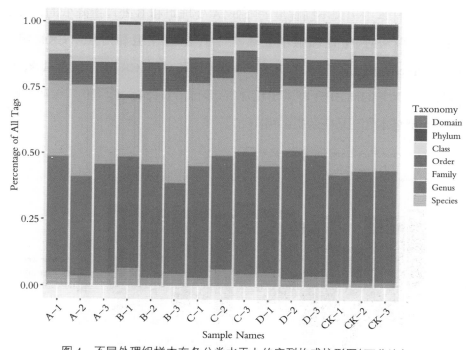

图4 不同处理组样本在各分类水平上的序列构成柱形图(百分比)
Fig.4 Histogram (percentages) of sequences of samples from different treatment groups at each classification level

优势物种很大程度上决定着微生物群落生态结构以及功能结构,了解群落在各个水平物种组成情况能有效地对群落结构的形成、改变以及生态影响等进行解读。我们统计了各个层级分类水平上处理样本的物种组成情况,然后用堆叠图的形式,直观展示不同的样品在各个分类层级水平上的物种丰度的变化情况。图5为不同处理样本目分类水平上的物种分布堆叠图,

Xanthomonadales 为各处理样本的第一优势菌群，在 A 处理组、C 处理组和对照(CK)组中较多，B 处理组和 D 处理组相对较低。Gemmatimonadales 在对照(CK)中和 A 处理组中最高，其次是 C 处理组，B 处理组和 D 处理组中最低。Sphingomonadales 在处理组中含量基本相等，对照(CK)含量最高，该菌种可能由于处理的原因出现降低，是重点关注的菌群。Planctomycetales 菌群在各处理样本的变化浮动不大，Tepidisphaerales 在对照(CK)中含量较高，在处理组样本中出现不同程度的降低，A 处理组和 B 处理组降低的最多。Bacillales 菌群在 D 处理组中显著高于其他处理组和对照（CK）组，Clostridiales 在 A 处理组、C 处理组、D 处理组和对照(CK)组中低含量，在 B 处理组中尤其的高，B-1 样本出现异常，该菌群占比超过了 30%，可能是取的该土壤样本含有某些杂质影响。

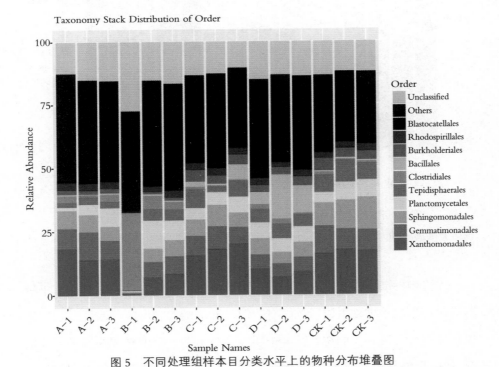

图 5 不同处理组样本目分类水平上的物种分布堆叠图
Fig. 5 Stacking plots of species distribution at the taxonomic level of sample orders of different treatment groups

图6为科水平上的物种分布堆叠图，Xanthomonadaceae 为各处理样本占比最高的菌群，为第一优势菌群，B处理组和D处理组较A处理组、C处理组和对照（CK）组较低，同目水平一致。Planctomycetaceae 在各处理样本的变化浮动不大，不是我们要关注的菌群。Tepidisphaeraceae 在对照（CK）组中最高，在各处理组中显著降低，其中A处理组最低，按照A、B、C、D处理组的顺序进行升高。Blastocatellaceae 在对照（CK）中较高，在各处理组中不同程度降低，有可能是该菌群的降低增加的连作障碍，可作为候选菌群进行研究。Comamonadaceae 在A处理组和B处理组中较CK组低，C处理组和D处理组较对照（CK）组不变，该菌群没有随着连作年限的增加而增加，在对照（CK）组中含量偏高，但在A处理组和B处理组含量很低，未能解释该原因，可能存在互作关系。

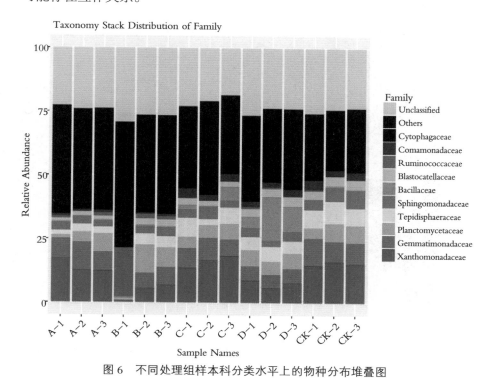

图6 不同处理组样本科分类水平上的物种分布堆叠图

FIg. 6 Species distribution stack diagram at the undergraduate classification level of different treatment groups

图7为不同处理组样本属水平上的物种分布堆叠图,可以看到Bacillus属在D处理组中尤为显著,在C处理组中略微升高,该菌属可以假定作为连作时间增加与菌种正相关的判断依据,根据功能特性进行判断是否和连作相关。RB41随着连作时间的增加逐渐增加,但对照(CK)组中依旧含量较高。

我们根据物种分类的表达谱数据,使用热图来展示不同的物种在各处理样本间的表达情况,同时根据热图上的聚类关系,也可以反映样本关系。为了减少噪音数据的影响,热图分析所挑选物种需要满足以下2个条件:(1)物种在至少一个样本的相对丰度(物种tags数/总tags数)达到0.1%以上;(2)选取满足条件(1)的相对总丰度在前25的物种进行热图分析。图8为不同处理样本属水平的物种分类热图,结果和图5、图6、图7不同处理样

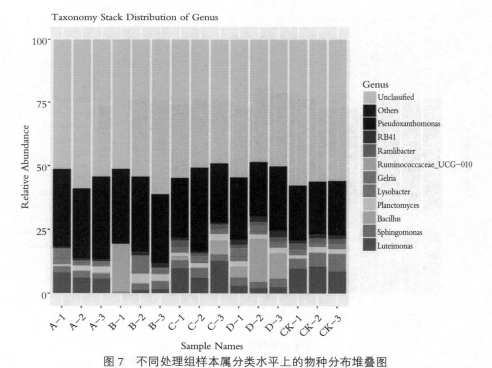

图7 不同处理组样本属分类水平上的物种分布堆叠图

Fig. 7 Stacking plots of species distribution at the level of genus classification in different treatment groups

本堆叠图一致，只是堆叠图显示前 10 的优势菌群，热图显示很多，并做了均一化的处理，以堆叠图为准即可。若进行详细数据挑选，也可用热图进行。

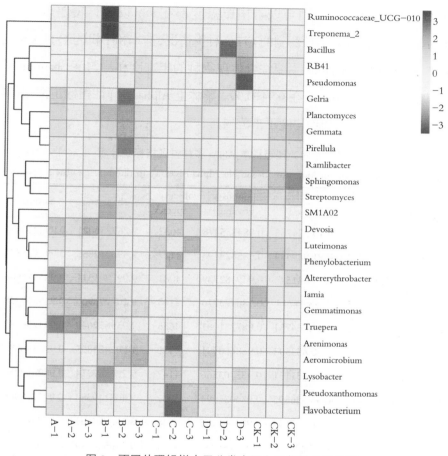

图 8　不同处理组样本属分类水平上的物种分类热图

Fig. 8 Heat maps of species classification at the genus classification level of samples from different treatment groups

2.1.4　不同处理组样本 Alpha 多样性分析

α 多样性是指特定生境或者生态系统内的多样性情况，它可以指示生境被物种隔离的程度，通常利用物种丰富度（种类情况）与物种均匀度（分布情况）2 个重要参数来计算。表 3 主要展示 Chao1，ACE，Shannon，observed species，Simpson 和 Good's Coverage 六大类常用的 α 多样性指数以及它们的相关分析结果。总体来说,Chao1 / ACE 指数主要关心样本的物种丰富度信息,各处理样本除 B-1 样本都在 4 500 以上;Good's Coverage 反映样本的低丰度 OTU 覆盖情况,从表中可以看出覆盖度都在 98% 以上,覆盖度很高;observed_species 表示能检测到的 OTU 种类情况,基本和下机的 OTUs 相差不多;Simpson / Shannon 主要综合体现物种的丰富度和均匀度,shannon 基本都在 9.2 以上,Simpson 在 0.94~0.99。整体上 Alpha 多样性指标正常,低丰度覆盖度高,丰富度和均匀度正常范围。

表 2 不同处理组样本 Alpha 多样性各指标统计表

Table 2 Statistical table of indicators of sample Alpha diversity in different treatment groups

处理样本	chao1	ace	goods_coverage	observed_species	shannon	simpson
A-1	4750.962141	4913.274755	0.984855478	4032	9.267256021	0.992216471
A-2	4777.008264	4891.989474	0.985824498	4127	9.596212856	0.994423375
A-3	4717.026882	4705.431546	0.989182739	4081	9.997991511	0.996817808
B-1	2549.979275	2788.41745	0.988773776	2143	6.107292966	0.945569278
B-2	5385.655172	5381.364751	0.987231081	4655	9.959652649	0.994730289
B-3	5143.129518	5160.853593	0.988660981	4512	10.02484865	0.996830097
C-1	5023.71875	5070.431151	0.985703566	4147	9.296587314	0.989632253
C-2	4667.825419	4795.986595	0.984057994	3932	9.408149384	0.993778086
C-3	4805.006729	4978.087552	0.983465386	3948	8.896874107	0.982940478
D-1	4979.969816	5087.422031	0.984785933	4331	9.746546452	0.995403273
D-2	5424.8	5519.514464	0.98420219	4569	9.084817256	0.982472722
D-3	5174.675749	5279.029111	0.983962972	4301	9.287055244	0.99111886
CK-1	4648.154867	4733.889657	0.985268847	3907	9.061163083	0.987071009
CK-2	4742.375546	4832.16038	0.984925944	3968	9.036444305	0.985847425
CK-3	4694.947137	4753.872161	0.986495611	4019	9.149981942	0.989199852

我们通过绘制稀释曲线(rarefaction curve)来评价测序量是否足以覆盖所有类群,并间接反映处理样本中物种的丰富程度。稀释曲线是利用已测得

序列中已知的 OTU 的相对比例,来计算抽取 n 个(n 小于测得 tags 序列总数)tags 时出现 OTU 数量的期望值,然后根据一组 n 值(一般为一组小于总序列数的等差数列)与其相对应的 OTU 数量的期望值做出曲线来。当曲线趋于平缓或者达到平台期时也就可以认为测序深度已经基本覆盖到样品中所有的物种。图 9 是各处理样本的稀释曲线,从图中可以看出,当测序的 Tags 数量到达 40 000 条时曲线趋于平缓到达平台期,说明测序的质量的合格的,深度和数据量足够。B-1 样本也在 40 000 条时到达平台期,前面数据已经说明,该样本中某类菌群的丰度很高,导致 OTUs 总数较低,属于特殊样本,但数据量角度来说,数据已经足够。

我们通过绘制 shannon 稀释曲线(shannon rarefaction curve)来评价测序量是否足够,并间接反映样品中物种的丰富程度。shannon 指数作为一个评价样本内物种多样性程度的指标,值越高代表多样性程度越高。shannon 稀释曲线是通过抽样 n 个 tags 来计算 shannon 指数的期望值,然后根据一组 n

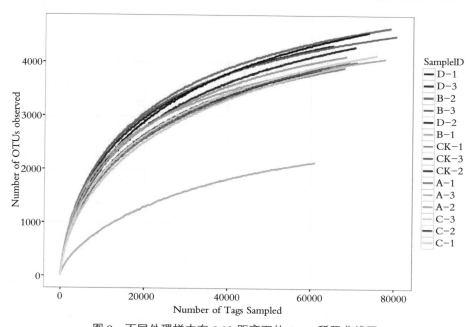

图 9　不同处理样本在 0.03 距离下的 OTU 稀释曲线图
Fig.9 OTU dilution curves of samples with different treatments at a distance of 0.03

值(一般为一组小于总序列数的等差数列)与其相对应的 shannon 的期望值做出曲线来,当曲线趋于平缓或者达到平台期时也就可以认为测序深度增加已经不影响物种多样性,测序量趋于饱和。图 10 为各样本的 shannon 稀释曲线,结论和图 9 的稀释曲线相同,测序数据量到达 40 000 条 Tags 后趋于稳定和平缓,数据足够。

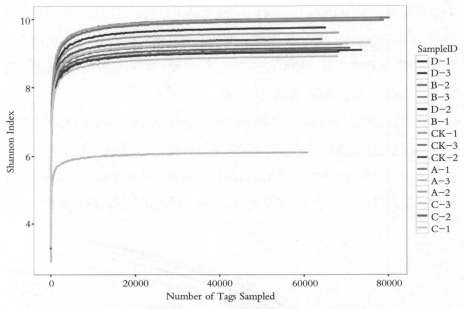

图 10　不同处理样本在 0.03 距离下的 shannon 稀释曲线图

Fig.10 Shannon dilution curves of samples with different treatments at a distance of 0.03

不同生境下的环境驱动因素能引起微生物 α 多样性差异。结合分组和采样信息,通过对 2 组或者多组间的 α 多样性进行假设检验,可以分析组间的物种多样性是否存在显著的差异,从而初步判断驱动群落多样性变化的潜在因素等。我们同时使用以下几种常见的假设检验方法来进行差异分析。(1)针对 2 个分组进行比较时,使用 T-test 检验和 wilcox 秩和检验;(2)针对 2 个以上分组进行比较时,使用 Tukey 检验和 Kruskal-Wallis 秩和检验。在进行差异分析的同时,我们还绘制了各个 Alpha 多样性指数的箱线图(下图展示了各个比较组的 shannon 指数箱线图)。箱形图可以直观的反应组内物种多样性的中位数、离散程度、最大值、最小值、异常值。图 11 是对各处理组样本汇总的 shannon 指数箱型图,图形展示有效数据分布的最大值(直线顶端)、最小值(直线底端)、中位数(盒子中线)、上四分位数(盒子顶边)、下四分位数(盒子底边)以及无效数据(直线外离群散点)。A 处理组最大值 9.997 9,最小值 9.267 2,B 处理组最大值 10.024 8,最小值 6.107 2,C 处理组最大值 9.408 1,最小值 8.896 8,D 处理组最大值 9.746 5,最小值 9.084 8,对照(CK)组最大值 9.149 9,最小值 9.036 4。B 处理组波动最大,和 B-1 样本特异性有关。

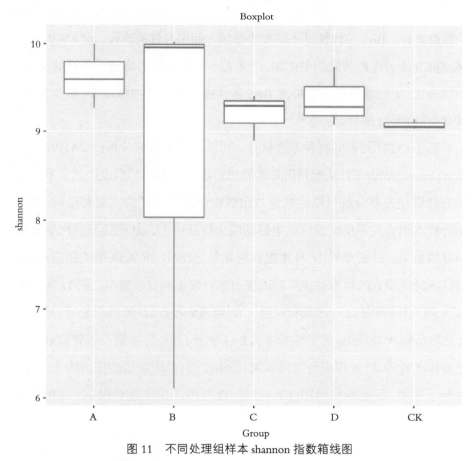

图 11　不同处理组样本 shannon 指数箱线图

Fig.11 Shannon index boxplot of samples from different treatment groups

2.1.5 不同处理组样本 Beta 多样性分析

Beta Diversity 是对不同样本的微生物群落构成进行比较。首先根据所有样本的 OTU 序列,使用软件 Muscle(v3.8.31)进行多序列比对,然后使用软件 TreeBeST(v1.9.2)构建 OTU 之间的系统发育树。再结合 OTU 的丰度信息,使用 R 语言中的 GUniFrac(v1.0)包来计算两两样本间的 Unweighted Unifrac 和 Weighted Unifrac 距离。最后,通过多变量统计学方法主坐标分析(PCoA,Principal Co-ordinates Analysis),NMDS,非加权组平均聚类分析(UPGMA,Unweighted Pair-group Method with Arithmetic Means)等分析,进一步从结果中挖掘各样品间微生物群落结构的差异和不同分类对样品间的贡献差异。Beta 多样性是不同生态系统之间多样性的比较,是物种组成沿环境梯度或者在群落间的变化率,用来表示生物种类对环境异质性的反应。一般来说,不同环境梯度下群落 Beta 多样性计算包括物种改变(多少)和物种产生(有无)2 部分。

基于 OTU 列表的物种丰度信息,可以开展主成分分析(PCA,Principal Component Analysis),从而利用降维的思想研究样本间的组成距离关系。这种方法借用方差分解可以有效地找出数据中最"主要"的元素和结构,将复杂的样本组成关系反映到横纵坐标的两个特征值上,从而达到简化数据复杂度的效果。分析结果中,样本组成越相似,反映在 PCA 图中的距离越近,而且不同环境间的样本往往可能表现出各自聚集的分布情况。图 12 为二维 PCA 图,可以看出 C 处理组和对照(CK)组最为接近,D 处理组最远,其次是 B 处理组和 A 处理组,各个样本除了 B-1 例外,其他样本聚类效果良好。跟进连作的特点,C 处理组与对照(CK)最为接近,说明该组的菌群由于连作,先趋于平缓,后逐渐和对照(CK)接近,D 处理组由于连作时间长,和对照(CK)距离远。

在微生物生态研究当中,UPGMA 分类树可以用于研究样本间的相似性,解答样本的分类学问题。利用 Mothur 软件,根据 weighted 和

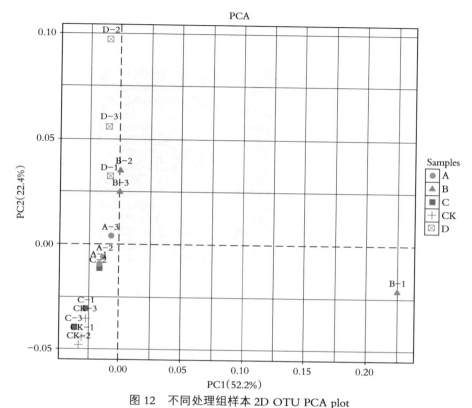

图 12　不同处理组样本 2D OTU PCA plot
Fig.12 2D OTU PCA plot of samples of different processing groups

unweighted Unifrac 矩阵信息,可以将样本进行 UPGMA 分类树分类。其中越相似的样本将拥有越短的共同分支。为了研究不同样本间的相似性,还可以通过对样本进行聚类分析,构建样本的聚类树。在环境生物学中,UPGMA25 (Unweighted Pair-group Method with Arithmetic Mean)是一种较为常用的聚类分析方法,它最早便是用来解决分类问题的。UPGMA 的基本思想是:首先将距离最小的 2 个样本聚在一起,并形成一个新的节点(新的样本),其分支点位于 2 个样本间距离的 1/2 处;然后计算新的"样本"与其他样本间的平均距离,再找出其中的最小 2 个样本进行聚类;如此反复,直到所有的样品都聚到一起,最终得到一个完整的聚类树。图 13 是基于 Unweighted Unifrac 距离的目水平 UPGMA 聚类树,从图中可以看出各组样品的聚类情况都很好,样本 B-1 单独一条。D 处理组和 C 处理组距离最近,其次是 B 处理组和对照(CK)组,A 处理组最远。

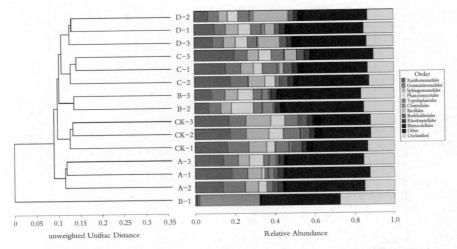

图 13　不同处理组样本基于 Unweighted Unifrac 距离的目水平 UPGMA 聚类树
Fig. 13　Horizontal UPGMA clustering tree for samples of different treatment groups based on Unweighted Unifrac distance

Analysis of Similarity（ANOSIM）分析是一种对微生物群落结构的非参数检验方法,用来检验组间的差异是否显著大于组内差异,从而判断分组是否有意义。下表是 Anosim 差异分析的结果表,其中 R 值表示差异程度,一般介于(0,1)之间,R>0,说明组间存在差异,一般 R>0.75:大差异;>0.5:中等差异,>0.25:小差异。R 等于 0 或在 0 附近(因为 R 偶尔也会<0,但一般认为是无效数据),说明组间没有差异。统计分析的可信度用 P-value 表示,P<0.05 表示统计具有显著性。表 3 是不同处理样本基于 unweighted unifrac 距离的 Anosim 分析结果表,A 处理组、B 处理组、C 处理组和 D 处理组和对照 CK 都存在差异,其中 A 处理组、C 处理组和 D 处理组与对照(CK)组是大差异,B 处理组与对照(CK)组是小差异。多组间比较 P 值为 0.001,具有显著差异。

表3 不同处理组基于 unweighted unifrac 距离的 Anosim 分析结果表

Table 3 Results of Anosim analysis of different treatment groups based on unweighted unifrac distance

Diffs	Rvalue	Pvalue	significant
CK-vs-A	1	0.1	
CK-vs-C	1	0.1	
CK-vs-D	1	0.1	
A-vs-B-vs-C-vs-D-vs-CK	0.752 6	0.001**	
A-vs-B-vs-C-vs-D	0.694 4	0.001**	

2.1.6 不同处理组样本群落功能分析

完整的微生物群落研究主要分为物种组成、多样性以及功能研究等几个重要方面。多种证据表明微生物的群落功能组成比物种组成与环境关系更为密切,随着分析技术发展,利用多样性测序数据进行群落功能预测已经成为群落研究的重要内容。我们将利用 FUNGuild、Tax4fun、FAPROTAX、BugBase 等多个预测软件,根据不同的数据类型有针对性地完成群落功能预测分析。其中 Tax4Fun 首先将 KEGG 数据库中已有基因组的原核生物 16S rRNA 序列与 SILVA 数据库中 16SrRNA 序列进行关联,然后将 KEGG 数据库已有的原核物种基因组进行序列打断,利用 UProC 对所有基因组的 KO 序列进行统计;最后利用 16S 的拷贝数对物种数目进行校正,最终实行 KEGG 预测以及 KO 丰度统计。根据 OTU 的物种注释和丰度信息,使用 Tax4Fun 软件可以进行 KEGG Pathway 的功能注释,并统计每个 Pathway 和 KOid 的丰度信息。图 14 是不同处理组样本丰度最大的 20 个 Pathway,颜色越深标明相关性越高,除去 B-1 组样本,可以看出 ko00500、ko00520、ko00680、ko02010 和 ko03440 在对照 CK 组中相关性较低,在 A 处理组、B 处理组和 D 处理组中相关性较高,可以作为重点关注的通路,推测出可能由于连作的原因,导致菌群丰度发生变化,影响到对应的代谢通路,导致辣椒产量产生改变。

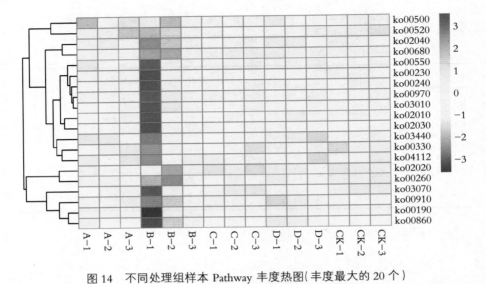

图 14 不同处理组样本 Pathway 丰度热图（丰度最大的 20 个）

Fig.14 Pathway abundance heat maps of samples from different treatment groups（the 20 samples with the largest abundance）

2.1.7 不同处理组样本物种差异分析

图 15 展示了不同组中丰度差异显著的物种，柱状图的长度代表差异物种的影响大小（即为 LDA Score）。从图中可以看出，A 处理组中显著差异的前 3 类菌群分别是 Erythrobacteraceae、Hyphomicrobiaceae 和 Devosia，B 处理组中差异显著的物种有 5 种，但前 3 类菌群差异显著，分别是 Clostridiales、Bacteroidia 和 Bacteroidales。C 处理组前三显著差异的菌群是 Caulobacterales、Caulobacteraceae 和 Mexicana，D 处理组前 3 类显著差异的菌群是 RB41、Thermoactinomyces 和 Chloroflexia，而对照（CK）组中前 3 类显著差异的菌群是 Luteimonas、Burkholderiales 和 Stenotrophobacter。从图中可以看出 B 处理组中的差异显著的菌群最少，只有 5 种。A 处理组的差异菌群最多，说明连作时间越短，差异的菌群越多，随着连作时间的增加，差异菌群趋于稳定和减少，通过筛选就能有效挑选各组的优势菌群。对照（CK）组的选取的是玉米地的样本，理论上玉米地也属于玉米一直连作的样本，对照（CK）组的差异菌群和连作时间较长处理的土壤类似，菌群也逐渐趋于稳定。C 处理组

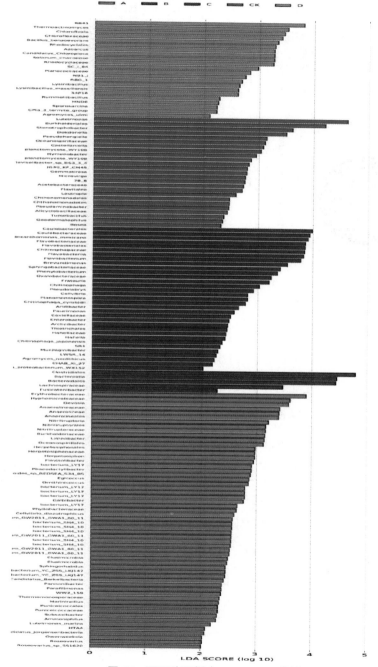

图 15　不同处理样本 LDA Score 分析

Fig. 15 LDA Score analysis of samples with different treatments

和 D 处理组前三的菌群 Caulobacterales、Caulobacteraceae、Mexicana、RB41、Thermoactinomyces 和 Chloroflexia 可以作为连作时间较长的参考标准,该 6 类菌群成为差异显著的菌群,可以推测连作时间较长,同理 A 处理组和 B 处理组前 3 类菌群的逐渐减少说明连作年代过长,也可以作为参考的指标。

图 16 不同处理组样本 LEFse 差异分析图

Fig. 16 LEFse difference analysis of samples in different treatment groups

2.2 ITS 数据分析

2.2.1 不同处理组样本数据处理分析

图 17 为数据预处理的分布图,reads QC filter 为低质量 reads,右边为红色图标,占比小于 1%;Non-overlap 为没有 overlap 的未组装 reads,右边为紫色图标,占比小于 5%;tag QC filter 为未通过"tag 过滤"的 tags,低质量、未组装和未通过的 tags 总量小于总数据量的 5%;Chimera 为嵌合体;effective tags 为有效用于后续分析的 tags,每个样品的占比都在 90%以上,有效 Tags 数量占据原始 PE Reads 的百分比从 A-1 到 CK-3 分别为:95.84%、94.49%、94.99%、96.57%、95.88%、95.24%、92.43%、96.26%、94.09%、93.87%、96.03%、95.32%、96.22%、96.21%和 95.24%。有效 Tags 数据量是足够进行后续分析,可靠数据。Tags 丰度统计显示经过 Tags 质控后的得到的高质量 Tags 数量最低的为 87 376 条,最高的为 97 055 条,Total length 最低的为 3 200 万,Max length 为 449,Min length 最低 202,最高 277,N50 为 380 左右,N90 为 335。从上述统计数据来看,下机数据足够,过滤后有效数据充足,为后续分析做

好了基础。

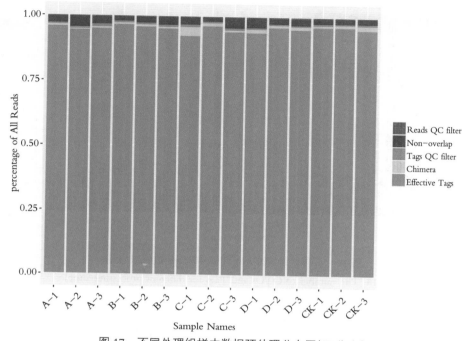

图 17 不同处理组样本数据预处理分布图（百分比）

Fig. 17 Distribution diagram of data preprocessing in different processing groups (percentage)

2.2.2 不同处理组样本 OTU 分析

为了研究样品的物种的组成多样性信息，用 Uparse 软件对所有样品的全部 Effective Tags 序列聚类，默认提供以 97%的一致性（Identity）将序列聚类成为 OTUs（Operational Taxonomic Units）结果，并计算出每个 OTU 在各个样品中的 Tags 绝对丰度和相对信息。图 18 是对不同处理组样本的 OTUs 和 Tags 数量的统计，总 Tags 数量平均 92162 条，有物种注释的 Tags 数量平均 91460 条，没有物种注释的 Tags 为 0 条，unique tags 的数量平均为 16677 条，总丰度为 1 的 OTUs 所对应的 Tags 会被过滤掉，平均为 701 条。

不同生境下的微生物群落，其物种分布存在一定程度的相似性和特异性。在多分组（或样本）情况下，为了解不同分组（或样本）之间的 OTU 差异情况，我们可以根据 OTU 丰度信息开展韦恩图分析，从而了解不同样本或

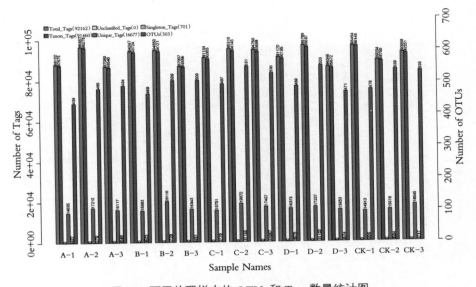

图 18　不同处理样本的 OTUs 和 Tags 数量统计图

Figure 18. The number of OTUs and Tags for different processing samples

者分组之间的 OTU 的共有或者特有信息。选取比较分组平均丰度大于 1 的所有 OTU(即高丰度并集 OTU)进行韦恩图分析。图 19 是各处理组样本之间共有和特有 OTUs 的韦恩图和各个集合的数量统计柱状图,对照(CK)vsA 处理:A 处理组特有 178 条 OTUs,对照(CK)特有 242 条 OTUs,两处理组共有 308 条 OTUs;对照(CK)vsB 处理:B 处理组特有 240 条 OTUs,对照(CK)特有 244 条 OTUs,两处理组共有 306 条 OTUs;对照(CK)vsC 处理:C 处理组特有 228 条 OTUs,CK 特有 216 条 OTUs,两处理组共有 334 条 OTUs;对照(CK)vsD 处理:D 处理组特有 222 条 OTUs,对照(CK)特有 232 条 OTUs,两处理组共有 318 条 OTUs。A 处理 vsB 处理 vsC 处理 vsD 处理 vs 对照(CK):A 处理组特有 65 条 OTUs,B 处理组最低,特有 98 条 OTUs,C 处理组特有 84 条 OTUs,D 处理组和对照(CK)特有 87 条 OTUs 和 134 条 OTUs,5 组相比(CK)组特有最多,也符合真菌和连作时间的增加菌种的变化特点。右图下方连线表示共有,单独一点表示特有。

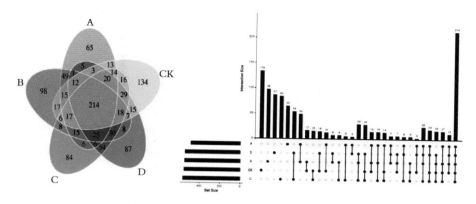

图 19　不同处理样本 OTUs 比较韦恩图
Fig. 19 Comparison of wynn diagram with OTUs of different processing samples

2.2.3　不同处理组样本物种分类分析

微生物物种分类一般分为界、门、纲、目、科、属、种 7 个等级,而每个 OTU 代表某类型分类水平集合。因此根据 OTU 的序列信息进行物种注释,能将分析结果与实际的生物学意义进行关联,从而研究群落中物种的变化关系等内容。

物种注释方法:Uparse 在构建 OTUs 的过程中会选取代表性序列(OTUs 中丰度最高的那条 Tag 序列),将这些代表性序列集合用 RDP Classifier 的 Naïve Bayesian assignment 算法,与 silva 数据库进行物种注释(设定置信度的阈值为 0.8~1)。表 4 为不同处理组样本物种注释表,每个样本纲水平注释最高 58 050 条 OTUs,最低 18 454 条 OTUs,从整体上来看,真菌在处理组样本中的 OTUs 显著低于对照(CK)组,A 处理组到 D 处理组,从纲、目、科和属的水平,OTUs 逐渐降低,符合预期随着连作时间的增加,真菌种类降低,趋于物种单一,和相关文章的报道的趋势相同,而对照(CK)组种类最多。

图 20 是每个处理样本在各分类水平上的序列构成柱形图,从图中可以看出门水平的注释率在 45%~75%,纲水平注释到的较少,门水平、科水平和种水平的注释率相比较高,种水平的注释率高也说明了真菌的研究较多,很

表 4 不同处理组样本物种注释 Tags 数量统计表

Table 4 Statistical table of quantity of Tags of sample species in different processing groups

样本	Domain	Phylum	Class	Order	Family	Genus	Species
A-1	87 675	39 169	30 754	29 717	27 716	18 024	7 726
A-2	96 278	47 567	32 567	32 342	28 705	17 006	7 809
A-3	86 540	43 492	29 296	29 201	24 301	12 738	6 813
B-1	93 724	53 880	28 441	28 350	27 409	17 200	3 715
B-2	94 121	35 368	25 583	25 500	24 331	17 946	12 668
B-3	86 556	41 866	25 143	24 953	23 294	14 889	7 184
C-1	90 355	61 892	48 737	45 817	37 336	18 630	10 272
C-2	94 145	62 312	28 995	28 658	24 282	5 952	3 959
C-3	93 699	56 011	26 672	25 873	21 419	14 771	11 311
D-1	90 195	57 748	30 443	28 606	22 703	11 154	4 875
D-2	95 697	55 254	18 454	17 434	15 344	11 777	8 959
D-3	85 672	55 235	27 481	25 092	21 651	10 342	5 183
CK-1	96 148	71 519	58 050	55 384	53 387	34 906	28 684
CK-2	88 750	69 243	51 855	51 680	50 581	29 579	19 337
CK-3	92 351	63 411	41 983	41 766	37 985	24 626	19 555

多被注释出来,但整体的注释率低于细菌,原因在于从种类上细菌远大于真菌,真菌的高级程度较细菌要高,部分土壤中不会存在过多的真菌类别。整体上处理组的真菌类别和注释低于对照(CK)组,说明连作导致物种趋于单一。

优势物种很大程度决定着微生物群落的生态结构以及功能结构,了解群落在各个水平的物种组成情况能有效地对群落结构的形成、改变以及生态影响等进行解读。我们统计了各个层级分类水平上的各处理样本的物种组成情况,然后用堆叠图的形式,直观展示不同的样本在各个分类层级水平上的物种丰度的变化情况。图 21 为目分类水平上的物种分布堆叠图,Sordariales 为各样本的第一优势菌群,在对照(CK)组中占比最多,达到了

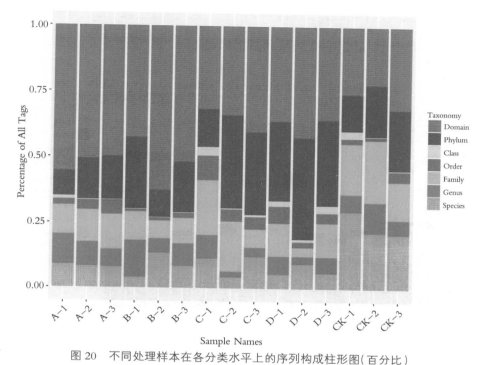

图 20 不同处理样本在各分类水平上的序列构成柱形图(百分比)
Fig. 20 Histogram of sequence composition of different processing samples at each classification level (percentage)

25%,其余处理组样本在 10%左右,处理组中无明显变化。Hypocreales 在对照(CK)、A 处理组和 B 处理组中相差不多,在随着连作时间的增加,C 处理组和 D 处理组出现明显降低,该真菌的变化可作为辣椒土壤连作的判定依据。Saccharomycetales 的丰度情况和 Hypocreales 的正好相反,在 C 处理组和 D 处理组中的丰度最高,在 A 处理组、B 处理和对照(CK)中的丰度显著降低,这两种真菌的变化可作为有效的判断依据。Eurotiales 的变化趋势和 Hypocreales 的相同,但相对丰度要小很多,也可作为参考标准。

图 22 为科水平上的物种分布堆叠图,Lasiosphaeriaceae 为科水平丰度最高的菌群,在对照(CK)组样本中占比达到了 25%,其余处理组样本中丰度平均在 5%左右。Nectriaceae 为主要关注菌群,在对照(CK)组、A 处理组和 B 处理组中的丰富较高,随着连作年限的增加含量显著降低,Pichiaceae 的丰度情

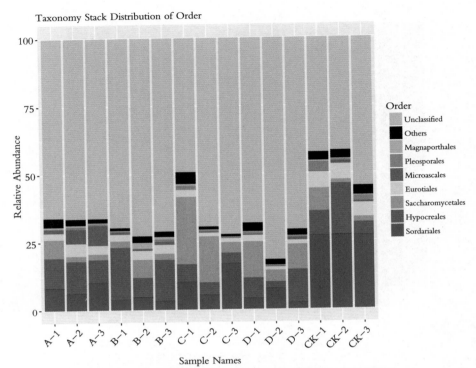

图 21 不同处理组样本目分类水平上的物种分布堆叠图

Fig. 21 Stacking plots of species distribution at the taxonomic level of different treatment groups

况和 Nectriaceae 的正好相反,在 C 处理组和 D 处理组中的丰度最高,在 A 处理组、B 处理组和对照(CK)中的丰度显著降低,该两组菌群可作为后期的主要研究对象,来对连作的菌群问题进行详细的解释。其余菌群的变化相比上述 3 种变化不明显。

我们根据物种分类的表达谱数据,使用热图来展示不同的物种在各样品间的表达情况,同时根据热图上的聚类关系,也可以反映样本关系。为了减少噪音数据的影响,热图分析所挑选物种需要满足以下 2 个条件:(1)物种在至少一个样本的相对丰度(物种 tags 数/总 tags 数)达到 0.1%以上;(2)选取满足条件(1)的相对总丰度在前 25 的物种进行热图分析。图 23 科水平的物种分类热图,结果和图堆叠图一致,只是堆叠图显示前 10 的优势菌群,热图显示很多,并做了均一化的处理,以堆叠图为准即可。

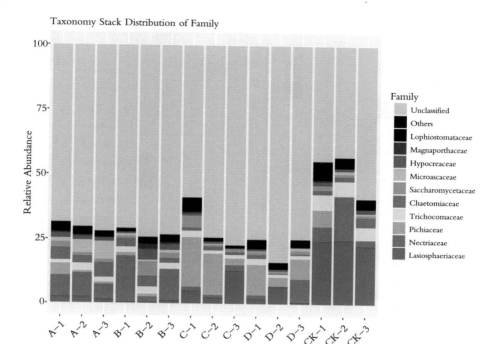

图22 不同处理组样本科分类水平上的物种分布堆叠图

Fig.22 Species distribution stack diagram at the undergraduate classification level of different treatment groups

2.2.4 不同处理组样本 Alpha 多样性分析

α 多样性是指特定生境或者生态系统内的多样性情况，它可以指示生境被物种隔离的程度，通常利用物种丰富度(种类情况)与物种均匀度(分布情况)2个重要参数来计算。表 5 主要展示 Chao1,ACE,Shannon,observed_species, Simpson 和 Good's Coverage 六大类常用的 α 多样性指数以及它们的相关分析结果。总体来说，Chao1/ACE 指数主要关心不同处理组样本的物种丰富度信息，平均值分别在 624 和 626；Good's Coverage 反映不同处理组样本的低丰度 OTU 覆盖情况，从表中可以看出覆盖度都在 99%以上，覆盖度很高；observed_species 表示能检测到的 OTU 种类情况，基本和下机的 OTUs 相差不多；Simpson/Shannon 主要综合体现物种的丰富度和均匀度,shannon 基本都在 4.3 以上，Simpson 平均值在 0.896。整体上不同处理组 Alpha 多样性

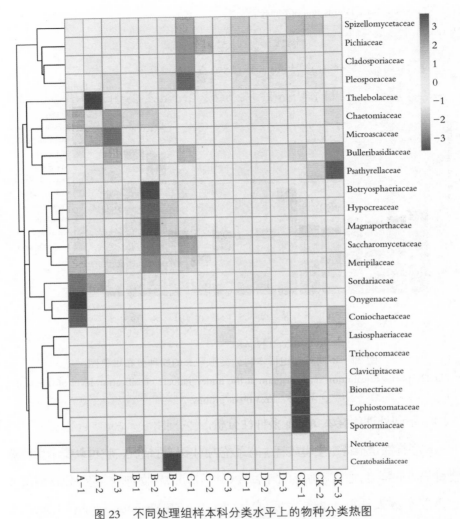

图 23 不同处理组样本科分类水平上的物种分类热图

Fig. 23 Heat maps of species classification at the undergraduate classification level of different treatment groups

指标正常,低丰度覆盖度高,丰富度和均匀度正常范围。

我们通过绘制稀释曲线(rarefaction curve)来评价测序量是否足以覆盖所有类群,并间接反映样品中物种的丰富程度。稀释曲线是利用已测得序列中已知的 OTU 的相对比例,来计算抽取 n 个(n 小于测得 tags 序列总数) tags 时出现 OTU 数量的期望值,然后根据一组 n 值(一般为一组小于总序

表5 不同处理组样本Alpha多样性各指标统计表

Table 5 Statistical table of indicators of sample Alpha diversity of different treatment groups

样本	chao1	ace	goods_coverage	observed_species	shannon	simpson
A-1	546.2040816	528.2000975	0.998825207	439	4.484292663	0.837594469
A-2	569.0857143	588.171497	0.998867862	485	5.157975484	0.913584468
A-3	611.9428571	621.0065746	0.99850936	494	4.857071557	0.902189192
B-1	604.0192308	591.5058069	0.998730315	469	4.388740665	0.88458407
B-2	610.4827586	610.8225456	0.998841916	509	4.588780692	0.876024515
B-3	638.0483871	650.4320617	0.998532742	509	5.001553327	0.911124842
C-1	668.3653846	635.7857182	0.998516961	497	5.357428181	0.94741821
C-2	667.1142857	683.6572756	0.998640395	551	4.307152313	0.85899222
C-3	640.915493	648.2162469	0.998655268	530	4.291443711	0.869399355
D-1	630.2307692	643.2119034	0.998492156	489	4.771696164	0.898069309
D-2	672.2207792	685.684555	0.998578848	553	3.967076499	0.821801031
D-3	586.328125	589.9159072	0.998575964	471	4.602788157	0.894178686
CK-1	579.7536232	598.8902672	0.998762325	478	5.530500616	0.950475121
CK-2	681.875	671.268105	0.998569014	539	5.255184495	0.936369708
CK-3	658.0166667	653.4968144	0.998678953	535	5.389172394	0.940116653

列数的等差数列)与其相对应的OTU数量的期望值做出曲线来。当曲线趋于平缓或者达到平台期时也就可以认为测序深度已经基本覆盖到样品中所有的物种。图24是不同处理组样本的稀释曲线,从图中可以看出,当测序的Tags数量到达75 000条时曲线趋于平缓到达平台期,说明测序的质量的合格的,深度和数据量足够。

我们通过绘制shannon稀释曲线(shannon rarefaction curve)来评价测序量是否足够,并间接反映样品中物种的丰富程度。图25为各样品的shannon稀释曲线,结论和上述的稀释曲线相同,测序数据量到达75 000条Tags后趋于稳定和平缓,数据足够,各个处理样本的浮动变化较小,D-2样本略微

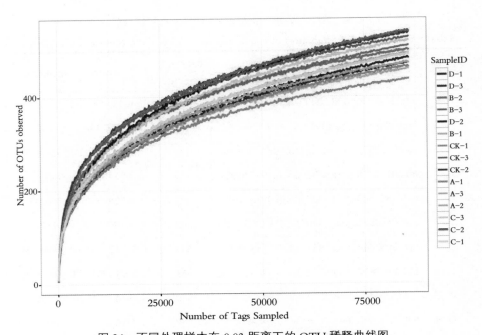

图 24 不同处理样本在 0.03 距离下的 OTU 稀释曲线图

Fig. 24 OTU dilution curves of samples with different treatments at a distance of 0.03

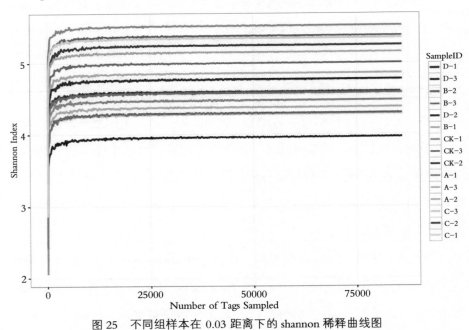

图 25 不同组样本在 0.03 距离下的 shannon 稀释曲线图

Fig. 25 Shannon dilution curves of samples from different groups at a distance of 0.03

偏低,也在正常范围之内。

不同生境下的环境驱动因素能引起微生物 α 多样性差异。结合分组和采样信息,通过对两组或者多组间的 α 多样性进行假设检验,可以分析组间的物种多样性是否存在显著的差异,从而初步判断驱动群落多样性变化的潜在因素等。我们同时使用以下几种常见的假设检验方法来进行差异分析。(1)针对 2 个分组进行比较时,使用 T-test 检验和 wilcox 秩和检验;(2)针对 2 个以上分组进行比较时,使用 Tukey 检验和 Kruskal-Wallis 秩和检验。在进行差异分析的同时,我们还绘制了各个 Alpha 多样性指数的箱线图(下图展示了各个比较组的 shannon 指数箱线图)。箱形图可以直观的反应组内物种多样性的中位数、离散程度、最大值、最小值、异常值。图 26 是对各组

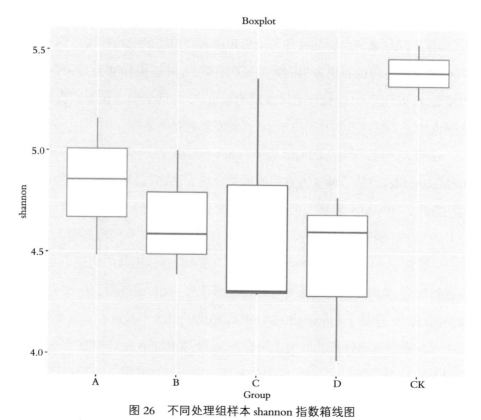

图 26 不同处理组样本 shannon 指数箱线图

Fig.26 Shannon index boxplot of samples from different treatment groups

样品汇总的 shannon 指数箱型图，A 处理组最大值 5.157，最小值 4.484，B 处理组最大值 5.001，最小值 4.388，C 处理组最大值 5.357，最小值 4.291，D 处理组最大值 4.771，最小值 3.967，对照 CK 组最大值 5.530，最小值 5.255。对照 CK 组的 shannon 值最高，从盒形图明显可以看出随着处理时间的增加 shannon 值越来越低，符合实验的预期变化趋势。

2.2.5　不同处理组样本 Beta 多样性分析

基于 OTU 列表的物种丰度信息，可以开展主成分分析（PCA，Principal Component Analysis），从而利用利用降维的思想研究样本间的组成距离关系。这种方法借用方差分解可以有效地找出数据中最"主要"的元素和结构，将复杂的样本组成关系反映到横纵坐标的 2 个特征值上，从而达到简化数据复杂度的效果。分析结果中，样品组成越相似，反映在 PCA 图中的距离越近，而且不同环境间的样品往往可能表现出各自聚集的分布情况。图 27 为二维 PCA 图，可以看出各组的聚类情况良好，A 处理组和 B 处理组聚类最为接近，C 处理组和 D 处理组聚类最为接近，C 处理组单独在图的右下方，该图也体现了随着连作时间的增加，真菌的聚类变化趋势。

在环境生物学中，UPGMA25（Unweighted Pair-group Method with Arithmetic Mean）是一种较为常用的聚类分析方法，它最早便是用来解决分类问题的。UPGMA 的基本思想是：首先将距离最小的 2 个样品聚在一起，并形成一个新的节点（新的样品），其分支点位于 2 个样品间距离的 1/2 处；然后计算新的"样品"与其他样品间的平均距离，再找出其中的最小 2 个样本进行聚类；如此反复，直到所有的样品都聚到一起，最终得到一个完整的聚类树。图 28 是基于 Unweighted Unifrac 距离的目水平 UPGMA 聚类树，从图中可以看出各组样品的聚类效果显著，A 处理组和 B 处理组聚为一类，C 处理组和 D 处理组聚为一类，对照（CK）组单独一类，说明 A 处理组和 B 处理组的处理时间接近且较短，C 处理组和 D 处理组较远，处理时间长，数据效果很好，能明显看出变化趋势。

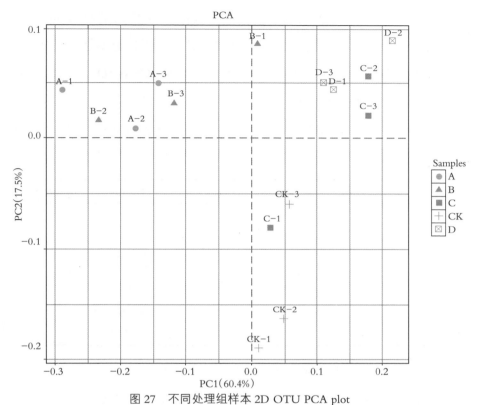

图 27 不同处理组样本 2D OTU PCA plot

Fig.27 2D OTU PCA plot of samples of different treatment groups

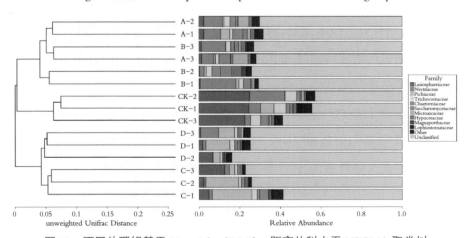

图 28 不同处理组基于 Unweighted Unifrac 距离的科水平 UPGMA 聚类树

Fig.28 Horizontal UPGMA clustering trees of different treatment groups based on Unweighted Unifrac distance

表 5 是基于 unweighted unifrac 距离的 Anosim 分析结果表,A 处理组、B 处理组、C 处理组和 D 处理组和对照(CK)都存在差异,且都是大差异。多组间 A-vs-B-vs-C-vs-D-vs-CK、A-vs-B-vs-C-vs-D 比较 P 值为 0.001,R 值大于 0.94,说明具有显著差异。

表 6 不同处理组基于 unweighted unifrac 距离的 Anosim 分析结果表
Table 6 Results of Anosim analysis of different treatment groups based on unweighted unifrac distance

Diffs	Rvalue	Pvalue	Significant
CK-vs-A	1	0.1	
CK-vs-B	1	0.1	
CK-vs-C	1	0.1	
CK-vs-D	1	0.1	
A-vs-B-vs-C-vs-D-vs-CK	0.9733	0.001**	
A-vs-B-vs-C-vs-D	0.9475	0.001**	

2.2.6 不同处理组样本群落功能分析

完整的微生物群落研究主要分为物种组成、多样性以及功能研究等几个重要方面。多种证据表明微生物的群落功能组成比物种组成与环境关系更为密切,随着分析技术发展,利用多样性测序数据进行群落功能预测已经成为群落研究的重要内容。我们将利用 FUN Guild、Tax 4 fun、FAPROTAX、Bug Base 等多个预测软件,根据不同的数据类型有针对性地完成群落功能预测分析。基于 OTU 丰度表格信息,利用 FUN Guild 开展真菌功能注释。由于缺乏真菌基因组数据,因此 FUN Guild 并不开展 KEGG 注释,而是通过整合已发表文章数据,进行一种称为 Guild 的功能分类预测。Guild 是一个生态学概念,用于描述物种资源利用吸收所进行的功能分类,FUN Guild 中的 Guild 分类包括动物病原菌、植物病原菌、木质腐生菌等 12 种。开展 Guild 预测,可以从其他生态角度研究真菌功能。图 29 为不包含没有注释的 OTU 的 Guilds 图,真菌的营养方式分为 2 种,分别是病理营养型(吸取宿主营养并伤害宿主)和病理营养型—腐生营养型(吸取宿主营养并伤害宿主—通过分解已死亡细胞获取营养物质),从图中可以看出 Animal Pathogen 在 CK 组、A 组和 B 组中相对较多,在 C 组和 D 组中的丰度显著降低,Endophyte、Clavicipitaceous Endophyte 和 Animal Pathogen 变化趋势相同,由于连作时间增加,该几类真菌都出现逐渐降低的变化,可以作为后续实验着重关注的真菌,也可以作为判断连作年限增加的初步检测依据,如果有对应几种真菌的检测试剂盒,可以简单有效的进行检测该几类真菌,结果是增加,就可以说明连作年限的变化。根据每立方单位的该真菌的含量,就可以对连作年限变化进行量化,为此类研究提供有力依据。

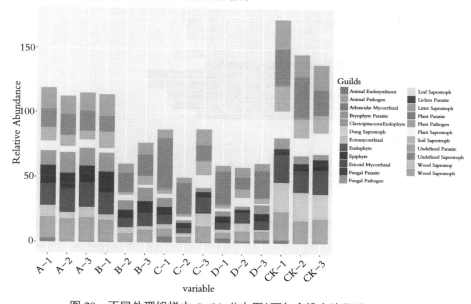

图 29 不同处理组样本 Guilds 分布图(不包含没有注释的 OTU)
Fig.29 Guilds distribution diagram of samples of different treatment groups (excluding unannotated OTU)

2.2.7 不同处理组样本物种差异分析

通过 LEFse 分析组间菌群差异,可以找出各组间特异的主要菌群,有助于开发 biomaker 等研究。利用 LEFse 软件对差异组间进行分析,LEFse 先对所有组样品间进行 kruskal-Wallis 秩和检验(一种多样本比较时 常用的检验方法），将筛选出的差异再通过 wilcoxon 秩和检验（一种两样本成组比较常用的检验方法）进行两两组间比较，最后筛选出的差异使用 LDA（Linear Discriminant Analysis）得出的结果进行排序得到左图，左图展示了不同组中丰度差异显著的物种，柱状图的长度代表差异物种的影响大小（即为 LDA

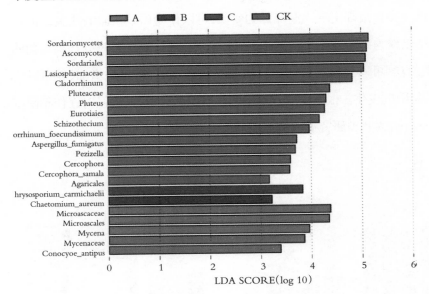

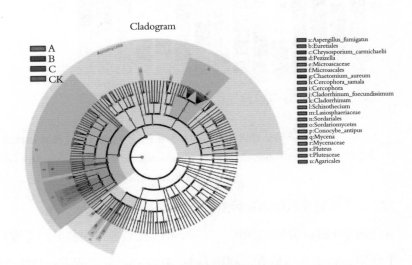

图 30　不同处理组样本 LEFse 差异分析图

Fig.30 LEFse difference analysis of samples in different treatment groups

Score)。随后通过将差异映射到已知层级结构的分类树上方式得到进化分支图(右图)。在进化分支图中,由内至外辐射的圆圈代表了由门至属(或种)的分类级别。在不同分类级别上的每一个小圆圈代表该水平下的一个分类,小圆圈直径大小与相对丰度大小呈正比。 着色原则:无显著差异的物种统一着色为黄色。详细见图30。从图中可以看出,对照(CK)组的差异菌群最多,其次是A处理组的样本,D处理组没有显著差异的菌群,C处理组和D处理组各有一类菌群,有效的说明连作时间越长,真菌的菌群越趋于单一,甚至于多组间比较没有差异菌群,B处理组和C处理组的Chaetomium_aureum和Hrysosporium_carmichaelii可作为有益菌群来关注,连作时间的增加,该两类菌群逐渐降低到无,同理A处理组中这五类真菌也可作为关注点,可优先关注得分较高的菌群。对照(CK)组的菌群种类是最多的,说明空白对照的菌群丰度最高,连作时间增加,菌群丰度降低,菌群趋于单一化,减少的菌群就可以作为我们改良土壤或者监控土壤变化的有力依据,可为土壤改良打下良好的数据基础。

3 讨论

土壤微生物是土壤生态系统变化的敏感指标之一,其活性和群落结构变化能敏感地反映出土壤生态系统的质量和健康状况(钟文辉等,2004)。如今,土壤微生物指标已被公认为土壤生态系统变化的预警及敏感指标(任天志等,2000)。其中,土壤细菌占土壤微生物总数的70%~90%,是土壤中最活跃的因素(曹志平,2007)。因此分析辣椒连作土壤的生物学指标和细菌与真菌群落结构变化或许有助于发现和找到克服辣椒连作障碍的方法。

在土壤微生物群落多样性特征分析中,Chao、Ace、Shannon指数值越大,Simpson指数值越小,说明样品的物种多样性越高(Grice et al,2009)。秦越等(2015)研究认为,连作会导致马铃薯根际土壤微生物群落结构组成失调,土壤从"细菌型"向"真菌型"土壤转化。本试验中,B组中的Chao、Ace、

Shannon 指数值最大,C 组和 D 组逐渐降低,B 组中 Simpson 指数值最小,随着连作时间的增加 C 组和 D 组逐渐升高,这表明随着连作时间的增加,物种多样性逐渐降低。同时,辣椒种植连作时间增加,细菌丰富度显著降低,物种的多样性也有所降低,而真菌丰富度增加,物种的多样性增加。这与 CARTWRIGHT 等(2016)和 FIERER 等(2006)认为土壤微生物多样性与种植时间呈负相关的结果一致。

土壤微生物数量受土壤养分含量、作物类型以及感病与否等理化及生态因素影响(杨尚东等,2013)。本文分析结果显示,细菌中 Bacillus 属在 D 处理组中尤为显著,在 C 处理组中略微升高,RB41 随着连作时间的增加逐渐增加,但对照(CK)组中依旧含量较高。真菌中 Saccharomycetales 的丰度情况在 C 处理组和 D 处理组中的丰度最高,在 A 处理组、B 处理和对照(CK)中的丰度显著降低。随着辣椒连作时间的增加,细菌和真菌的种类逐渐趋于单一化,而且出现细菌型土壤向真菌型转变的趋势。这与许多前人的研究结果(Li et al,2012;周宝利等,2010;杜茜等,2012)相一致。同时这一现象亦表明,辣椒连作土壤中微生物群落结构发生显著变化,使土壤由多样细菌型土壤向单一性真菌型转变,土壤生态系统失调可能是辣椒连作土壤容易发生连作障碍的主要原因。

土壤微生物是土壤环境质量的主要指标之一(曹志平,2007)。本文对辣椒连作和相邻玉米土壤细菌和真菌群落结构的分析结果显示,辣椒连作时间较长的土壤中主要的细菌菌群为 Caulobacterales、Caulobacteraceae、Mexicana、RB41、Thermoactinomyces 和 Chloroflexia,可以作为连作时间较长的参考标准,该 6 类菌群成为差异显著的菌群,可以推测连作时间较长,同理连作时间较短的土壤前 3 类菌群的逐渐减少说明连作年代过长,也可以作为参考的指标,A 处理组的差异菌群最多,说明连作时间越短,差异的菌群越多,随着连作时间的增加,差异菌群趋于稳定和减少。真菌中 Animal Pathogen 在 CK 组、A 组和 B 组中相对较多,在 C 组和 D 组中的丰度显著降低，Endophyte、Clavicipitaceous Endophyte 和 Animal Pathogen 变化趋势相同，由于连作时间增加，该几类真菌都出现逐渐降低的变化。本文的分析结果与岳冰冰（2012）在烤烟连作土壤的研究结果相类似。此外，这一结果亦证实了辣椒连作容易发生连作障碍降低产量的根本原因，即连作导致表征土壤肥力的生物学性状劣化的同时，土壤微生物群落结构亦发生显著变化，多样性下降，导致土壤中速效养分含量降低，使得缺乏养分的作物长势偏弱，连作土壤中较为单一的微生物群落结构对病原菌的拮抗能力下降，从而使产量降低。

4 结论

本实验结果表明，随着辣椒连作时间的增加，土壤微生物群落结构发生显著变化，细菌菌群多样性逐渐降低，导致土壤微生物群落由多种细菌型向单一真菌型转变，最终导致辣椒产量的降低。可将 Caulobacterales、Caulobacteraceae、Mexicana、RB41、Thermoactinomyces 和 Chloroflexia 这 6 类细菌菌群成为差异显著的菌群，作为连作时间过长的参考。可将真菌 Animal Pathogen、Endophyte 和 Clavicipitaceous Endophyte 这 3 类真菌的降低作为连作时间增加，物种多样性单一化的参考，为后续进一步土壤改良工作研究打下来良好的基础。

参考文献

[1] CARTWRIGHT J,DZANTOR E K,MOMEN B. Soil microbial community profiles and functional diversity in limestone cedar glades[J]. Catena,2016,147:216-224.

[2] FIERER N,JACKSON R B. The diversity and biogeography of soil bacterial communities [J]. Proceedings of the National Academy of Sciences of the United States of America,2006,103(3):626-631.

[3] Grice E A,Kong H H,Conlan S,et al.Topographical and temporal diversity of the human skin microbiome[J]. Science,2009,324(5931):1190-1192.

[4] LI P D, DAI C C, WANG X X, et al. 2012. Variation of soil enzyme activities and microbial community structure in pennut monocropping system in substropical China [J]. African Journal of Agriculture Research, 7(12): 1970-1879.

[5] 曹志平. 土壤生态学[M]. 北京：化学工业出版社, 2007, 211-222.

[6] 杜茜, 卢迪, 马琨. 马铃薯连作对土壤微生物群落结构和功能的影响[J]. 生态环境学报, 2012, 21(7): 1252-1256.

[7] 秦越, 马琨, 刘萍. 马铃薯连作栽培对土壤微生物多样性的影响[J]. 中国生态农业学报, 2015, 23(2): 225-232.

[8] 任天志, Grego S. 持续农业中的土壤生物指标研究[J]. 中国农业科学, 2000, 33(1): 68-75.

[9] 杨尚东, 吴俊, 赵久成, 等. 番茄青枯病罹病植株和健康植株根际土壤理化性状及生物学特性的比较[J]. 中国蔬菜, 2013, 22: 64-69.

[10] 岳冰冰. 烤烟连作改变了根际土壤微生物的多样性[D]. 哈尔滨：东北林业大学, 2012.

[11] 周宝利, 徐妍, 尹玉玲, 等. 不同连作年限土壤对茄子土壤生物学活性的影响及嫁接调节[J]. 生态学杂志, 2010, 29(2): 290-294.

[12] 钟文辉, 蔡祖聪. 土壤管理措施及环境因素对土壤微生物多样性影响研究进展[J]. 生物多样性, 2004, 12(4): 456-465.

微生物菌剂对拱棚连作辣椒生长、产量及品质的影响

高晶霞[1],牛勇琴[2],吴雪梅[2],王学梅[1],谢 华[1]

(1.宁夏农林科学院种质资源研究所,宁夏银川 750002;
2.宁夏回族自治区彭阳县蔬菜产业发展服务中心,宁夏彭阳 756500)

摘 要:本研究通过不同微生物菌剂施用对拱棚连作辣椒生长发育、产量及品质的影响,结果表明:处理1、处理4辣椒株高、茎粗、开展度、果长、果粗、单株结果数、单果重均高于其他3个处理及空白对照(CK)。处理4小区总产量、折合亩产量、折合亩产值均高于其他4个处理及空白对照(CK),分别为202.3 kg、5 767.4 kg、8 074.4 元,分别比处理1、处理2、处理3、处理4、处理5及空白对照(CK)亩产量增幅9.1%、7.0%、4.5%、5.0%、10.7%;在辣椒采收盛期,处理5连作辣椒维生素C含量最高,为110.7 mg·g^{-1},处理1拱棚连作辣椒维生素C含量次之,且连作辣椒可溶性糖含量最高为19.58 mg·g^{-1},处理1及空白对照(CK),连作辣椒可溶性蛋白含量最高,分别为1.81 mg·g^{-1}、1.80 mg·g^{-1}。

关键词:微生物菌剂;连作;辣椒;生长发育;产量;品质

Effects of MicrobialInoculants on Growth、Yield and Quality of Capsicum Continuous Cropping

GAO Jing-xia[1], NIU Yong-qin[2], WU Xue-mei[2], WANG Xue-mei[1], XIE Hua[1]

(1. Ningxia Academy of Agriculture and Forestry Plant Resources, Ningxia, Yinchuan, 750002; 2. The Ningxia Hui Autonomous Region, Pengyang County AgriculturalTechnology Extension and Service Center, Pengyang, Ningxia 756500)

Abstract: In this experiment, the growth, yield and quality of Capsicum were studied by different microbial inoculants. The results show that: The plant height, stem diameter, development degree, fruit length, fruit size, single plant result and single fruit weight of continuous cropping pepper, treatment 1 and treatment 4 were higher than those of the other 3 treatments and blank control (CK), The total yield, yield and output value of treatment 4 were higher than those of the other 4 treatments and the blank control (CK), 202.3 kg, 5767.4 kg and 8 074.4 yuan respectively, The yield increased by 9.1%, 7%, 4.5%, 5%, 10.7%, respectively, In the hot pepper harvest period, treatment 5 was treated with the highest VC content in continuous cropping pepper, For 110.7 mg·g^{-1}, The content of VC in the capsicum of the continuous cropping of the arch shed was the second, The content of soluble sugar in the continuous cropping of continuous cropping pepper was 19.58 mg·g^{-1}, Treatment 1 and blank control (CK) showed that the soluble protein content of continuous cropping pepper was the highest, which was 1.81 and 1.80 mg·g^{-1} respectively

Key words: organic fertilizer; straw; continuous cropping; capsicum; growth and development

辣椒(*Capsicum annuum* L.)为中国栽培面积较大的蔬菜,种植面积达142万hm^2,年产干、青椒1 300亿kg[1]。但随着种植年限增加及不合理的水肥管理等原因,辣椒连作障碍十分普遍,主要表现为病害加重、产量降低、品质下降、土壤理化性状恶化、土壤微生物区系改变等[2-5]。防治连作障碍,最好的方法就是在提高管理水平的基础上,改善土壤及土壤微生物。目前微生物菌剂产品种类繁多,效果不一[6-10]。

近年来,宁夏南部山区拱棚辣椒发展非常迅速,栽培面积逐年增加,生产模式因区域不同而各具特色。但因为过分追求经济利益,化肥和农药超量盲目使用,造成土壤生态环境恶化,作物产量和品质下降,引发了严重的连作障碍,对食品安全及农业的可持续发展产生不良影响[11,12]。目前,宁夏南部山区连作障碍方面的研究仍很薄弱,对高效设施农业的发展不能提供有效的技术支撑。这些问题严重阻碍了蔬菜生产的发展和农民收益的提高,制约了宁夏南部山区蔬菜产业的可持续健康发展。

本试验以牛角椒为研究对象,通过选用4种单一微生物菌剂对拱棚连作辣椒生长发育、产量及品质的研究,为创造良好的设施土壤生态环境、解决辣椒连作障碍及获得优质辣椒提供一定的理论依据。

1 材料与方法

1.1 试验地点

试验地点位于宁夏彭阳县新集乡拱棚辣椒示范基地（连作辣椒3 a以上,全氮1.05 g·kg^{-1},全磷0.97 g·kg^{-1},全钾23.0 g·kg^{-1},有机质23.0 g·kg^{-1},碱解氮63.8 g·kg^{-1},速效磷30.4 g·kg^{-1},速效钾195 g·kg^{-1},pH 7.63,全盐1.17 g·kg^{-1}）。

1.2 供试材料

试验选用微生物菌剂为地菌净、荧光假单孢杆菌、巨大芽孢杆菌、枯草芽孢杆菌(购自广州市微元生物科技有限公司)。

1.3 试验方法

试验设6个处理。

处理1:地菌净施用处理,5 kg/667 m^2,500倍液稀释,每小区175 g,用水87.5 kg,每株0.99 kg水溶液。

处理2:荧光假单孢杆菌施用处理,150 g/667 m^2,1 000倍液稀释,每小区5.3 g,用水5.3 kg,每株0.06 kg水溶液。

处理3:巨大芽孢杆菌施用处理,2 kg/667 m^2,500倍液稀释,每小区70 g,用水35 kg,每株0.4 kg水溶液。

处理4:枯草芽孢杆菌施用处理,1.43 kg/667 m^2,500倍液稀释,每小区50 g,用水25 kg,每株0.285 kg水溶液。

处理5:荧光假单孢杆菌+巨大芽孢杆菌+枯草芽孢杆菌三种按(1:1:2)混合施用处理,每小区125.3 g,用水62.65 kg,每株水溶液0.71 kg水溶液。

处理 6：以不施微生物菌剂为空白对照（CK）。

微生物菌剂稀释后灌根，3 次重复，每个小区 23.4 m²，保证辣椒株数 100 株左右，分别在定植后 10 d、苗期（定植后 30 d）、开花坐果期（定植后 50 d）、开花盛期（定植后 70 d）、结果盛期（定植后 90 d）灌根。

1.4 调查项目

1.4.1 辣椒植株性状的测定

每次灌根 5 d 后调查辣椒冠幅、株高、茎粗（每小区 5 株）。

1.4.2 辣椒果实性状及品质的测定

辣椒采收盛期，每个小区取 50 个长势均匀的果实，测量单果重、果长、果粗，之后送宁夏农林科学院测量果实品质。用钼蓝比色法测定辣椒果实中维生素 C 含量，苯酚硫酸比色法测定可溶性糖含量，紫外分光光度法测定可溶性蛋白质含量，PRO-101 型糖度计测定可溶性固形物含量。

1.4.3 产量统计

每个小区随机取 10 株测辣椒单株坐果数、单株产量、小区产量。

1.5 数据统计及分析

试验数据采用 Excel 2010 进行统计分析和 SPSS 17.0 软件 Duncan 方差分析。

2 结果分析

2.1 微生物菌剂对拱棚连作辣椒生长的影响

2.1.1 微生物菌剂对拱棚连作辣椒株高的影响

从图 1 中可知，微生物菌剂对拱棚连作辣椒株高在辣椒定植后 15 d、35 d、75 d、95 d 连作辣椒株高存在差异性（$P<0.05$），处理 1 株高最高，分别在 26.33 cm、45.0 cm、70.13 cm，85.07 cm、155.73 cm，空白对照（CK）株高最矮。

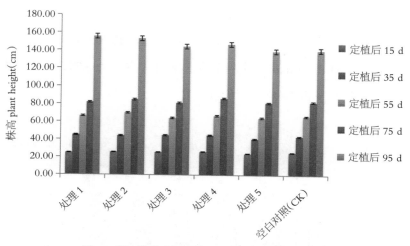

图 1 微生物菌剂对拱棚连作辣椒株高的影响

Figure. 1 Effect of microbial inoculants on the height of continuous cropping pepper

2.1.2 微生物菌剂对拱棚连作辣椒茎粗的影响

从图 2 中可知,微生物菌剂对拱棚连作辣椒茎粗在辣椒定植后 15 d、35 d 差异不显著,分别在 5.17~5.33 mm,8.07~8.79 mm,定植后 55 d、75 d、95 d 辣椒茎粗存在差异性($P<0.05$),空白对照(CK)明显高于其他 5 个处理。

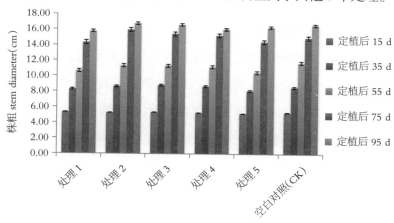

图 2 微生物菌剂对拱棚连作辣椒茎粗的影响

Figure. 2 Effect of microbial inoculants on the stem diameter of continuous cropping pepper

2.1.3 微生物菌剂对拱棚连作辣椒冠幅的影响

从图3可知,微生物菌剂对拱棚连作辣椒冠幅定植后15 d、35 d差异不显著。定植后55 d、75 d、95 d冠幅存在显著性差异(P<0.05),处理3、处理5冠幅大于其他几个处理及空白对照(CK)。

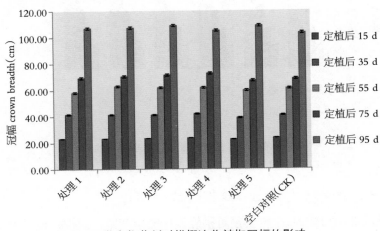

图3 微生物菌剂对拱棚连作辣椒冠幅的影响

Figure. 3 Effect of microbial inoculants on the development of continuous cropping pepper

2.2 微生物菌剂对拱棚连作辣椒果实性状的影响

2.2.1 微生物菌剂对拱棚连作辣椒果长影响

从图4可知,微生物菌剂对拱棚连作辣椒果长有影响。在辣椒结果盛期,处理2连作辣椒果长最长,为26.35 cm,处理1、处理5辣椒果长无差异,分别为25.37 cm、25.35 cm,处理3和空白对照(CK)辣椒果长无差异,分别为24.41 cm和24.37 cm,处理4辣椒果长最短,为23.26 cm。

2.2.2 微生物菌剂对拱棚连作辣椒果粗的影响

从图5可知,微生物菌剂对拱棚连作辣椒果粗差异显著。在辣椒结果盛期,处理2连作辣椒果粗最粗,为40.28 mm;处理1、空白对照(CK)连作辣椒果粗次之;处理3、处理4、处理5拱棚连作辣椒果粗均最细。

2.2.3 微生物菌剂对连作辣椒单株结果数的影响

从图6可知,微生物菌剂对拱棚连作辣椒单株结果数差异显著（P<

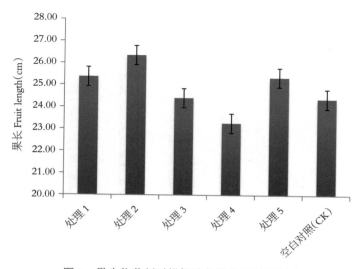

图 4 微生物菌剂对拱棚连作辣椒果长的影响

Figure. 4 Effect of microbial inoculants on the length of continuous cropping pepper

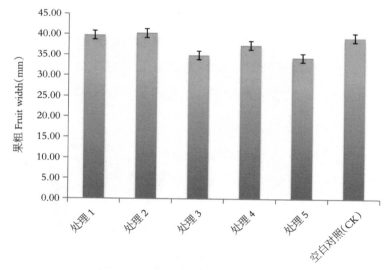

图 5 微生物菌剂对拱棚连作辣椒果粗的影响

Figure.5 Effect of microbial inoculants on the Roughage of continuous cropping Capsicum

0.05)。在辣椒结果盛期,处理 4 拱棚连作辣椒单株结果数最多,为 26.67 个;处理 2、处理 3、处理 5,连作辣椒单株结果数次之;处理 1,空白对照(CK)辣椒单株结果数最少。

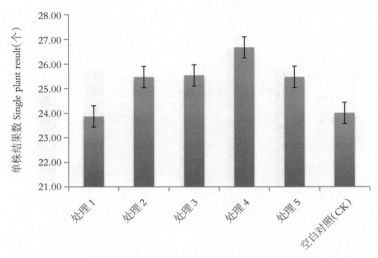

图 6 微生物菌剂对拱棚连作辣椒单株结果数的影响

Figure. 6 Effect of microbial inoculants on the single plant results of continuous cropping pepper

2.2.4 微生物菌剂对连作辣椒单果重的影响

从图 7 可知,微生物菌剂对拱棚连作辣椒单果重有差异。在辣椒结果盛期,处理 1 拱棚连作辣椒单果重最大,为 92.16 g;处理 3、处理 4、处理 5 及空白对照(CK)连作辣椒单果重差异不显著,处理 2,拱棚连作辣椒单果重最低。

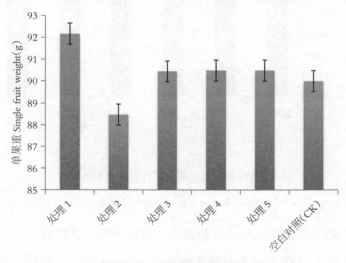

图 7 微生物菌剂对拱棚连作辣椒单果重的影响

Figure .7 Effect of microbial inoculants on single fruit weight of continuous cropping pepper

2.2.5 微生物菌剂对拱棚连作辣椒产量的影响

从图 8 和表 1 中可知,微生物菌剂对拱棚连作辣椒产量存在显著性差异(P<0.05),处理 4 小区总产量、折合亩产量、折合亩产值均高于其他 4 个处理及空白对照(CK),分别为 202.3 kg、5 767.4 kg、8 074.4 元,分别比处理 1、处理 2、处理 3、处理 5 及空白对照(CK)亩产量增幅 9.1%、7.0%、4.5%、5.0%、10.7%;处理 3、处理 5 小区总产量、折合亩产量、折合亩产值差异不显著(P<0.05),空白对照(CK)小区总产量、折合亩产量、折合亩产值最低。

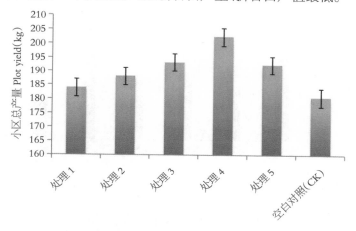

图 8 微生物菌剂对连作辣椒小区产量的影响

Figure .8 Effect of microbial inoculants on the plot yield of continuous cropping pepper

表 1 微生物菌剂对连作辣椒产量的影响

Table 1. Effect of microbial inoculants on the yield of continuous cropping pepper

处理 Handle	折合总产量 Aggregateyield (kg/667 m^2)	折合总产值 Aggregateout- put value(元)	比对照 Comparative control±(kg)	显著性分析 Saliency analysis 5%	位次 Precedence
处理 1	5 241.9	7 338.7	94.0	cd	5
处理 2	5 361.7	7 506.4	213.8	bc	4
处理 3	5 507.0	7 709.8	359.1	b	2
处理 4	5 767.4	8 074.4	619.5	a	1
处理 5	5 481.3	7 673.8	333.4	b	3
空白对照(CK)	5 147.9	7 207.1		d	6

2.3 微生物菌剂对连作辣椒品质的影响

从图9、图10、图11可知,微生物菌剂对拱棚连作辣椒品质存在显著性差异($P<0.05$),在辣椒采收盛期,处理5,辣椒果实维生素C含量最高,为110.7 mg·g^{-1},处理1拱棚连作辣椒维生素C含量次之,其他3个处理果实维生素C:处理2>处理3>处理4,空白对照(CK)辣椒维生素C含量最低,为40.0 mg·g^{-1};微生物菌剂对拱棚连作辣椒可溶性糖含量有差异,处理1连

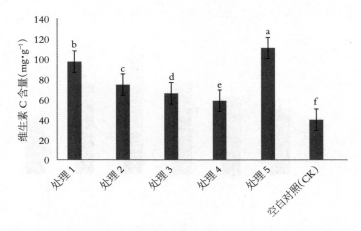

图9 微生物菌剂对拱棚连作辣椒维生素C含量的影响

Figure .9 Effect of microbial inoculants on vitamin C content in continuous cropping pepper

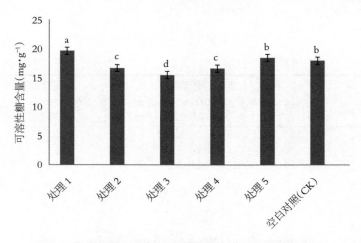

图10 微生物菌剂对拱棚连作辣椒可溶性糖含量的影响

Figure.10 Effect of microbial inoculants on soluble sugar content in continuous cropping pepper

作辣椒果实可溶性糖含量最高,为19.58 mg·g⁻¹,处理3可溶性糖含量最低,为15.45 mg·g⁻¹,其他3个处理及空白对照(CK)连作辣椒可溶性糖含量均在15.58~18.41 mg·g⁻¹;微生物菌剂对拱棚连作辣椒可溶性蛋白含量有差异,处理1、空白对照(CK)连作辣椒可溶性蛋白含量最高,分别为1.81 mg·g⁻¹、1.80 mg·g⁻¹,其他4个处理可溶性蛋白含量分别在1.45~1.71 mg·g⁻¹。

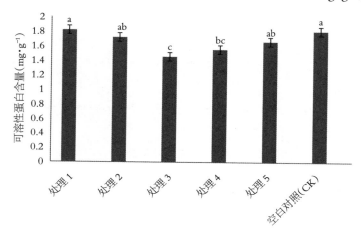

图11 微生物菌剂对拱棚连作辣椒可溶性蛋白的影响

Figure.11 Effect of microbial inoculants on soluble protein of continuous cropping pepper

3 结论

本试验通过不同微生物菌剂施用对拱棚连作辣椒生长发育、产量及品质的研究,结果表明:处理1、处理4辣椒株高、茎粗、开展度、果长、果粗、单株结果数、单果重,均高于其他3个处理及空白对照(CK)。处理4小区总产量、折合亩产量、折合亩产值均高于其他4个处理及空白对照(CK),分别为202.3 kg、5 767.4 kg、8 074.4元,分别比处理1、处理2、处理3、处理5及空白对照(CK)亩产量增幅9.1%、7.0%、4.5%、5.0%、10.7%;处理1、处理2小区总产量、折合亩产量、折合亩产值差异不显著($P<0.05$),空白对照(CK)小区总产量、折合亩产量、折合亩产值最低。在辣椒采收盛期,处理5连作辣椒维生素C含量最高,为110.7 mg·g⁻¹,处理1拱棚连作辣椒维生素C含量次之,

处理1连作辣椒可溶性糖含量最高为19.58 mg·g^{-1}，处理1及空白对照（CK），连作辣椒可溶性蛋白含量最高，分别为1.81 mg·g^{-1}、1.80 mg·g^{-1}。

综合分析结果：处理4、处理5，在一定程度上提高了连作辣椒产量及品质，为进一步明确微生物菌剂对辣椒连作障碍缓解效果，建立切实可行的缓解技术提供科学依据。

参考文献

[1] 杨定清,雷绍荣,谢永红.有机食品与农业可持续发展[J].农业环境与发展,2007,(2):43-45.

[2] 夏振远,李云华,杨树军.微生物菌肥对烤烟生产效应的研究[J].中国烟草科学,2002(3):28-30.

[3] 黄科,刘明月,蔡雁平,等.氮磷钾施用量与辣椒品质的相关性研究[J].西南农业大学学报,2002,24(4):349-352.

[4] 刘来,黄保健,孙锦,等.大棚辣椒连作土壤微生物数量、酶活性与土壤肥力的关系[J].中国土壤与肥料,2013(2):5-10.

[5] 何文寿.设施农业中存在的土壤障碍及其对策研究进展[J].土壤,2004,36(3):235-242.

[6] 孙继民,邹学校,罗尊长,等.辣椒连作研究进展[J].辣椒杂志,2011(2):1-7,27.

[7] 郭红伟,郭世荣,刘来,等.辣椒连作对土壤理化性状、植株生理抗性及离子吸收的影响[J].土壤,2012,44(6):1041-1047.

[8] 赵晓玲.生物菌剂处理不同有机肥对温室辣椒产量和品质的影响[J].热带农业科学,2015,35(3):12-14.

[9] 田俊岭,彭桂香,李永涛,等.一种高效光合菌剂对辣椒生长及土壤微生物的影响[J].生物技术进展,2014,4(3):197-200.

[10] 文吉辉,黄志农,徐志德.光合细菌菌剂对辣椒产量及其病虫害的影响[J].辣椒杂志,2013,11(1):19-22.

[11] 李玉利,杨忠兴,仇璇,等.土壤改良剂对大棚辣椒连作土壤理化性质的影响[J].安徽农业科学,2014,42(33):11676-11677,11739.

[12] 常彦莉.施用BGA土壤调理剂对辣椒及土壤理化性质的影响[J].农业科技与信息,2015(1):28-30.

辣椒连作土壤微生物群落及土壤离子对微生物菌剂的响应

高晶霞[1],吴雪梅[2],牛勇琴[2],裴红霞[1],王学梅[1],谢 华[1]

（1. 宁夏农林科学院种质资源研究所,宁夏银川 750002；
2. 宁夏回族自治区彭阳县蔬菜产业发展服务中心,宁夏彭阳 756500）

摘 要：【目的】筛选1~2种微生物配比,以减缓辣椒连作障碍。【方法】以辣椒连作8年大棚为对象,通过不同微生物菌剂配比的施用,研究连作辣椒土壤对不同微生物菌剂用量的影响特征,测定辣椒连作土壤中细菌、真菌和放线菌数量、土壤酶活性、土壤离子变化。【结果】胶质芽孢杆菌:巨大芽孢杆菌:枯草芽孢杆菌=2:1:2,土壤中细菌数量最多,为 $8.1×10^6$,真菌数量最少,为 $1.2×10^6$,可显著增加土壤中细菌数量,降低土壤中有害真菌和放线菌数量,防止土壤由细菌型向真菌型转变;胶质芽孢杆菌:巨大芽孢杆菌:枯草芽孢杆菌=2:1:2,蔗糖酶活性、脲酶活性最强,分别为777.07、201.53 mg/g·24 h。【结论】巨大芽孢杆菌:枯草芽孢杆菌=2:1:2处理,提高了土壤酶活性,减少了土壤中离子的富集,能有效减缓辣椒土壤连作障碍。

关键词：微生物菌剂；连作；辣椒；土壤；微生物群落

Response of Soil Microbial Community and Soil Ions to Microbial Agents in Pepper Continuous Cropping

GAO Jing-xia[1], GAO Yu[2], WU Xue-mei[2], NIU Yong-qin[2],
PEI Hong-xia[1], XIE Hua[1]

（1. *Institute of Germplasm Resources, Ningxia Academy of Agriculture and Forestry Plant Resources, Ningxia Yinchuan 750002, China*; 2. *The Ningxia Hui Autonomous Region, Pengyang County Agricultural Technology Extension and Service Center, NingxiaPengyang 756500, China*）

Abstract:【Objective】The proportion of 1 to 2 microorganisms was screened to slow down the barrier of continuous cropping.【Method】Taking the greenhouse of pepper continuous cropping for 8 years as the object, through the application of different proportions of microbial agents, the influence characteristics of pepper soil on the dosage of different microbial agents were studied, and the quantity of bacteria, fungi and actinomycetes, soil enzyme activity and soil ion changes in pepper continuous cropping soil were measured.【Result】The number of bacteria in the soil is $8.1×10^6$, and the number of fungi is $1.2×10^6$, which can significantly increase the number of bacteria in the soil, reduce the number of harmful fungi and actinomycetes in the soil, and prevent the soil from changing from bacterial type to fungal type.The highest activities of sucrase and urease were 777.07 mg/g and 201.53 mg/g·24 h, respectively.【Conclusion】Bacillus macrospore :bacillus subtilis =2∶1∶2 treatment can improve the soil enzyme activity, reduce the accumulation of ions in the soil, and effectively slow down the barrier of continuous cropping in the soil of pepper.

Key words: Compound microbial agents; Continuous cropping; Pepper; Soil; Microbial community

辣椒(*Capsicum annuum* L.)又名番椒,原产于拉丁美洲热带地区,茄科辣椒属,为一年生草本植物,在我国普遍栽培[1]。辣椒营养丰富,维生素C含量在蔬菜中居第一位[2]。

【研究意义】辣椒是宁夏南部山区栽培面积较大的蔬菜之一,但受耕地的限制,蔬菜连作现象十分普遍,长期连作导致土壤微生物菌群失衡,有害病原微生物大量繁殖,病虫害发生频繁,土传病害逐年加重[1-4],同时辣椒生产中由于长期大量使用化肥、农药和频繁的耕作等造成辣椒土壤生产力降低、病害频发、农药残留超标等一系列难题,并由此引发了一系列具有区域特点的生态与环境问题,致使辣椒区生态环境恶化,资源紧缺,辣椒质量和生产效益下降,已成为制约辣椒丰产栽培可持续发展的瓶颈[5-9]。微生物菌肥是由一种或数种有益微生物细菌经发酵而成的无毒害无污染的生物性肥料。已在多种作物上广泛应用[4-7],并且证明有提高作物产量品质及增强抗病性的作用。但多为单一菌剂。

【前人研究进展】陈雪丽[10]、王茹[11]等研究表明,植物对土传病害的抗性与根际土壤微生物关系密切。而土壤中微生物类群与土壤酶的活性密切相关[12-15],其活性的大小可以较敏感地反映土壤中生化反应的方向和强度;耿丽平[16]等研究发现玉米秸秆还田配施微生物菌剂在一定程度上提高了土壤速效钾、土壤微生物量碳氮的含量和土壤纤维素酶活性,且土壤微生物量碳、氮含量表层明显高于下层;张丽荣[17]等研究表明微生物菌剂施入土壤后在作物根系周围形成了有益微生物优势菌群,提高了土壤微生物的活性,增加了土壤微生物数量;钱海燕[18]等研究发现秸秆还田配施微生物菌剂对转化酶活性以及真菌、氨化细菌、好气自生固氮菌、磷细菌、纤维素分解菌数量的增长效果最为显著,而对过氧化氢酶、脲酶活性以及土壤微生物细菌、放线菌、硝化细菌数量增长的效果不及秸秆还田配施 N、P、K;雷先德[19]等研究发现微生物菌剂能够保持和提高土壤微生物丰富度指数和香农－威尔(Shannon-Wierner)多样性指数,稳定农田土壤环境中微生物种群多样性。

【本研究切入点】通过单一菌剂在拱棚辣椒上试验研究结果筛选出的枯草芽孢杆菌对辣椒连作土壤有明显减缓作用,但未开展复配菌剂在连作辣椒土壤的相关试验,能否减缓土壤连作障碍尚待研究。

【拟解决的关键技术问题】本试验以辣椒连作 8 a 塑料大棚为研究对象,通过不同微生物菌剂的配比施用,研究连作辣椒土壤对不同微生物菌剂用量的影响特征,测定了辣椒连作土壤中细菌、真菌和放线菌数量、土壤酶活性、土壤离子变化特点,以为筛选出 1~2 种不同微生物菌剂配比比例,为能够有效缓解拱棚辣椒土壤连作障碍提供科学依据。

1 材料与方法

1.1 试验地点

试验地点位于宁夏南部山区彭阳县新集乡拱棚辣椒示范基地(连作辣椒

8 a),实施地点土壤状况(全氮 1.05 g·kg^{-1},全磷 0.97 g·kg^{-1},全钾 23.0 g·kg^{-1},有机质 23.0 g·kg^{-1},碱解氮 63.8 g·kg^{-1},速效磷 30.4 g·kg^{-1},速效钾 195 g·kg^{-1},pH 7.63,全盐 1.17 g·kg^{-1})。

1.2 供试材料

微生物菌剂(胶质芽孢杆菌、巨大芽孢杆菌和枯草芽孢杆菌)均为宁夏尚博农植物保护科技有限公司提供。

1.3 试验方法

试验设 5 个处理。

处理 1:胶质芽孢杆菌:巨大芽孢杆菌:枯草芽孢杆菌=1:1:1;

处理 2:胶质芽孢杆菌:巨大芽孢杆菌:枯草芽孢杆菌=2:1:2;

处理 3:胶质芽孢杆菌:巨大芽孢杆菌:枯草芽孢杆菌=1:2:2;

处理 4:胶质芽孢杆菌:巨大芽孢杆菌:枯草芽孢杆菌=2:3:3;

处理 5:胶质芽孢杆菌:巨大芽孢杆菌:枯草芽孢杆菌=3:2:3;

对照(CK):不施菌肥。

微生物菌剂按照不同比例混合后,均以 500 倍稀释灌根,3 次重复,每个小区保证辣椒株数 100 株以上,分别在幼苗生长期、营养生长期、开花坐果期、结果盛期灌根,每株灌 100 mL。

1.4 调查项目

1.4.1 土壤取样

结果盛期取根际土壤,采样深度为 30 cm,每个处理分散 5 点取样 1 000 g,取 100 g 保持鲜样,4℃冷藏,其余风干待测。

1.4.2 土壤微生物数量测定

采用稀释平板法测定新鲜土壤中细菌、真菌、放线菌数量,细菌采用牛肉膏蛋白胨琼脂培养基;真菌采用马丁氏琼脂培养基;放线菌采用改良高氏 1 号培养基。实验步骤:称取 10 g 土样分别加入备好盛有 90 mL 无菌水的三角瓶中,置于振荡器上振荡 10 min,从所制悬浮液中吸取 1 mL 移入 9 mL 无菌水的试管中,依次类推稀释至所需浓度。真菌稀释 10^3 倍、放线菌稀释 10^4 倍、细菌稀释 10^5 倍,3 类微生物测定分别在培养 3、7、9 d 后统计菌落数。

1.4.3 土壤八大离子测定

钾离子和钠离子:火焰光度法,钙离子和镁离子:EDTA 滴定法或者 ICP、原子吸收光谱法,氯离子:硝酸银滴定法,硫酸根离子:EDTA 滴定法或者硫酸钡比浊法,碳酸根和重碳酸根离子:双指示剂滴定法。

1.4.4 土壤酶活性测定

土壤脲酶采用苯酚钠-次氯酸钠比色法,土壤蔗糖酶、纤维素酶采用 3,5 二硝基水杨酸比色法,过氧化氢酶采用高锰酸钾滴定法。

1.4.5 叶绿素含量

第 3 次灌根后测量辣椒叶绿素(每小区 5 株)。

2 结果与分析

2.1 连作辣椒土壤酶活性对微生物菌剂的响应

酶是土壤或基质中生物活性最强的部分,反映了土壤或基质中各种生化过程的强度,可以作为评价土壤或基质肥力状况的生物活性指标,酶活性越高,土壤或基质性状越好[23]。从图 1~4 可知,不同处理连作辣椒土壤酶活性有差异,土壤蔗糖酶是影响土壤碳代谢的关键酶。处理 2 蔗糖酶活性最强,为 777.07 mg/g·24 h,处理 3、5 蔗糖酶活性次之,分别为 575.95 mg/g·24 h、593.51 mg/g·24 h,处理 1 与对照(CK)蔗糖酶活性最小,分别为 520.61 mg/g·24 h、518.29 mg/g·24 h;土壤脲酶是表征土壤氮素转化的关键酶。处理 2 脲酶活性最强,为 201.53 mg/g·24 h,处理 3、5 及对照(CK)脲酶活性次之,处理 1、4 脲酶活性最弱,分别为 141.61 mg/g·24 h、137.29 mg/g·24 h;处理 2~4 纤维素酶活性最强,分别为 0.18 mg/g·min、0.17 mg/g·min、0.18 mg/g·min,处理 1、5 纤维素酶活性次之,对照(CK)纤维素酶活性最弱,为 0.08 mg/g·min;过氧化氢酶是一种分解土壤中过氧化氢进而减小其对植物毒害作用的氧化还原酶,可以用来表征土壤的生化活性。处理 2、4、对照(CK)过氧化氢酶活性差异不显著,均为 7.38 mg/g·72 h,处理 1、3 过氧化氢酶活性次之,处理 5 过氧化氢酶活性最弱,为 7.22 mg/g·72 h。

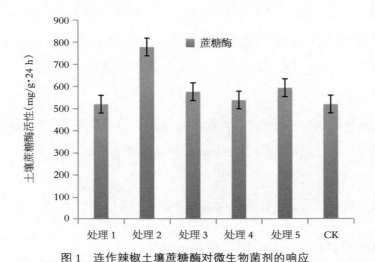

图 1 连作辣椒土壤蔗糖酶对微生物菌剂的响应

Fig. 1 Response of sucrase in continuous cropping pepper soil to microbial agents

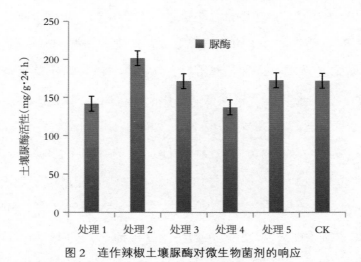

图 2 连作辣椒土壤脲酶对微生物菌剂的响应

Fig. 2 Response of soil urease of continuous cropping pepper to microbial agents

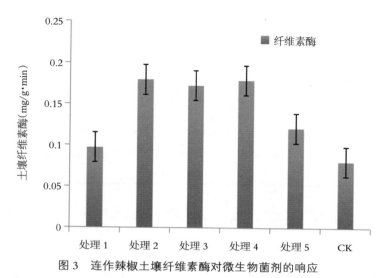

图 3 连作辣椒土壤纤维素酶对微生物菌剂的响应

Fig. 3 Response of soil cellulase in continuous cropping pepper to microbial agents

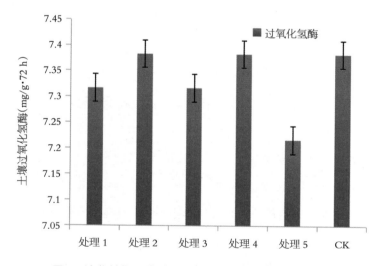

图 4 连作辣椒土壤过氧化氢酶对微生物菌剂的响应

Fig. 4 Response of catalase in soil of continuous cropping pepper to microbial agents

2.2 连作辣椒土壤微生物数量对微生物菌剂的响应

从表 1 可以看出,不同处理连作辣椒微生物数量差异显著,处理 2 土壤中细菌数量最高,为 $8.1×10^6$,处理 5、对照(CK)土壤中细菌数量次之,处理 1、3、4 土壤中细菌数量最低,分别为 $5.6×10^6$、$5.3×10^6$、$5.4×10^6$;处理 3、处理 4

土壤中真菌数量最高，分别为 $6.4×10^6$、$8.2×10^6$，处理 1、5 及对照（CK）土壤真菌数量次之，处理 2 土壤中真菌数量最低，为 $1.2×10^6$；处理 2、3 土壤中放线菌数量最低，分别为 $1.9×10^6$、$1.6×10^6$；处理 4、对照（CK）土壤中放线菌数量最高，分别为 $6.6×10^6$、$7.2×10^6$，处理 1、处理 5 土壤中放线菌数量次之。

表 1 连作辣椒土壤微生物数量对微生物菌剂的响应

Table 1 Response of soil microbial quantity of continuous cropping pepper to microbial microbial agent

处理 Treatment	细菌/真菌 Bacteria/Fungi	细菌 Bacteria CFU/g	真菌 Fungi CFU/g	放线菌 Actinomycetes CFU/g
处理 1	1.44	$5.6×10^6$	$3.9×10^6$	$4.2×10^6$
处理 2	6.75	$8.1×10^6$	$1.2×10^6$	$1.9×10^6$
处理 3	4.42	$5.3×10^6$	$6.4×10^6$	$1.6×10^6$
处理 4	0.66	$5.4×10^6$	$8.2×10^6$	$6.6×10^6$
处理 5	2.57	$7.7×10^6$	$3.0×10^6$	$2.5×10^6$
CK	1.24	$6.8×10^6$	$5.5×10^6$	$7.2×10^6$

2.3 连作辣椒土壤离子对微生物菌剂的响应

从表 2 可以看出，不同处理土壤中碳酸氢根离子含量均低于对照（CK），且差异显著（$P<0.05$），分别比对照减少 33.3%、20.5%、30.8%、28.2%、43.6%；处理 3、4 土壤中氯离子含量高于对照（CK）均比对照增加 14.3 %，说明氯离子在土壤中富集；处理 1、2、5 土壤中氯离子含量与对照（CK）无明显差异，说明氯离子在土壤中并未富集；处理 1、3 土壤中硫酸根离子含量均低于（CK），处理 2、4、5 土壤中硫酸根离子含量均高于对照（CK），分别比对照增加 22.4%、6.35%、9.23%，说明硫酸根离子在土壤中大量富集；处理 1、3、5 土壤中钙离子含量均高于对照（CK），分别比对照增加 11.1%、36%、50%，说明土壤中钙离子在土壤中大量富集，处理 2、4 钙离子含量均低于对照（CK）；处理 2、4 土壤中镁离子含量均高于对照（CK），分别比对照增加 50%、18.1%，说明镁离子在土壤中大量富集，处理 1、3、5 土壤中镁离子含量均低于对照（CK）；处理 2、4 土壤中钾离子均低于对照（CK），处理 1、3、5 土壤中钾离子含量均高于对照

(CK),分别比对照增加18.3%、12.5%、43.7%,说明钾离子在土壤中大量富集;处理2、3、4、5土壤中钠离子含量均低于对照(CK),说明钠离子在土壤中并未富集,处理1土壤中钠离子含量高于对照(CK),比对照增加1.5%,说明钠离子在土壤中稍有富集,但并不明显。

表2 连作辣椒土壤离子对微生物菌剂的响应
Table2 Response of soil ions in continuous cropping pepper to microbial agents

处理 Treatment	碳酸氢根离子 Bicarbonateion (g/kg)	氯离子 Chlorideion (g/kg)	硫酸根离子 Sulfateion (g/kg)	钙离子 Calciumion (g/kg)	镁离子 Magnesiumion (g/kg)	钾离子 Potassiumion (g/kg)	钠离子 Sodiumion (g/kg)
处理1	0.26	0.12	0.50	0.18	0.08	0.060	0.066
处理2	0.31	0.12	0.76	0.14	0.18	0.040	0.043
处理3	0.27	0.14	0.37	0.25	0.07	0.056	0.051
处理4	0.28	0.11	0.63	0.15	0.11	0.036	0.057
处理5	0.22	0.14	0.65	0.32	0.08	0.087	0.036
CK	0.39	0.12	0.59	0.16	0.09	0.049	0.065

2.4 连作辣椒叶绿素含量对微生物菌剂的响应

从图5可以看出,不同处理对连作辣椒叶绿素含量有影响,处理2、5叶

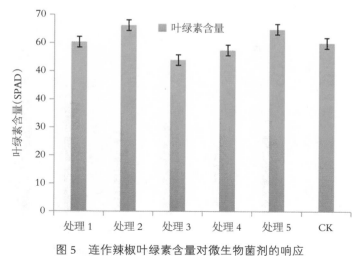

图5 连作辣椒叶绿素含量对微生物菌剂的响应
Fig.5 Response of chlorophyll content of continuous cropping capsicum to microbial agents

绿素含量最高,分别为 66.277 SPAD、64.87 SPAD,处理 1、对照(CK)次之,分别为 60.33 SPAD、60.0 SPAD,处理 3、4 叶绿素含量最低,分别为 53.94 SPAD、57.44SPAD。

2.5 连作辣椒抗病性对微生物菌剂的响应

从图 6~7 可以看出,不同处理连作辣椒白粉病发病率及病情指数有差

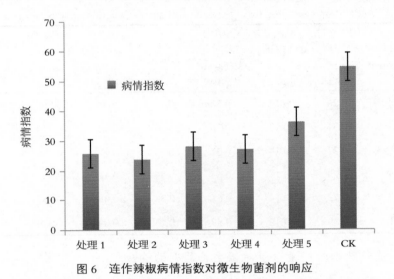

图 6 连作辣椒病情指数对微生物菌剂的响应

Fig.6 Response of condition index of continuous cropping pepper to microbial agents

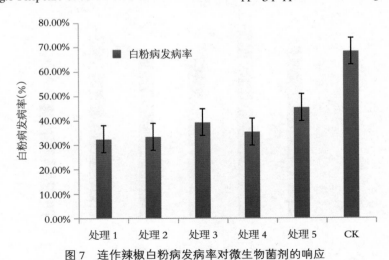

图 7 连作辣椒白粉病发病率对微生物菌剂的响应

Fig. 7 Response of powdery mildew incidence of continuous cropping pepper to microbial agents

异，对照（CK）辣椒白粉病发病率、病情指数最高，分别为68.2%、54.51，处理5辣椒白粉病发病率、病情指数次之，分别为45.33%、36.23，处理1~4辣椒白粉病发病率、病情指数较低，分别在32.4%~39.33%，23.8~28.11。

3 讨论

微生物是维持土壤质量的重要组成部分，能够帮助植物适应养分胁迫的环境，改善土壤养分的吸收和利用[20,21]。以微生物为核心研发的新型复合菌剂、生物肥料、新型生物农药的使用，不仅可以减少化肥、农药的施用，有效提高土壤生产力，而且还可以起到对作物健株、控害的作用[22-24]。土壤酶是土壤有机质分解与养分转化和循环的驱动力，是土壤质量和生态稳定性的重要指标[5-7]，土壤酶活性的高低能反映土壤生物活性和土壤生化反应强度，土壤过氧化氢酶、脲酶、蔗糖酶等关键酶的共同作用促进了作物的生长发育和根系活力的提高[8-9]。研究发现，接种微生物菌剂能够改善土壤结构和营养状况，调节土壤酶的活性，提高土壤肥力和养分利用率[20]。芽孢杆菌菌剂能够显著改善根际土壤微生态，提高土壤脲酶、蔗糖酶、蛋白酶、过氧化物酶和脱氢酶活性[21]。本试验结果基本与前人研究结果一致。

4 结论

本研究中，处理2、5叶绿素含量最高，分别为66.27 SPAD、64.87SPAD；对照（CK）辣椒白粉病发病率、病情指数最高，分别为68.2%、54.51；处理2蔗糖酶活性、脲酶活性最强，处理2、3、4纤维素酶活性最强；处理2土壤中细菌数量最高，为$8.1×10^6$，处理2、3土壤中放线菌数量最少，分别为$1.9×10^6$、$1.6×10^6$；不同处理土壤中离子含量与对照（CK）相比，均有不同程度富集，处理3、处理4土壤中氯离子含量高于对照（CK）均比对照增加14.3%，处理2、4、5土壤中硫酸根离子含量均高于对照（CK），分别比对照增加22.4%、6.35%、9.23%，说明硫酸根离子在土壤中大量富集；处理2、3、4、5土壤中钠离子含量均低于对照(CK),说明钠离子在土壤中并未富集,处理1土壤中钠离子含量高于对照(CK),比对照增加1.5%,说明钠离子在土壤中稍有富集,但并不明显。

综上所述,处理2(胶质芽孢杆菌:巨大芽孢杆菌:枯草芽孢杆菌=2:1:2)可以显著降低土壤中有害真菌和放线菌数量,防止土壤由细菌型向真菌型转变;可以为辣椒生长提供足够的养分,提高土壤酶活性,减轻病害发生程度,减少土壤中离子的富集,并能有效缓解辣椒土壤连作障碍。

参考文献

[1] 吴凤芝,赵凤艳,刘元英.设施蔬菜连作障碍原因综合分析与防治措施[J].东北农业大学学报,2000,31(3):241-247.

[2] 吴凤芝,孟立君,王学征.设施蔬菜轮作和连作土壤酶活性的研究.植物营养与肥料学报,2006,12(4):554-558.

[3] 邢宇俊,程智慧,周艳丽,等.保护地蔬菜连作障碍原因及其调控[J].西北农业学报,2004,13(1):120-123.

[4] 郑军辉,叶素芬,喻景权.蔬菜连作障碍产生原因及生物防治.中国蔬菜,2004(3):56-58.

[5] 解媛媛,谷洁,高华,等.微生物菌剂酶制剂化肥不同配比对秸秆还田后土壤酶活性的影响[J].水土保持研究,201,17(2):238-243.

[6] 蔡燕飞,廖宗文,章家恩,等.生态有机肥对番茄青枯病及土壤微生物多样性的影响[J].应用生态学报,2003,14(3):349-353.

[7] 王明友,杨秀凤,郑宪和,等.复合微生物菌剂对番茄的光合特性及产量品质的影响[J].土壤肥料,2004(4):37-39.

[8] 王明友,李光忠,杨秀凤,等.微生物菌肥对保护地黄瓜生育及产量、品质的影响研究初报[J].土壤肥料,2003(3):38-40.

[9] 吉牛拉惹.生物菌肥对土壤肥力影响的研究初报[J].西昌学院学报:自然科学版,2005,19(4):25-26.

[10] 陈雪丽,王光华,金剑,等.两株芽孢杆菌对黄瓜和番茄根际土壤微生物群落结构影响[J].生态学杂志,2008,27(11):1895-1900.

[11] 王茹华,周宝利,张启发,等.嫁接对茄子根际微生物种群数量的影响[J].园艺学报,2005,32(1),124-126.

[12] 赵兰坡,姜岩.土壤磷酸酶活性测定方法的探讨[J].土壤通报,1986,6(3):138-141.

[13] 郭明,尹亚梅,何良荣,等.农用化学物质对土壤脲酶活性的影响[J].农业环境保护,2000,19(2):68-71.

[14] 郭瑞英,陈清,李晓林.土壤微生物抑病性与土壤健康[J].中国蔬菜,2005,S1:78-82.

[15] 张晶,张惠文,李新宇,等.土壤微生物生态过程与微生物功能基因多样性[J].应用生态学报,2006,17(6):1129-1132.

[16] 耿丽平,李小磊,赵全利,等.添加微生物菌剂对小麦产量及土壤生物学性状的影响[J].江苏农业科学,2017,45(5):50-54.

[17] 张丽荣,陈杭,康萍芝,沈瑞清.不同微生物菌剂对番茄产量及土壤微生物数量的影响[J].湖北农业科学,2013,52(22):5452-5454+5518.

[18] 钱海燕,杨滨娟,黄国勤,等.秸秆还田配施化肥及微生物菌剂对水田土壤酶活性和微生物数量的影响[J].生态环境学报,2012,21(3):440-445.

[19] 雷先德,李金文,徐秀玲,等.微生物菌剂对菠菜生长特性及土壤微生物多样性的影响[J].中国生态农业学报,2012,20(4):488-494.

[20] 李自刚,王新民,刘太宇,等.复合微生物菌肥对怀地黄连作障碍修复机制研究[J].湖南农业科学,2008(5):62-65.

[21] 曹齐卫,张卫华,李利斌,等.济南地区日光温室土壤养分的分布状况和累积规律[J].应用生态学报,2012,23(1):115-124.

[22] 王立刚,李维炯,邱建军,等.生物有机肥对作物生长、土壤肥力及产量的效应研究.土壤肥料,2004(5):12-16.

[23] 曲再红,杜相革.土壤添加物、土壤微生物和番茄苗期生长相互关系的研究.中国农学通报,2004,20(4):84-86.

[24] 赵贞,杨延杰,林多,等.微生物菌肥对日光温室黄瓜生长发育及产量品质的影响[J].中国蔬菜,2012(18):149-153.

[23] Zou C J, Zhang Y Y, Zhang Y M, et al. Regulation of biochar on matrix en-zyme activities and microorganisms around cucumber roots under continuous cropping[J]. Chinese Journal of Ap-plied Ecology,2015,26(6):1772-1778.

有机肥与玉米秸秆配施对连作辣椒生长发育、产量及品质的影响

高晶霞[1]，牛勇琴[2]，吴雪梅[2]，谢 华[1]，王学梅[1]，裴红霞[1]

（1. 宁夏农林科学院种质资源研究所，宁夏银川　750002；
2. 宁夏回族自治区彭阳县蔬菜产业发展服务中心，宁夏彭阳　756500）

摘　要：通过有机肥与玉米秸秆配施对拱棚连作辣椒生长发育、产量及品质的研究，结果表明：处理3、处理4、处理6连作辣椒株高、茎粗、开展度、单株结果数、单果重、净光合速率，均高于处理2、处理1及空白对照（CK）。处理6小区总产量、折合亩产量、折合亩产值均最高，分别为237.4 kg、6 766.9 kg、744.6元，分别比处理1、处理2、处理3、处理4、处理5及空白对照（CK）亩产量增幅5.6%、18.9%、2.2%、4.4%、18.1%、26.7%。在辣椒采收盛期，处理1、处理6维生素C含量最高，分别为105 mg/g、101 mg/g。

关键词：有机肥；玉米秸秆；连作；辣椒；生长发育；产量

Effects of Combined Application of Organic Fertilizer and Corn Stalk on Growth, Yield and Quality of Continuous Cropping Pepper

GAO Jing-xia[1], NIUYong-qin[2], WUXue-mei[2], WANG Xue-mei[1], XIE Hua[1], PEI Hong-xia[1]

（1. Ningxia Academy of Agriculture and Forestry Plant Resources, Ningxia, Yinchuan, 750002; 2. The Ningxia Hui Autonomous Region, Pengyang County Vegetable Industry Development Service, Ningxia 756500）

Abstract: The effects of combination of organic manure and straw on the growth index, yield and quality of Capsicum in continuous cropping of arch shedint, The results show that: The plant height, stem diameter, development degree, single plant number, single fruit weight and net photosynthetic rate of treated 3, 4 and 6 continuous cropping pepper were treated, all of them were higher than treatment 2, treatment 1and blank control. treatment 6, the total output of the plot, the yield per mu, and the output value of Mu are the highest. 237.4 kg, 6 766.9 kg and 744.6 yuan, respectively, the yield per mu increased by 5.6%, 18.9%, 2.2%, 4.4%, 18.1% and 26.7% respectively compared with treatment 1, treatment 2, treatment 3, treatment 4, treatment 5 and blank control (CK). In the harvest period, Treatment 1, treatment 6 Vitamin C content of the highest, 105 mg/g and 10 1mg/g, respectively.

Key words: organic fertilizer; straw; continuous cropping; capsicum;growth and development;yield

辣椒是人们日常生活中必不可少的重要食品，也是农民务农收入的重要经济来源。随着辣椒生产日趋规模化和专业化，辣椒连作障碍日益加重，成为辣椒栽培上亟待解决的一大难题，严重制约辣椒生产的可持续发展，备受学者关注[1]。目前普遍认为辣椒连作障碍的主要原因有病虫害、营养失调和自毒作用，这些因素导致了土壤理化性状恶化及土壤酶活性降低，从而直接或间接地影响了辣椒的生长[2-4]。施用化肥虽可以显著提高农作物的产量，但随着施肥量的增加，肥效递减的问题日益突出。过量的化肥投入不仅使生产成本增加，同时也带来了农产品品质下降、环境污染等问题。精制有机肥是集有机营养与无机营养元素于一体的全营养型肥料[5]，有关施用有机肥可以提高蔬菜产量和品质的研究已有报道[6,7]。许多研究报道指出：微生物肥料与有机肥(或无机肥)配施营养合理，肥效持久，可以提高产量，增加收益，改善品质，减轻病害，改良土壤环境，增强土壤可持续利用能力，抗旱保水，消除污染[8,9]。本试验通过鸡粪、羊粪、牛粪与玉米秸秆配施对连作辣椒生长发育、产量及品质的影响，以期为进一步明确有机肥对辣椒连作障碍缓解效果，建立切实可行的缓解技术提供科学依据。

1 材料与方法

1.1 试验地点

试验地点位于宁夏彭阳县新集乡拱棚辣椒核心试验示范基地(连作辣椒 3 a 以上,全氮 1.07 g·kg^{-1},全磷 0.95 g·kg^{-1},全钾 22.0 g·kg^{-1},有机质 24.0 g·kg^{-1},碱解氮 63.0 g·kg^{-1},速效磷 29.4 g·kg^{-1},速效钾 192 g·kg^{-1},pH 7.50,全盐 1.10 g·kg^{-1})。

1.2 供试材料

羊粪、牛粪、鸡粪、玉米秸秆,使用前,处理 1~6 分别进行腐熟处理。

1.3 试验方法

试验设 6 个处理。

处理 1:底肥施腐熟羊粪,每 667 m² 用量 4 000 kg,约 6.4 m³ 左右;

处理 2:底肥施腐熟牛粪,每 667 m² 用量 5 000 kg,约 8 m³ 左右;

处理 3:底肥施腐熟消毒鸡粪,每 667 m² 用量 1 500 kg;

处理 4:底肥施腐熟羊粪+玉米秸秆,每 667 m² 用量羊粪 4 000 kg+玉米秸秆 2 000 kg;

处理 5:底肥施腐熟牛粪+玉米秸秆,每 667 m² 用量牛粪 5 000 kg+玉米秸秆 2 000 kg;

处理 6:底肥施腐熟鸡粪+玉米秸秆,每 667 m² 用量鸡粪 1 500 kg+玉米秸秆 2 000 kg;

空白对照(CK):底肥不施有机肥。

每个处理 3 次重复,随机排列。

1.4 调查项目

1.4.1 植株

在辣椒开花期、结果盛期、采收中后期调查株高、茎粗、开展度、单株结果数、单果重。

1.4.2 产量

每个处理随机取 10 株辣椒调查单株产量、小区产量及折合 667 m² 产量。

1.4.3 叶绿素

用叶绿素仪（CSPAD502）测定相对叶绿素含量，并以 SPAD 值表示。每片叶重复 3 次，取平均值。

1.4.3 品质

用铝蓝比色法测定辣椒果实中维生素 C 含量，苯酚硫酸比色法测定可溶性糖含量，紫外分光光度法测定可溶性蛋白质含量，PRO-101 型糖度计测定可溶性固形物含量。

1.4.4 净光合速率

用美国产 Li-6400 型光合仪测定净光合速率（Pn），每小区选有代表性的 5 片叶上午 10:00-11:00 进行测定，每叶重复测定 3 次，取其平均值。

1.5 数据处理

数据用 EXCEL 和 SPSS 11.0 软件进行单因素方差分析，多重比较采用 LSR 法（Duncan's 法），显著水平 $P<0.05$。

2 结果分析

2.1 有机肥与秸秆配施对连作辣椒株高的影响

从图 1 中可知，有机肥与秸秆配施对拱棚连作辣椒株高之间差异不显著。在辣椒生长的 3 个时期，处理 3、处理 4 株高均高于其他有机肥与秸秆的辣椒株高，牛粪、牛粪+秸秆株高最低。

2.2 有机肥与秸秆配施对连作辣椒茎粗的影响

从图 2 可知，有机肥与秸秆配施对拱棚连作辣椒茎粗有差异。在辣椒生长的 3 个时期，处理 3、处理 6 茎粗均高于其他有机肥与秸秆，分别为 17.44 mm、17.34 mm，处理 1、处理 4 茎粗次之，处理 2、处理 5 及空白对照（CK）植

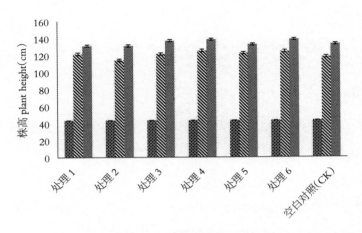

图 1 不同处理对连作辣椒株高的影响

Fig.1 Effects of different treatments on plant height of continuous cropping pepper

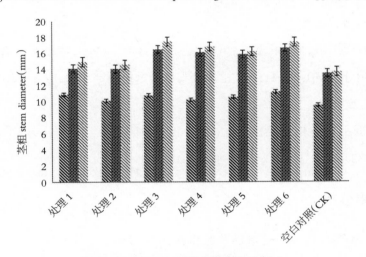

图 2 不同处理对连作辣椒茎粗的影响

Fig.2 Effects of different treatments on stem thickness of continuous cropping pepper

株茎粗,在辣椒生长的 3 个时期最低。

2.3 有机肥与秸秆配施对连作辣椒开展度的影响

从图 3 可知,有机肥与秸秆配施对拱棚连作辣椒开展度差异不显著。在辣椒生长的 3 个时期,处理 3、处理 6 开展度均高于其他有机肥与秸秆,处理 1、处理 4 茎粗次之,处理 2、处理 5、空白对照(CK)植株开展度在辣椒生长的三个时期最低。

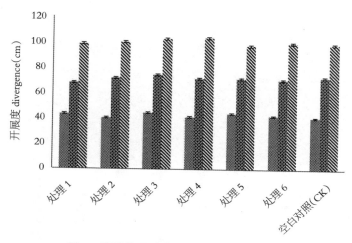

图 3 不同处理对连作辣椒开展度的影响
FIG.3 Effects of different treatments on the development degree of continuous cropping pepper

2.4 有机肥与秸秆配施对连作辣椒叶绿素含量的影响

从图 4 可知,有机肥与秸秆配施对拱棚连作辣椒叶绿素含量有差异。在辣椒生长盛期,处理 3、处理 1 叶绿素含量均高于其他处理,分别为 66.02 SPAD、64.81 SPAD,处理 2 叶绿素含量次之,处理 4、处理 5、空白对照(CK)叶绿素含量之间差异不显著,处理 5 叶绿素含量最低为 56.77 SPAD。

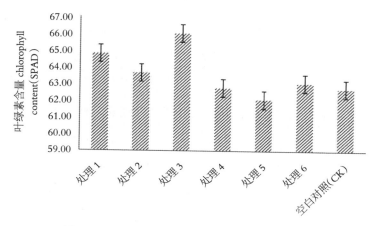

图 4 不同处理对连作辣椒叶绿素含量的影响
Fig.4 Effects of different treatments on chlorophyll content of continuous cropping pepper

2.5 有机肥与秸秆配施对连作辣椒净光合速率(PN)的影响

从图5中可知,有机肥与秸秆配施对拱棚连作辣椒净光合速率有差异。在辣椒采收盛期,处理1、处理3净光合速率均高于其他有机肥及空白对照(CK),分别为26.76 $\mu mol \cdot m^{-2} \cdot s^{-1}$、27.27 $\mu mol \cdot m^{-2} \cdot s^{-1}$,处理4、处理5净光合速率次之,处理2、处理6及空白对照(CK)净光合速率最低。

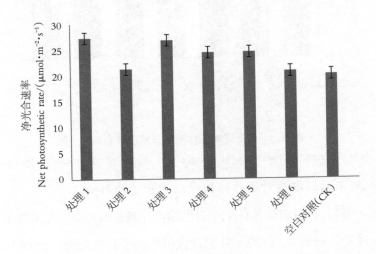

图5 不同处理对连作辣椒净光合速率(PN)的影响
Fig.5 Effects of different treatments on net photosynthetic rate (PN) of continuously cultivated pepper

2.6 有机肥与秸秆配施对连作辣椒单株结果数的影响

从图6可知,有机肥与秸秆配施对拱棚连作辣椒单株结果数有差异。在辣椒采收盛期,处理1、处理5单株结果数高于其他有机肥及秸秆单株结果数,分别为23.73个、23.2个,处理3、处理4、处理6单株结果数次之,处理4及空白对照(CK)单株结果数最少。

2.7 有机肥与秸秆配施对连作辣椒单果重的影响

从图7可知,有机肥与秸秆配施对拱棚连作辣椒单果重有差异。在辣椒采收盛期,处理1、处理3、处理6单果重高于其他有机肥及秸秆,分别为115.9g、115.8g、115.6g,处理2、处理4及空白对照(CK)单果重次之,处理5

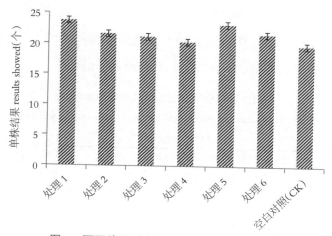

图 6 不同处理对连作辣椒单株结果数的影响

Fig.6 Effects of different treatments on the number of pepper per plant

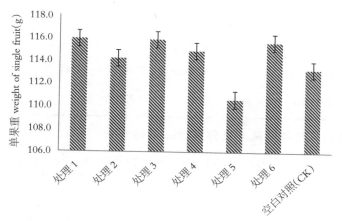

图 7 不同处理对连作辣椒单果重的影响

Fig.7 Effects of different treatments on single fruit weight of continuously cultivated pepper

单果重最低。

2.8 有机肥与秸秆配施对连作辣椒产量的影响

从表 1 可以看出,有机肥与秸秆配施对连作辣椒产量存在显著性差异($P<5\%$、$P<1\%$)。处理 3、处理 6、处理 4 均高于牛粪、羊粪、牛粪+秸秆及空白对照。处理 5 小区总产量、折合亩产量、折合亩产值均最高,分别为 237.4 kg、6 766.9 kg、744.6 元,分别比处理 1、处理 2、处理 3、处理 4、处理 5 及空白对

照(CK)亩产量增幅 5.6%、18.9%、2.2%、4.4%、18.1%、26.7%。处理 1、处理 3、处理 4 小区总产量、折合亩产量、折合亩产值差异不显著(P<5%)，处理 2、处理 5 及空白对照小区总产量、折合亩产量、折合亩产值最低。

表 1 不同处理对连作辣椒产量的影响

Table 1. Effect of combination of organic manure and straw on the yield of continuous cropping pepper

处理 Handle	小区总产量 Total yield of small area(kg/23.4 m²)				折合总产量 Aggregate yield (kg/667 m²)	折合总产值 Aggregate output value (元)	比对照 Comparative control± (kg)	显著性分析 Saliency analysis 5%、1%		位次
	I	II	III	平均						
处理 1	219.6	226.5	228.7	224.9	6 410.6	7 051.7	1 071.7	c	C	4
处理 2	200.4	203	195.2	199.5	5 686.6	6 474.7	547.2	d	C	6
处理 3	232.1	229.5	235.7	232.4	6 624.4	7 286.8	1 285.5	ab	AB	2
处理 4	228	230.5	224	227.5	6 484.7	7 133.2	1 145.8	bc	B	3
处理 5	200.2	204.9	198	201	5 729.4	6 553.1	618.5	d	C	5
处理 6	240.5	234.8	237	237.4	6 766.9	7 443.6	1 428.0	a	A	1
空白对照(CK)	184.4	187.6	190	187.3	5 338.9	5 872.8		e	D	7

2.9 有机肥与秸秆配施对连作辣椒品质的影响

从图 8-10 可知，有机肥与秸秆配施辣椒品质存在显著性差异，在辣椒

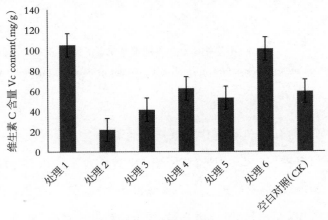

图 8 不同处理对连作辣椒维生素 C 含量的影响

Fig.8 Effects of different treatments on vitamin C content of continuously cultivated pepper

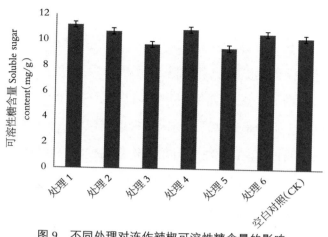

图 9 不同处理对连作辣椒可溶性糖含量的影响

Fig.9 Effects of different treatments on soluble sugar content of continuous cropping pepper

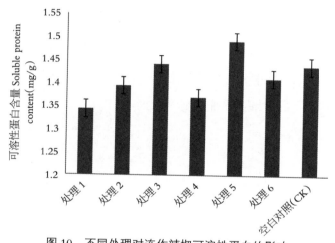

图 10 不同处理对连作辣椒可溶性蛋白的影响

Fig.10 Effects of different treatments on soluble protein of continuous cropping pepper

采收盛期,处理 1、处理 6 维生素 C 含量最高,分别为 105 mg/g、101 mg/g,处理 3、处理 5 及空白对照(CK)维生素 C 含量不显著,处理 2 维生素 C 含量最低,为 21.8 mg/g。处理 1 可溶性糖含量最高为 11.2 mg/g,处理 2、处理 4、处理 6 可溶性糖含量次之,处理 3、处理 5 可溶性糖含量最低。处理 5 辣椒可溶性蛋白含量最高为 1.49 mg/g,处理 3 及空白对照(CK)可溶性蛋白含量次之,处理 2 可溶性蛋白含量最低。

3 结论

前人研究,微生物肥料与有机肥(或无机肥)配施营养合理,肥效持久,可以提高产量,增加收益,改善品质[8,9],本试验研究结果与前人研究结果基本一致,通过有机肥与玉米秸秆配施对拱棚连作辣椒生长指标、产量及品质的研究,结果表明:处理3、处理5、处理4连作辣椒株高、茎粗、开展度、单株结果数、单果重、净光合速率均高于处理2、处理1、处理5及空白对照。处理6小区总产量、折合亩产量、折合亩产值均最高,分别比处理1、处理2、处理3、处理4、处理5及空白对照(CK)亩产量增幅5.6%、18.9%、2.2%、4.4%、18.1%、26.7%。在辣椒采收盛期,处理1、处理6维生素C含量最高,分别为105 mg/g、101 mg/g,处理5维生素C含量最低,为21.8 mg/g。

综合分析结果:处理1、处理3、处理4、处理6,在一定程度上提高了连作辣椒产量及品质,为进一步明确有机肥及秸秆配施对辣椒连作障碍缓解效果,建立切实可行的缓解技术提供科学依据。

参考文献

[1] 朱玲玲.有机肥对花椰菜产量和经济性状的影响[J].中国瓜菜,2006,(2):20-21

[2] 李松龄.有机—无机肥料配施对番茄产量及品质的影响[J].北方园艺,2006,(3):3-4.

[3] 耿士均,刘刊,商海燕,等.专用微生物肥对连作辣椒和番茄生长的影响[J].浙江农业科学,2012(5):651-656.

[4] 毕军,夏光利,张昌爱,等.有机生物活性肥料对冬小麦生长及土壤活性质量影响的试验研究[J].土壤通报,2005,36(2):230-233.

[5] 王立刚,李维炯,邱建军,等.生物有机肥对作物生长、土壤肥力及产量的效应研究[J].土壤肥料,2004(5):12-16.

[6] 孙艺文,吴凤芝.小麦、燕麦残茬对连作黄瓜生长及土壤酶活性的影响[J].中国蔬菜,2013,(4):46-51.

[7] Yu X Y, Wang X F, Zhang W Q, et al. Antisense suppression of an acid invertase

gene (MAI1) in muskmelon alters plant growth and fruit development [J]. Journal of Experimental Botany, 2008, 59(11):2969-2977.

[8] Nguyen M T, Ranamukhaarachchi S L. Soil-borne antagonists for biological control of bacterial wilt disease caused by Ralstonia solanacearum in tomato and pepper [J]. Journal of Plant Pathology, 2010, 92(2):395-406.

[9] Camprubi A, Estaun V, Nogales A, et al. Response of the grapevine rootstock Richter 110 to inoculation with native and selected arbuscular mycorrhizal fungi and growth performance in a replant vineyard[J]. Mycorrhiza, 2007, 18(4):211-216.

Study on Growth and Photosynthetic Characteristics of Pepper under Different Fertilization Modes

Jingxia GAO[1], Lihua CAO[2], Xuemei WU[3], Shoucai MA[3], Hua XIE[1]

(1. Institute of Plant Germplasm Resources, Ningxia Academy of Agro-Forestry Sciences, Yinchuan 750002, China; 2. Ningxia Rural Science and Technology Development Center, Yinchuan 750001, China; 3. Pengyang County Agricultural Technology Extension and Service Center of the Ningxia Hui Autonomous Region, Pengyang 756500, China)

摘 要：以新冠龙辣椒为供试材料,研究了不同施肥种类对拱棚辣椒生长发育、产量及光合特性的影响,结果表明：T_4辣椒植株的株高、开展度最大,分别为104.3、90.6 cm，T_1辣椒单果重、果长、果粗均最大,分别为101.6、25.49、4.86 cm。T_1辣椒叶片总叶绿素含量最高为66.7 mg/g,根系活力最强为100.6 μg/(g·FW·h)。T_5辣椒叶片净光合速率下降幅度最高为6.6 μmol/(m²·s)，T_1、T_2辣椒叶片净光合速率下降幅度最低,波动于1.3~2.5 μmol/(m²·s)。T_3辣椒叶片气孔导度下降幅度最高为2 658.1 mol/(m²·s)，T_2、T_5辣椒叶片气孔导度下降幅度最低,波动于386.7~428.7 mol/(m²·s)。T_1辣椒叶片蒸腾速率下降幅度最高为4.7 mmol/(m²·s)，T_5辣椒叶片蒸腾速率下降幅度最低为1.1 mmol/(m²·s)，辣椒叶片胞间CO_2浓度下降幅度最高为436.8 μmol/mol。T_1辣椒的小区产量、产量、效益均最高,分别为340.1 kg、96 448.5 kg/hm²、162 865.6 Yuan/hm²，T_3辣椒的小区产量、产量、效益最低,分别为260.8 kg、74 338.5 kg/hm²、124 888.5 Yuan/hm²。

关键词：施肥种类;拱棚;辣椒;生长发育;光合特性

Abstract：With the new pepper cultivar Guanlong as a test material, effects of different fertilizer application modes on growth and development, yield and photosynthetic characteristics of pepper were studied. The results showed that the plant height and plant expansion of pepper in T_4 were the largest, of 104.3 and 90.6 cm, respectively. T_1 showed the largest weight of single fruit, fruit length and fruit diameter, respectively of 101.6 g, 25.49 cm and 4.86 cm. The content of total chlorophyll in the leaves of T_1 was the highest, of 66.7 mg/g, and the root activity of T_1 was also the highest, of 100.6 g/(g·FW·h). The net photosynthetic rate of pepper leaves in T_5 was the highest, of 6. μmol/(m²·s), while the net photosynthetic rates of T_1 and T_2 were the lowest, fluctuating in the range of 1.3~2.5 μmol/(m²·s). T_3 showed the highest decrease amplitude of stomatal conductance in pepper leaves, of 2 658.1 mol/(m²·s), while the decrease amplitudes of stomatal conductance of pepper leaves in treatments T_2 and T_5 were the lowest, fluctuating in the range of 386.7~428.7mol/(m²·s). T_1 showed the highest decrease amplitude of transpiration rate of pepper leaves, of 4.7 mmol/(m²·s), while treatment T_5 showed the lowest decrease amplitude of transpiration rate of pepper leaves, of 1.1 mmol/(m²·s). The intercellular CO_2 concentration was the highest in the leaves of T_5, of 436.8 μmol/mol. T_1 showed the highest plot output, yield per hectare and benefit, of 340.1 kg, 96 448.5 kg/hm² and 162 865.6 Yuan/hm², respectively, while T_3 exhibited the lowest plot output, yield per hectare and benefit, of 260.8 kg, 74 338.5 kg/hm² and 124 888.5 Yuan/hm², respectively.

Key words: Fertilizer types; Arch shed; Pepper; Growth and development; Photosynthetic characteristics

Pengyang County is located at the south border of the Ningxia Hui Autonomous Region, the eastern foot of Liupan Mountain, 106°32′~106°58′ east longitude, 35°41′~36°17′ north latitude. In this area, the annual average air temperature is 7.4~8.5℃, the frost-free sea-son is 140~170 d, and the precipitation is 350~550 mm. This area has a typical temperate continental semiarid monsoon climate, which is especially suitable for cultivation of pepper[1-2]. However, with the years of facility cultivation increasing, continuous seasonal or allyear coverage over many years has changed ecological balance under natural condition. Especially, all-year multicropping leads to too-high soil output capacity, farmers improve fertilizer input blindly, and therefore, nutritional imbalance is aggravated, resulting in problems including soil deterioration, physiological disease, successive cropping obstacle, and decrease of yield and quality[3-4]. Directing at re quirement for arch greenhouse continuous high-yield pepper cultivation, inaccurate experience type topdressing, imbalanced N, P and K supplement, low fertilizer efficiency, and high labor consumption, main drip irrigation fertilizers were selected, a gradient experiment was designed to investigate the effects on growth and development, yield and photosynthetic traits, so as to select a formula fertilization method integrating water and fertilizer suitable for arch greenhouse pepper cultivation with good cost performance and easy extension and application.

Materials and Methods

Time, location and scale

This experiment was carried out from May 20 to September 30, 2015, in 5 arch greenhouses in Henan Arch Greenhouse Pepper Garden in Baihe Village, Xinji Township, Penyang County.

Material selection

The tested pepper variety was Guanlong. The fertilizers were selected from those with good reliable quality commonly applied in this area; for topdressing in middle and later stages, higher P and K contents were required; the fertilizers should have outstanding solubility; and the fertilizers should have high price/performance ratio.

Experimental methods Dosage of different fertilizers

Single-factor treatments were designed according to recommended dosages of various fertilizers, and fertilization was performed for 10 times together with drip irrigation, with natural condition as control. Plants were selected randomly for long-term fixed-point monitoring, root activity was determined in laboratory after sampling, chlorophyll contents and photosynthetic rates at different stages were monitored with movable instruments in field, and plant height, door pepper time, average weight of single fruit and yield were conventional measured in field.

①Root activity was determined by TTC method [5]. ②Leaf chlorophyll con-tent: Chlorophyll content in the 5th-8th functional leaves was determined by a portable SPAD-502 chlorophyll meter in the morning of a sunny day, each plant was determined for 3 times, and 5~7 plants in each treatment were determined. ③Light intensity: Photo-synthetic rates of pepper plants in different treatments were measured at 10:00 on a sunny day. The used instrument was a TPS-2 portable photo-synthesis system produced by PP-systems. Five plants with good uniform growth were selected from each treatment, and the fifth functional leaf on each plant was determined.

Results and Analysis

Effects of different fertilizer types on growth of pepper plants and fruit traits

It could be seen from Table 3 that, the application of the 5 kinds of fertilizers affected plant traits of pepper. The harvest time of door pepper in T_4, T_5 and CK was the earliest, all on June 17, the harvest time of door pepper in treatments T_2 and T_3 was the second earliest, and treatment T_1 showed the latest harvest time of door pepper, on June 26. Under the same monitoring time, there were differences in plant height, plant expansion, weight of single fruit, fruit length, fruit thickness, total chlorophyll content and root activity between the 5 treatments, T_4 showed the largest plant height and plant expansion, of 104.3 and 90.6 cm, respectively, T_2, T_3 and T_5 showed plantheights in the range of 93.7~97.9 cm and plant expansion value in the range of 86.5 −90.6 cm, while pepper in the CK exhibited the smallest plant height and plant expansion, of 79.5 and 72.3 cm, respectively. There were differences in fruit traits between the 5 treatments, pepper in T_1 showed the largest weight of single fruit, fruit length and fruit diameter, of 101.6 g, 25.49 cm and 4.86 cm, respectively, and pepper in other treatments exhibited weights of single fruit in the range of 88.2~92.3 g, fruit length in the range of 19.98~21.87 cm, and fruit diameter in the range of 4.67~4.70 cm. There were differences in chlorophyll content between the 5 treatments, T_1 showed the highest total chlorophyll content of 66.7 mg/g, T_3 exhibited the second highest chlorophyll content, while no remarkable differences were observed between T_2, T_4 and T_5 in total chlorophyll content. The 5 fertilizers had different effects on root activity of pepper, pepper in T_1 showed the highest root activity 100.6 μg/(g·FW·h), while T_3 exhibited the

Table 1 Application of fertilizers

Type	Place of production	Active ingredient	Recommended dosage	Specifications	Unit Price
Luxi punching fertilizer(T_1)	Luxi Group Co.,Ltd.	$N-P_2O_5-K_2O$ 16−5−30⩾51% (potassium sulfate type)	Diluted with clear water according to 1:300 ~500, and applied at a rate in the range of 75~150 kg/hm²	20 kg (5 kg×4 bags)	12.825 Yuan/kg 64.125 Yuan/time
Major−element water−soluble fertilizer(T_2)	Luxi Group Co.,Ltd.	$N-P_2O_5-K_2O$ 15−10−30⩾55% (containing nitrate nitrogen)	Diluted with clear water according to 1:300 ~500, and applied at a rate in the range of 75~150 kg/hm²	20 kg (5 kg×4 bags)	14.250 Yuan/kg 71.25 Yuan/time
Mosaic crystal potassium(T_3)	Mosaic Company Potassium Fertilizer Esther Hatch Limited Partnership	Potassium oxide⩾60%		50 kg	4.6 Yuan/kg 23 Yuan/time
Lvjuneng major−element water−soluble fertilizer(T_4)	Jiangsu Zhongdong Fertilizer Co., Ltd.	$N-P_2O_5-K_2O$ 15−7−30⩾50%	600−1 000 times diluent	25 kg (5 kg×5 bags)	14.775 Yuan/kg 73.875 Yuan/time
Compound fertilizer potassium sulfate type(T_5)	E′zhong Chemical Co., Ltd.	$N-P_2O_5-K_2O$ 18−5−22⩾45%			3.9 Yuan/kg 19.5 Yuan/time

weakest root activity of 76.4 $\mu g/(g \cdot FW \cdot h)$.

Table 2 Dosage of different fertilizers

Item	Fertilizer		
	Applying time(month/date)	Applying amount/kg	Investment//Yuan/hm^2
T_1	6/10	75	11 875.5
T_2	6/10	75	13 195.5
T_3	6/10	75	4 260.0
T_4	6/10	75	13 681.5
T_5	6/10	75	2 925.0
Control(CK)	–	–	3 510.0

Table 3 Effects of different fertilizers on plant growth traits

Fertilizer	Harvest time of door pepper month/date	Monitoring time month/date	Plant height//cm	Plant expansion cm	Weight of single fruit//g	Fruit length cm	Fruit diameter cm	Total chlorophyll content mg/g	Root activity $\mu g/(g \cdot FW \cdot h)$
T_1	6/26	7/23	88.5	84.8	101.6	25.49	4.86	66.7	100.6
T_2	6/25	7/23	95.9	88.3	88.5	21.49	4.68	59.7	82.7
T_3	6/25	7/23	93.4	86.5	88.2	21.27	4.70	62.2	76.4
T_4	6/17	7/23	104.3	92.7	92.3	19.98	4.67	58.4	91.6
T_5	6/17	7/23	97.9	90.6	89.1	21.86	4.70	58.2	88.3
CK	6/17	7/23	79.5	72.3	90.2	20.92	4.69	59.4	78.8

Effect of different fertilizers on decrease amplitude of net photosynthetic rate (Pn) of pepper leaves

It could be seen from Fig. 1 that there were differences in decrease amplitude of net photosynthetic rate of pepper leaves between the 5 fertilizers. T5 showed the highest decrease amplitude of net photosynthetic rate of pepper leaves, of 6.6 $\mu mol/(m^2 \cdot s)$, T_1 and T_2 exhibited the lowest decreaseamplitudes of net photosynthetic rate of pepper leaves, fluctuating in the range of 1.3~2.5 $\mu mol/(m^2 \cdot s)$, and T_3 and T_4 showed non-significant decrease amplitudes of net photosynthetic rate of pepper leaves, in the range of 4.8~5.3 $\mu mol/(m^2 \cdot s)$. The decrease of net photosynthetic rate would inevitably lead to decrease in accumulation of assimilate, thereby resulting in weakening of plant growth vigor and decrease in dry matter accumulation.

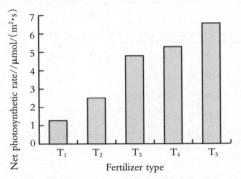

Fig. 1 Effect of different fertilizer types onthe decrease of net photosynthetic rate (Pn)of pepper

Effect of different fertilizers on decrease amplitude of stomatal conductance (Gs) of pepper leaves

As shown in Fig. 2, there were differences in decrease amplitude ofstomatal conductance of pepperleaves between the 5 fertilizers. T_3 showed the highest decrease amplitude of stomatal conductance of pepper leaves, of 2 658.1 mol/($m^2 \cdot s$), T_1 and T_4 exhibited the second highestdecrease amplitudes of stomatal conductance of pepper leaves, and thedecrease amplitudes of stomatal conductance of pepper leaves in treatments T_2 and T_5 were the lowest,

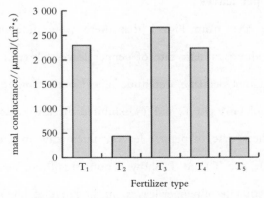

Fig. 2 Effect of different fertilizer typesonthe decrease of stomatal conductance (Gs) of pepper leaves

fluctuating in the range of 386.7 ~428.7 mol/(m² ·s). The decrease in stomatalconductance would lead to reductionof conduction capacity of photosynthetic substrate, and photosynthesiswould thus be affected inevitably.

Effect of different fertilizers on decrease amplitude of transpirationrate (Tr) of pepper leaves

It could be seen from Fig. 3 that the application of the 5 different fertilizers caused differences in decrease amplitude of transpiration rate of pepper leaves. T_1 showed the highest decrease amplitude of transpiration rate of pepper leaves, of 4.7 mmol/(m² ·s), T_3 exhibited the second highest decrease amplitude of transpiration rate of pepper leaves, the transpiration rates of pepper leaves of treatments T_2 and T_4 decreased non-significantly, and treatment T_5 showed the lowest decrease amplitude of transpiration rate of pepper leaves, of 1.1 mmol/(m²·s). The decrease of transpiration rate would inevitably cause reduction of transpiration pull, and the absorption and transportation rate of moisture and mineral nutrients would thus be slowed.

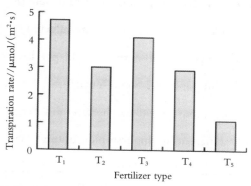

Fig. 3 Effect of different fertilization typeson the decrease of the transpiration rate(Tr) of pepper leaves

Effect of different fertilizers on decrease amplitude of intercellular CO_2 concentration of pepper leaves

It could be seen from Fig. 4 that there were differences in decrease amplitude of intercellular CO_2 concentration of pepper leaves between the 5 fertilizers. T_5 showed the highest decrease amplitude of intercellular CO_2 concentration of pepper leaves, of 436.8 μmol/mol, T_1 exhibited the second highest decrease amplitude of intercellular CO_2 concentration of pepper leaves, and the intercellular CO_2 concentrations of pepper leaves of treatments T_2, T_3 and T_4 were not remarkable, in the range of 390.86 ~399.82 μmol/mol.

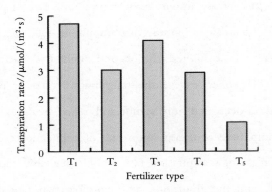

Fig. 4 Effect of different fertilizer types on CO_2 concentration (Ci) of pepper leaves

Relationship between fertilizer typeand yield and benefit

As shown in Table 4, different fertilizers affected yield and benefit of pepper. T_1 showed the highest plot output, yield per hectare and benefit, of 340.1 kg, 96 448.5 kg/hm² and 162 865.6 Yuan/hm², respectively, T_4 showed the second highest plot output, yield per hectare and benefit, T_2, T_5 and CK were not remarkably different in highest plot output, yield per hectare and

Table 4 Relationship between fertilizer type and yield benefit

Item	Treatment					
	T_1	T_2	T_3	T_4	T_5	CK
Plot output kg	340.1	278.0	260.8	313.1	295.2	300.0
Yield kg/hm²	96 448.5	79 242.0	74 338.5	89 247.0	84 144.0	85 524.0
Benefit Yuan/hm²	162 865.5	133 126.5	124 888.8	149 935.5	141 361.5	143 679.0

benefit, while T_3 exhibited the lowest plot output, yield per hectare and benefit, of 260.8 kg, 74 338.5 kg/hm^2 and 124 888.5 Yuan/hm^2, respectively.

Conclusions

In this experiment, drip fertigation of T_1 had greater effects on growth and development and yield of pepper grown in arch greenhouse. This formula fertilization integrating water and fertilizer is suitable for arch greenhouse pepper cultivation with good cost performance and easy extension and application.

References

[1] MAR(马瑞), HUIHJ(惠浩剑), MASC(马守才), et al. Introduction test of greenhouse pepper varieties (大棚辣椒品种引进观察试验)[J]. Mod Agric Sci Technol (现代农业科技), 2012(7):149-150.

[2] ZHOUG(周刚), LIUQ(刘琪), XUGY(徐国友), et al. Variety comparative test of greenhouse early-maturing pepper cultivars (大棚早熟栽培辣椒品比试验)[J]. Journal of Changjiang Vegetables (长江蔬菜), 2008,(3):48-49.

[3] FENGY(冯义). Green comprehensive control technique for diseases and pests of greenhouse pepper (大棚辣椒病虫害绿色综合防控技术)[J]. Journal of Changjiang Vegetables (长江蔬菜)[J], 2014(13):48-49.

[4] SUNHX(孙红霞), WUQ(武琴), ZHENGGX(郑国祥), et al. Effectiveness of EM in promoting soil microorganism activity and resisting eggplant and cucumber obstructions in succession cropping (EM对茄子、黄瓜抗连作障碍和增强土壤生物活性的效果)[J]. Soil(土壤), 2001,(5):264-267.

[5] ZOUQ (邹琦). Laboratory guide for plant physiology (植物生理学实验指导) [M]. Beijing: Chinese Agricultural Press (北京:中国农业出版社), 1998.

Combined effects of organic fertilizer and straw on microbial community and functions of pepper continuous cropping soil

Gao Jingxia[1], Peihongxia[1], Qinxiaojun[1], Dailijun[2], Zhao yunxia[1], Wang xuemei[1], Xie Hua[1]

(1. *Institute of Germplasm Resources, Ningxia Academy of Agriculture and Forestry Sciences, Yinchuan 750002, Ningxia*; 2.*Science and Technology Bureau of PengYang County, Pengyang 756500, China*)

Abstract: Continuous cropping is common in China because of its large population, limited arable land and environmental factors. A long-term continuous cropping, it cause serious soil-borne disease obstacles, yield and quality decline, severely restricted the development of pepper industry. In order to figure out the effects of pepper continuous cropping, to improve yield and reduce soil-borne disease incidence and to examine the effects on the soil microbial community composition and abundance. In this study, we examined seven treatments as follows: (A) cow dung 5 000 kg/667 m^2 and corn straw 2 000 kg/667 m^2, and (B) fowl dung 1 500 kg/667 m^2 and corn straw 2 000 kg/667 m^2, and (C) sheep Dung 4 000 kg/667 m^2 and corn straw 2 000 kg/667 m^2, and (D) fowl dung 1 500 kg/667 m^2, and (E) cow dung 5 000 kg/667 m^2, and (F) sheep Dung 4 000 kg/667 m^2, and control was no organic fertilizer. The results of pepper yield showed that there were significantly higher in the treatments than CK. The diversity of soil microbial community composition and abundancedemonstrate that Proteobacteriaand Ascomycota were the most abundance in Phylums, which generally agreed with many previous articles that Proteobacteria were the most common Phylums. In summary, compared with CK there were significant in the seven treatments, the treatment of fowl dung with corn straw was a best fertilizer combination for improving pepper yield and output value.

Key words:Pepper ;Continuous cropping;Microbial community;Soil -borne disease

INTRODUCTION

Many previous studi es have shown that soil microbial community composition and abundance can be affected by various factors, such as continuous cropping, fertilizing method and irrigation pattern [1,2]. Continuous cropping is common in China because of its large population, limited arable land and environmental factors[3,4]. However, pathogenic bacteria were produced by continuous cropping can inhibit plant growth and thus reduce crop yields[5,6].

Pepper is one of the most important Cash crops and it has a large planting area with high economic value.Especially in the Province of Ningxia, it has the annual average temperature is 7.5℃, the frost-free period is 158 days and the precipitation is 442.7 mm, which are very suitable conditions for the growth of pepper. Pepper grow in arch shed is very popular at this place,planting area is approaching twenty Ten thousand yield. Because of long -term continuous cropping, soil degradation, it cause serious soil-borne disease obstacles[7-9], yield and quality decline[10], severely restricted the development of pepper industry.

Recently, increasing numbers of studies have speculated that the disruption of the soil microbial community also contributes to the continuous cropping obstacles after long-term continuous cropping[1,7,11]. The results showed that there are many factors such as continuous cropping, Soil enzyme activity [12,13], who can lead to serious soil-borne disease and yield and quality decline. But the effects of continuous cropping on soil microbial community composition and abundance remain unclear. According to the province studies, there were few investigative about continuous cropping damage was evaluated in pepper [1]. In order to explore which fertilization treatment was suitable to improve soil microbial community composition and structure, decline serious soil-borne disease obstacles cased by continuous cropping. The next-generation sequencing technologies, an efficient, fast and accurate technique, providing an opportunity for investigative soil microbial community.

In this study, 16sRNA and ITS2 gene were selected to investigate the effect of long-term continuous cropping on soil bacterial and fungus abundance, community structures.

Through our research, we can find the effects that can effectively improve continuous cropping, and provide a better theoretical basis of soil bacterial and fungus communities on the pepper growth under continuous cropping system.

Materials and Methods

Sampling Sites and Experimental Design

The experiment was carried out in pepper demonstration base (35°45′N~36°14′N, 106°52′E~106°21′E), Xinji Township, Pengyang County, Ningxia Hui Autonomous Region, China. The mean annual temperature, frost-free period and total precipitation in this region are 7.5°C, 158d and 442.7 mm, respectively. The soil type is a yellow loam, five years or more were selected for continuous cropping. The initial values of soil physical and chemical properties as follows: total N 1.07 g·kg^{-1}, total P 0.95 g·kg^{-1}, total K 22.00 g·kg^{-1}, organic matter 24.00 g·kg^{-1}, alkali-hydrolyzed nitrogen 63.00 g·kg^{-1}, available

Phosphorus P 29.4 g·kg^{-1}, available Potassium K 192.0 g·kg^{-1}, pH 7.5, and total salt 1.10 g·kg^{-1}.

The field experiment was carried out in an arched shed. The organic fertilizer was decomposed before the experiment. In this study, there were seven treatments as follows: (A) cow dung 5 000 kg/667 m^2 and corn straw 2 000 kg/667 m^2, and(B) fowl dung 1 500 kg/667 m^2 and corn straw 2 000 kg/667 m^2, and (C) sheep Dung 4 000 kg/667 m^2 and corn straw 2 000 kg/667 m^2, and(D) fowl dung 1 500 kg/667 m^2, and(E) cow dung 5 000 kg/667 m^2, and (F)sheep Dung 4 000 kg/667 m^2, and control was no organic fertilizer.

Five sampling points were determined using the Z-line method. Soil samples from the top 30cm of the pepper rhizosphere soil were collected. After mixing the soil at each sample site, take about 100 g of soil samples by quartering. All samples were collected using sterile containers and transported on dry ice to the laboratory, then samples were stored at −70°C.

DNA Extraction, 16s and ITS gene amplification, and Sequencing

DNA was extracted from 500 mg soil samples using CTAB method[14]. The DNA quality and concentration were measured by a NanoDrop Spectrophotometer (Thermo Fisher Scientific Inc., United States), then DNA was stored at −20°C.

High-throughput sequencing of the bacterial 16S rRNA and fungus ITS2 genes present in the soil samples was performed by GeneDenovo Bioinformatics Technology Co. Ltd (Guangzhou, China)using the Illumina HiSeq 2500 platform. The 16S V3−V4 region was amplified using the following primers: 341F(CCTACGGGNGGCWGCAG) and 806R (GGACTACHVGGGTATCTAAT). The ITS2 region was amplified using the forward primer KYO2F (GATGAAGAACGYAGYRAA) and reverse primer ITS4RT(CCTCCGCTT

ATTGATATGC). All PCR reactions were carried out with 15 μL of Phusion ® High-Fidelity PCR Master Mix (New England Biolabs); 0.2 μM of forward and reverse primers, and about 10 ng template DNA. Thermal cycling consisted of initial denaturation at 98℃ for 1 min, followed by 30 cycles of denaturation at 98℃ for 10 s, annealing at 50℃ for 30 s, and elongation at 72℃ for 30 s. Finally 72℃ for 5 min. Mix same volume of 1X loading buffer (contained SYB green) with PCR products and operate electrophoresis on 2% agarose gel for detection. PCR products was mixed in equidensity ratios. Then, mixture PCR products was purified with Qiagen Gel Extraction Kit (Qiagen, Germany). Sequencing libraries were generated using TruSeq® DNA PCR-Free Sample Preparation Kit (Illumina, USA) following manufacturer's recommendations and index codes were added. The library quality was assessed on the Qubit 2.0 Fluorometer (Thermo Scientific) and Agilent Bioanalyzer 2100 system.

All clean sequences data were accessible in NCBI SRA database under the accession number AAA.

Data Analysis

All statistical differences were calculated by analysis of variance, and the means were segregated by the LSD multiple comparison test at $P < 0.05$ using the IBM SPSS Statistics software (SPSS, v.19). Shannon and Simpson diversity indices and the Chao1 and abundance-based coverage estimator (ACE) richness estimators, all calculated with QIIME software (version 1.7.0)[15].

Results

Pepper Yield Analysis

The mean of pepper yield and output value for three years under our experimental treatment showed as followed: six treatments mentioned in our experimental design compared with CK had strong significant under 5% and 1% level (Tab.1). In the treatment of fowl dung and fowl dung with corn straw, cow dung and cow dung with corn straw, sheep dung and sheep dung with corn straw, there were no significant under 5% level respectively. There were strong significant among cow dung, sheep dung and fowl dung treatment, but there were no significant between cow dung and sheep dung under 1% level. In summary, six kinds of combinations fertilization methods are effective for increasing pepper production and output value.Compare with other treatments, the treatment of fowl dung with corn straw was a best fertilizer combination for improving pepper yield and output value.

Tab.1 The yield of pepper analysis under each treatment in an three years

Treatment	Total yield of small area (kg/23.4 m²)				Yield (kg/667 m²)	Output value (CNY)	Comparative control±(kg)	Saliency analysis 5%、1%
	I	II	III	Average				
Sd	219.6	226.5	228.7	224.9	6 410.6	7 051.7	1 071.7	c C
Cd	200.4	203.0	195.2	199.5	5 686.6	6 474.7	547.2	d C
Fd	232.1	229.5	235.7	232.4	6 624.4	7 286.8	1 285.5	ab AB
Sd+Cs	228.0	230.5	224.0	227.5	6 484.7	7 133.2	1 145.8	bc B
Cd+Cs	200.2	204.9	198.0	201.0	5 729.4	6 553.1	618.5	d C
Fd+Cs	240.5	234.8	237.0	237.4	6 766.9	7 443.6	1 428.0	a A
Ck	184.4	187.6	190.0	187.3	5 338.9	5 872.8		e D

† Sd, sheep dung; Cd, Cow dung; Fd, fowl dung; Cs, corn straw; significant under 5% and 1% level labeled a, b, c, d and A, B, C,D respectively.

Diversity of Soil Bacterial and Fungus Communities and Relative Abundance

The number of OTUs (at a 97% similarity level) in the seven treatments ranged from 55 336 to 99 329 (Tab.2). According to OTUs, ACE and Chao 1, it reflected similar trends among the treatments. there were significant between

CK and other treatments. In terms of the Shannon, there were no significant among Sd, Fd+Cs and CK. Simpson diversity index declared that there were significant among Cd, Fd and CK at 5% level.

There were nine Phylums whose relative abundance were more then 2% (Fig.1). *Proteobacteria* was most abundance phylum in the seven treatments. The second most abundant phylum was *Planctomycetes* of the total sequences. Compare with CK, there were significantly difference in the other treatments, it reflected similar trends showed in richness and diversity(Chao1 and ACE). In term of Shannon diversity, the treatment of sheep dung (F) there was no significantly between sheep dung and CK. The treatment of cow dung, fowl

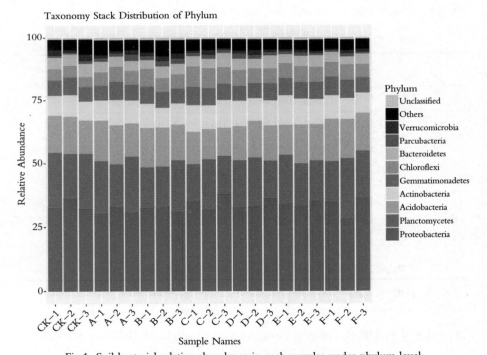

Fig.1 Soil bacterial relative abundance in each samples under phylum level

† CK, no organic fertilizer (CK); A, Cow dung and corn straw (Cd+Cs); B, fowl dung and corn straw(Fd+Cs); C, sheep dung and corn straw(Sd+Cs); D, fowl dung(Fd); E, Cow dung(Cd); F, sheep dung(Sd); There were three biological replicates in each treatment.

dung, sheep dung with cow straw, cow dung with cow straw, showed significantly compared with CK. In Simpson diversity, we can see there were significantly between cow dung and CK, as well as fowl and CK(Tab. 2).

Tab.2 The 16sRNA gene alpha diversity of seven treatments

Treatment	OTUs	Chao1	ACE	Shannon	Simpson
Sd	97655±6876.898b	6560.579±192.0780b	6567.663±375.558b	10.279±0.076a	0.997±0.00039a
Cd	94550±16252.092b	6514.343±25.127b	6414.842±72.141b	10.461±0.090b	0.998±0.00032b
Fd	99329±3818.049b	7310.679±88.269b	7324.22±207.751b	10.663±0.089b	0.998±0.00035b
Sd+Cs	92330±4050.551b	7030.774±41.844b	7053.968±60.868b	10.36±0.121b	0.998±0.00033a
Cd+Cs	87677±9267.735b	6837.624±115.852b	6812.933±171.164b	10.47±0.102b	0.998±0.00018a
Fd+Cs	82214±14919.157b	6789.442±236.232b	6762.135±232.413b	10.244±0.054a	0.997±0.00009a
Ck	55336±22734.977a	5856.616±410.120a	5903.657±369.507a	10.189±0.103a	0.997±0.00025a

† Sd, sheep dung; Cd, Cow dung; Fd, fowl dung; Cs, corn straw; significant under 5% level labeled a, b, respectively. Values are means (n=3) with standard errors. Different lowercase letters in a column indicate differences between treatments according to a LSD multiple comparison test.

The ITS2 gene alpha diversity showed in Table 3. Under a 97% similarity level, count the number of OTUs in the seven treatments ranged from 88 613.667 to 105 448.667. There were no differences among the treatments in terms of the OTUs. Fungus richness and diversity were calculated in seven treatments with Chao1 and ACE. Compare with the treatment of Fd and Fd+Cs, there was significant between Fd and Fd+Cs at 5% level. In terms of the Shannon diversity, we also analyze the difference in the treatment Sd and Fd+Cs, as well as Cd and Fd+Cs, there was no differences between Sd and Cd. To analyze the Simpson diversity, the difference we can found between Sd, Cd, Ck and Fd+Cs, respectively(Tab.3).

There were five Phylums whose relative abundance were more then 2%. The abundance of Ascomycota was most phylum in the seven treatments. The second most abundant phylum was *Mortierellomycota*. Compare with CK, there were significantly difference in the other treatments of the relative abundance of *Ascomycota*, especially in the treatments of E (Cow dung) and F(sheep dung). The relative abundance of *Mortierellomycota* and *Basidiomycota* in the treatment of E (Cow dung) and F (sheep dung) less than other treatments. *Mucoromycota* phylum only distributed in treatment of Cow dung and corn straw (Cd+Cs) and Cow dung (Cd), and *Chytriomycota* detected in individual samples named CK-3 and A-3, but in other samples of the same group was not detected. Therefore, the *Chytriomycota* phylum detected in our research, its existed or not depend on further experimental verification. In summary, different fertilization methods in this study play a key role in changing fungus microbial community abundance and composition, making it more suitable for the growth and development of pepper, decline soil -borne disease cased by long -term continuous cropping activity(Fig.2).

Fungal Functional Gene Analysis

Based on the abundance of OTUs, FUNGuild was used to perform fungal function annotation. The relative abundance of animal pathogen, dung saprotroph, endophyte, epiphyte, plant saprotroph, soil saprotroph and Wood saprotroph in the treatment fowl dung with corn straw(Fd+Cs) and Cow dung (Cs) has significant compare with other treatments. In this study, we found that the relative abundance of lichen parasite in the other treatment reduced a lot, especially in the treatment cow dung(Cd) and sheep dung(Cd), there was no lichen parasite detected (Fig.3). Compared with traditional fertilizer application, base fertilizer with organic fertilizer could reduce damage fungus, improving the relative abundance which suitable for pepper growing.

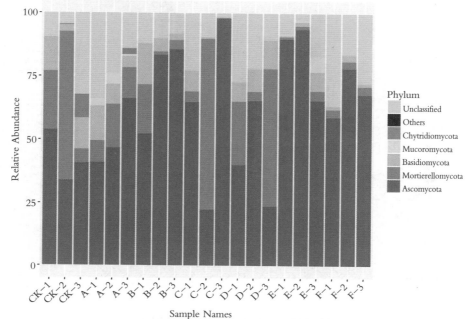

Fig.2 Soil fungus relative abundance in each samples under phylum level

† CK, no organic fertilizer (CK); A, Cow dung and corn straw (Cd+Cs); B, fowl dung and corn straw(Fd+Cs); C, sheep dung and corn straw(Sd+Cs); D, fowl dung(Fd); E, Cow dung(Cd); F, sheep dung(Sd); There were three biological replicates in each treatment.

Tab.3 The ITS2 gene alpha diversity of seven treatments

Treatment	Chao1	Ace	Shannon	Simpson
Sd	88 613.667±581.085	224.101±10.213ab	4.623±0.828b	0.902±0.088b
Cd	98 553±8 469.918	222.443±10.250ab	4.936±0.145b	0.945±0.010b
Fd	102 510±12 155.308	216.658±4.691b	4.290±0.775ab	0.848±0.120ab
Sd+Cs	90 915.667±11 748.480	221.020±23.829ab	3.919±1.515ab	0.782±0.209ab
Cd+Cs	87 728.333±12 840.783	225.567±24.265ab	4.260±0.868ab	0.860±0.109ab
Fd+Cs	105 448.667±11 472.812	248.631±21.884a	3.034±1.099a	0.644±0.199a
Ck	104 939.667±8 456.383	229.647±8.543ab	4.523±0.350ab	0.911±0.024b

† Sd, sheep dung; Cd, Cow dung; Fd, fowl dung; Cs, corn straw; significant under 5% level labeled a, b, respectively. Values are means (n=3) with standard errors. Different lowercase letters in a column indicate differences between treatments according to a LSD multiple comparison test.

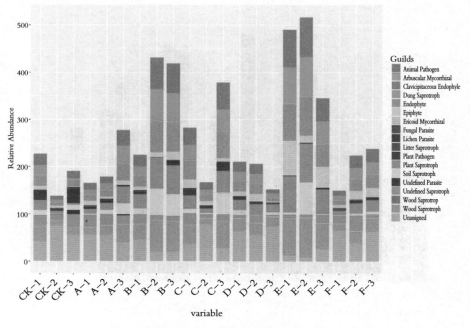

Fig.3 Functional gene analysis of fungus

Relationships between Bacterial and fungus Community Structure and the total yield of pepper

A redundancy analysis (RDA) was performed to identify the correlations between the soil bacterial and fungus community structure and the total yield of pepper (Fig.4). The relationship between bacterial and the yield of pepper showed in Fig.4a could explain 68.55% of the variation, and 75.92% of the variation detected between fungus and the total yield of pepper (Fig.4b). It reflected similar trends in relationships between Bacterial and funguscommunity structure and the total yield of pepper.The Sphingomonas, Planctomyces, Pirellula, Lysobacter, RB41, H16 and Iamiacorrelated closely and negatively correlationwith there years yield of pepper.Cladorrhinum and chaetomium showed closely and negatively correlation with I and II years yield of pepper, but we found closely and positively correlation with III yield of pepper.

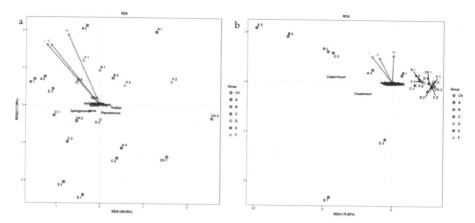

Fig.5 Correlations between Bacterial and fungus community structure and the total yield of pepper determined by RDA analysis. (a) correlations between bacterial communities and the total yield of pepper detected by RDA analysis. (b) correlations between funguscommunities and the total yield of pepper detected by RDA analysis.I, II andIIIlabeled for the total yield of pepper three years.The direction of an arrow indicates the steepest increase in the variable and the length indicates the strength relative to the other variables.

DISCUSSION

Learning the diversity of soil bacterial and fungus communities and relative abundanceinpepper continuous cropping systems could be helpful to provide a better understanding of soilproductivity in long −term continuous cropping. Long −term continuous cropping lead to lower organic matter content and serious soil−borne disease and yield and quality decline, caused by unscientific fertilization and farming system. Compared with ourresults, previous study already reported continuous cropping caused soilorganic matterdecline and serious soil−borne disease[16]. With the peppercropping years increasing, soil organic matter significantly increased, which couldbe also attributed to long −term oversupply of chemical fertilizer [17]. Soil diversity of soil bacterial and fungus communities and relative abundance areconsidered to be an key points of soil quality.In our study, soil diversity of soil bacterial and fungus communities significantly decreased with three years of peppercropping, the results showed that increase soil organic matter content, scientific fertilization can help increase pepper soil microbial community compositionabundance.

Pepper long-term continuous croppingwas carried out in Ningxia Hui Autonomous Region, the soil type is a yellow loam, five years or more were selected for continuous cropping.In our study, pepper yield of seven treatments revealed that six treatments mentioned in our experimental design compared with CK had strong significant under 5% and 1% level (Tab.1). Compared with many previous studies, such as black pepper [1], cotton [18], watermelon[6], cucumber [19], the growths of which were significantly in microbial community diversity by continuous cropping.

In this study, diversity of soil bacterial and fungus communities and relative abundance showed that

Proteobacteriaand Ascomycota were the most abundance in Phylums level, which generally agreed with many previous articles that *Proteobacteria* were the most common Phylums [20,21]. The relative abundance of *Ascomycota* significantly showed in soil fungus relative abundance in each samples under phylum level (Fig.2), which caused by long-term continuous cropping system. The relative abundances of *Ascomycota* phyla increased with long-term continuous cropping of pepper,which agreed with previous observations [22-25]. In addition, previous study has found that *Proteobacteriaand Ascomycota* were most abundance in increased in continuous cropping system [1,26-29]. In conclusion, continuous cropping in pepper cultivationcaused serious soil-borne disease obstacles, yield and quality decline. Bacterial and fungus communities from thepepper revealed that the soil fungus communitydiversity, was significantly affected by the pepper continuous cropping system. All these changes mightfinally result inpepper yield decline, and serious soil-borne disease. Therefore, a suitable fertilization and farming system was a great due to improved long-term continuous cropping.

CONCLUSION

Pepper yield analysis, Six treatments in our experimental design compared with CK had strong significant under 5% and 1% level. There were significant among cow dung, sheep dung and fowl dung treatment, except the treatment of cow dung and sheep dung under 1% level. Compare with other treatment, we can find the treatments of fowl dung with corn straw was a best choose for increasing pepper production and output value, to reduce soil-borne disease obstacles, make a pepper in a favorable growth environment.

Diversity of soil bacterial, our results demonstrate that there were fine Phylums whose relative abundance were more then 2%, it reflected similar trends among the treatments judging with ACE and Chao 1, Simpson diversity index declared that there were significant among Cd, Fd and CK at 5% level. *Proteobacteria* was most abundance phylum in the seven treatments. Compare with seven treatmentsbacterial communities and relative abundance, there were no significant in phylum. Fungus richness and diversity were calculated in seven treatments with Chao1 and ACE, the results showed that there was significant between Fd and Fd+Cs at 5% level. *Ascomycotawas* most abundance phylum in the seven treatments.In terms of the Shannon and simpson diversity, the difference we can found between Sd, Cd, Ck and Fd+Cs, respectively.

In conclusion, the treatment of fowl fung with corn straw (Fd+Cs) is more effective than the other treatments in promoting fungus microbial community abundance and composition, making it more suitable for the growth and development of pepper, decline soil-borne disease cased by long-term continuous cropping activity.Thetreatment of fowl fung with corn straw could be an ideal crop residue management mode for reducing soil-borne disease cased by long-term continuous cropping activity.

Acknowledgments

This work was supported by the whole industrial chain innovation demonstration project (QCYL-2018-03) and National technical system of bulk vegetable industry(CARS-23-g24). We thank the Genedenovo Biotechnology Co., Ltd (Guangzhou, China) for providing sequencing and data analysis.

References

[1] Xiong W, Li Z, Liu H, et al. The Effect of Long-Term Continuous Cropping of Black Pepper on Soil Bacterial Communities as Determined by 454 Pyrosequencing[J]. Plos One, 2015,10(8):e0136946.

[2] Wu L, Chen J, Wu H, et al. Effects of consecutive monoculture of Pseudostellaria heterophylla on soil fungal community as determined by pyrosequencing [J]. Scientific Reports, 2016, 6:26601.

[3] CHEN Youqi, Peter H, XU Bin. Spatial Modeling of Land Use and Its Effects in China[J]. Progress in Geography, 2000.

[4] HUANG, P, ZHU, L. G, ZHAI, X. J,et,al. Hepatitis C virus infection and risk factors in the general population: a large community-based study in eastern China, 2011 - 2012[J]. Epidemiology & Infection:1-10.

[5] Li H, Wang J, Liu Q, et al. Effects of consecutive monoculture of sweet potato on soil bacterial community as determined by pyrosequencing [J]. Journal of Basic Microbiology, 2019,59(2).

[6] Ruiping Y, Yanling M, Changming L, et al. The Effects of Cattle Manure and Garlic Rotation on Soil under Continuous Cropping of Watermelon (Citrullus lanatus L.)[J]. Plos One,2016, 11(6):e0156515

[7] Zu C, Li Z, Yang J, Yu H, Sun Y, Tang H, et al. Acid Soil Is Associated with Reduced Yield, Root Growthand Nutrient Uptake in Black Pepper (Piper nigrum L.).

Agric Sci. 2014; 5: 466−473.

[8] W. Mao, J.A. Lewis, R.D. Lumsden, & K.P. Hebbar. . Biocontrol of selected soilborne diseases of tomato and pepper plants. Crop Protection, 17(6), 0−542.

[9] Wang Xiaobing, Luo Yongming, Li. EFFECTS OF LONG-TERM STATIONARY FERTILIZATION EXPERIMENT ON INCIDENCE OF SOIL-BORNE DISEASES AND BIOLOGICAL CHARACTERISTICS OF PEANUT IN CONTINUOUS MONOCROPPING SYSTEM IN RED SOIL AREA [J]. Acta Pedologica Sinica, 2011.

[10] Ting-Gang W , Yu-Qing J , Xiao-Feng D U , et al. Effects of Different Fertilizer Treatments on Yield and Quality of Pepper[J]. Acta Agriculturae Jiangxi, 2013.

[11] Huaiying Yao, Xiaodan Jiao, Fengzhi Wu. Effects of continuous cucumber cropping and alternative rotations under protected cultivation on soil microbial community diversity[J]. Plant & Soil, 284(1−2):195−203.

[12] MA Yunhua, WEI Min, WANG Xiufeng. Variation of microflora and enzyme activity in continuous cropping cucumber soil in solar greenhouse [J]. Chinese Journal of Applied Ecology, 2004, 15(6):1005.

[13] Li-Na Z , Yong Y U , Bing H , et al. Effect of Melon Continuous Cropping on Soil Fertility and Soil Enzyme Activity[J]. heilongjiang agricultural sciences, 2016.

[14] K Reanwarakorn, S Klinkong, J Porsoongnurn. First report of natural infection of Pepper chat fruit viroid in tomato plants in Thailand [J]. New Disease Reports, 2011, 24(24).

[15] Bokulich, N.A., S. Subramanian, J.J. Faith, D. Gevers, J.I. Gordon, R. Knight, et al. 2013. Quality −filtering vastly improves diversity estimates from Illumina amplicon sequencing. Nat. Methods 10:57 − 59. doi:10.1038/nmeth.2276

[16] Reeves DW. The role of soil organic matter in maintaining soil quality in continuous cropping systems. Soil Tillage Res. 1997; 43: 131−167.

[17] Zu C, Li Z, Yang J, Yu H, Sun Y, Tang H, et al. Acid Soil Is Associated with Reduced Yield, Root Growthand Nutrient Uptake in Black Pepper (Piper nigrum L.). Agric Sci. 2014; 5: 466−473.

[18] Yang L, Tan L, Zhang F, et al. Duration of continuous cropping with straw return affects the composition and structure of soil bacterial communities in cotton fields[J].

Canadian Journal of Microbiology, 2017:cjm-2017-0443.

[19] Feng-Zhi WU.Effect of different base fertilizers on soil microbial community diversity in cucumber rhizosphere[J]. plant nutrition and fertilizer science, 2008, 14(3):576-580.

[20] Nacke H, Thürmer A, Wollherr A, Will C, Hodac L, Herold N, et al. Pyrosequencing-based assessmentof bacterial community structure along different management types in German forest and grasslandsoils. PloS One. 2011; 6: e17000. doi: 10.1371/journal.pone.0017000 PMID: 21359220.

[21] Shen Z, Wang D, Ruan Y, Xue C, Zhang J, Li R, et al. Deep 16S rRNA Pyrosequencing Reveals a Bacterial Community Associated with Banana FusariumWilt Disease Suppression Induced by Bio-OrganicFertilizer Application. PloS One. 2014; 9: e98420. doi: 10.1371/journal.pone.0098420 PMID: 24871319

[22] Ling, Ning, Deng, Kaiying, Song, Yang, et al. Variation of rhizosphere bacterial community in watermelon continuous mono-cropping soil by long-term application of a novel bioorganic fertilizer[J]. Microbiological Research, 169(7-8):570-578.

[23] YANG Yong, ZHANG Xuejun, LI Meihua, et al.Effects of microbiological fertilizer on rhizosphere soil fungus communities under long-term continuous cropping of protected Hami melon[J]. Chinese Journal of Applied & Environmental Biology, 2018.

[24] Ruqiang Cui, Hegui Wang, Yajing Zeng, et al. Application of Deep Pyrosequencing to the Analysis of Soil Microbial Communities in Different Lotus Fields [J]. Journal of Residuals Science & Technology, 2016, 13(S2):S235-S241.

[25] Zhu Siyuan, Wang Yanzhou, Xu Xiaomin, et al. Potential use of high-throughput sequencing of soil microbial communities for estimating the adverse effects of continuous cropping on ramie (Boehmeria nivea L. Gaud)[J]. Plos One, 13(5):e0197095.

[26] Yang, Lei, Tan, Lanlan, Zhang, Fenghua, et al. Duration of continuous cropping with straw return affects the composition and structure of soil bacterial communities in cotton fields[J]. Canadian Journal of Microbiology:cjm-2017-0443.

[27] Rodrigues Richard R, Pineda Rosana P, Barney Jacob N, et al. Plant Invasions Associated with Change in Root-Zone Microbial Community Structure and Diversity [J]. Plos One, 10(10):e0141424.

[28] Peng Dang, Xuan Yu, Hien Le, et al. Effects of stand age and soil properties on soil bacterial and fungal community composition in Chinese pine plantations on the Loess

Plateau[J]. Plos One, 2017, 12(10):e0186501.

[29] Shiwen, Wang, Jing, Ren, Ting, Huang, et al. Evaluation of soil enzyme activities and microbial communities in tomato continuous cropping soil treated with jerusalem artichoke residues[J]. Communications in Soil Science & Plant Analysis, 49(22):2727-2740.

辣根素水乳剂对连作辣椒光合特性及土壤酶活性的影响

高晶霞[1]，吴雪梅[2]，牛勇琴[2]，高 昱[2]，谢 华[1]

（1. 宁夏农林科学院种质资源研究所，宁夏银川 750002；
2. 宁夏回族自治区彭阳县蔬菜产业发展服务中心，宁夏彭阳 756500）

摘 要：辣椒连作容易导致土壤的活性降低、菌群失调以及病虫害增加，从而导致土壤的整体生产能力下降。本研究通过对20%的辣根素水乳剂进行不同剂量的调制，分析不同药量处理对连作辣椒的光合特性以及土壤酶活性的影响。结果显示：20%的辣根素水乳剂对连作辣椒的生长指标、光合作用特性以及土壤酶的活性都有一定程度的改善，不同剂量的药量处理产生的改善效应也都不相同。研究结果为我国连作辣椒土壤的修复及病虫害防治提供重要的理论及实践参考，同时对种植地区的生态保护及环境防治提供有益的帮助。

关键词：辣根素；连作；辣椒；光合特性；土壤酶活性

Effect of Horseradish Water Emulsion on the Biological Community of Pepper Continuous Cropping Soil

Gao Jingxia[1], Wu Xuemei[2], Niu Yongqin[2], Gao Yu[2], Xie Hua[1]

(1. Institute of Germplasm Resources, Ningxia Academy of Agriculture and Forestry Sciences, Yinchuan 750002, Ningxia; 2. Vegetable Industry Development Service Center of Pengyang County, Ningxia Hui Autonomous Region, Pingyang 756500, China)

Abstract: Continuous cropping of peppers can easily lead to a decrease in soil activity, an imbalance in flora, and an increase in pests and diseases, which can lead to a decline in overall soil productivity. In this study, the effects of different doses on photosynthetic characteristics and soil enzyme properties of continuous cropping pepper were analyzed by adjusting different doses of 20% capsaicin water emulsion. The research results show that 20% capsaicin water emulsion has improved the growth index, photosynthesis characteristics and soil enzyme activity of continuous cropped pepper to a certain extent, and the improvement effects produced by different doses of drug treatment are also different. The research results provide important theoretical and practical references for the continuous cropping of pepper soil in China and the prevention and control of plant diseases and insect pests. At the same time, they provide useful help to the ecological protection and environmental control in the planting areas.

Key words: horseradish; continuous cropping; pepper; photosynthetic characteristics; soil enzyme activity

在我国农业经济作物种植中,辣椒是一种重要的经济作物,辣椒原产于巴西等拉美地区,属于茄科植物,一般为一年生经济作物,由于辣椒的维生素含量特别丰富,并且能够抵抗潮湿及阴冷,搭配其他蔬菜口感很好,因此在我国的种植面积较广,目前主要在宁夏和江西等区域,在目前蔬菜种植中,困扰农户的主要问题是辣椒连作的问题,辣椒连作容易导致土壤的活性降低、菌群失调以及病虫害增加,从而导致土壤的整体生产能力下降,与此同时,过多的农药以及化肥在土壤中的残留,进一步破坏了土壤的修复能力,造成了严重的环境生态问题,这些因素都是辣椒连作带来的连锁反应,严重影响了辣椒作物的生长发育状态[1-3]。针对这种情况,目前常用的做法是对土壤进行消毒,恢复土壤的自我修复能力,学者研究发现土壤的消毒措施可以提升土壤的活性,降低了土壤的病虫害,为部分经济作物的连作提供了基础保障[4-6]。也有学者通过混合多种农药,调节农药成分,尽可能修复土壤的活性,降低土壤的病虫害等影响,同时研究混合农药对土壤活性的影响,进行了大量实验研究[7]。

综合目前的文献研究可以发现,目前针对辣椒等连作经济作物的土壤修复的文献主要针对混合各种农药成本,进行土壤修复及连作经济作物防御病虫害的影响研究,较少涉及连作经济作物本身的生长特性的分析,同时没有考虑植物性农药对连作经济作物土壤的修复能力,为此,本研究以改善土壤活性为研究视角,选择辣根素为研究对象,辣根素作为一种植物性农药,区别于化学合成性农药,辣根素对自然环境的伤害效应最低,环保特性最高,同时作为土壤处理剂能够有效杀灭多种土壤中的多种病菌,同时农药的残留度很低,副作用最小[8,9]。本研究通过对20%的辣根素水乳剂进行不同剂量的调制,分析不同药量处理对连作辣椒的光合特性以及土壤酶活性的影响,研究结果为我国连作辣椒土壤的修复及病虫害防治提供重要的理论及实践参考,同时对种植地区的生态保护及环境防治提供有益的帮助。

1 材料与方法

1.1 供试材料

本研究选用的辣椒品种为"巨峰1号",供试的试验田选取自宁夏彭阳县新集乡辣椒核心试验基地,土壤面积为0.15亩,分为4个小区,每小区面积25 m²,土壤的基本参数如下:pH值为8.37,碱解氮78.2 mg·kg^{-1},有效磷18.3 mg·kg^{-1},速效钾406 mg·kg^{-1},有机质38.7 g·kg^{-1}。所用的试验田地势平缓,相关的浇灌设施齐全,土壤的养分能力及病虫害水平处于平均值,试验田的土壤管理参照正常管理。20%辣根素水乳剂由中农齐民(北京)科技发展有限公司提供。对照组采用的50%多菌灵可湿性粉剂选自宁夏尚博农植物保护有限公司,所用的肥料为经济作物专用的果蔬有机肥,主要购买自宁夏银川市益力特种苗有限公司。植株随机排列,同时在测试过程中,不进行药剂喷洒。

1.2 试验方法

在试验前,对试验土壤进行工地整备,清除土壤中的杂草及植株残体,进行正常的施肥灌溉,对试验田做畦,铺设滴灌系统,滴灌测试,检查滴灌管是否有堵塞。施药前2~3 d,每天使用滴灌设备清水滴灌,将畦浸透,调节土壤湿度处于65%水平,同时对作物覆盖塑料膜,整体或单垄覆盖,4丝以上塑料膜效果最佳,四周压实,尽量使塑料膜紧贴地面,检查塑料膜是否有破损,如有及时修补。施药前,先用清水滴灌15~30 min,使水分在滴灌点附近扩展开。

根据本次试验的要求,分成5组进行处理,利用文丘里系统滴灌施药,滴灌时间约30~40 min(沙土时间短,黏土时间长;浅根系作物时间短,深根系作物时间长),确保施药均匀。滴药结束后,继续用清水滴灌15~30 min。前3组的辣根素水乳剂的施药浓度分别为7 L/667 m²、9 L/667 m²、11 L/667 m²,也即处理方式1,2,3。处理方式4和5为对照组,分别采用50%多菌灵可湿性粉剂以及无措施空白对照组,试验中各药剂按照不同的施药浓度进行精准喷灌施药,并进行土壤表面覆膜,覆膜3~5 d后进行种植,50%多菌灵可湿性粉剂按照2 kg/667 m²用量拌土撒施,同样进行土壤表面覆膜,覆膜3~5 d后进行种植,空白组仅对辣椒植株进行覆膜,不做其他处理,覆膜3~5 d后进行种植。

1.3 指标测定

1.3.1 生长指标

在种植30 d后,测量辣椒植株的各项生长指标,其中辣椒的株高采用杜克ls-p激光测距仪进行测量,茎粗选用实验室的游标卡尺测量,植株生物量采用梅特勒托乐多ME204/02物理电子天平进行测量。

1.3.2 辣椒的叶绿素及光合作用指标测量

辣椒植株的叶绿素含量采用经典文献中的丙酮乙醇混合液法进行测量,随机选择5株种植30 d的辣椒植株,在每株上面选取位置一致的地方对辣椒植株的复叶进行展开,利用实验室购买的美国OPTI-SCIENCES OS-5p+便携式脉冲调制叶绿素荧光仪对相关参数进行测量,并在间隔5 min后进行数据采集。

1.3.3 土壤指标

对于连作辣椒的土壤测量采用 pH 测量仪、细菌培养基以及比色法等方法,分别测量连作辣椒土壤的 pH、电导率、土壤微生物以及酶活性等参数。

1.4 数据处理

数据用 EXCEL 和 SPSS 11.0 软件进行统计分析,显著水平为 $P<0.05$。

2 结果分析

2.1 不同处理方式对连作辣椒植株生长状态的影响

图 1 为不同辣根素处理方式后,种植 30 d 的辣椒植株的株高和茎粗生长情况,从图中可以发现,辣椒植株的株高和茎粗在经过不同辣根素方式处理后,均高于不处理的对照情况(处理方式 5),特别是处理方式 3 的 11 L/667 m² 浓度的辣根素水乳剂,对辣椒植株的茎粗和株高影响最为显著,两个生长指标增长明显,辣椒在定植 30 d 的株高和茎粗分别增加了 15.73%、6.67%,同时从图 1 中可以看到,处理方式 1 和处理方式 2 相比处理方式 4 也即 50%多菌灵可湿性粉剂,对辣椒植株的株高和茎粗的促进作用低于处理方式 4,也即 50%多菌灵可湿性粉剂对辣椒植株生长状态的影响高于处理方式 1 和 2,显示了辣根素水乳剂浓度配比对促进辣椒植株生长状态有显著影响。

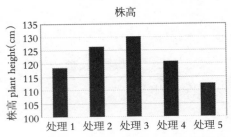
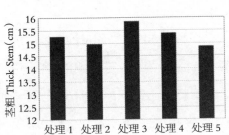

注:处理方式 1 为 7 L/667 m² 浓度的辣根素水乳剂,处理方式 2 为 9 L/667 m² 浓度的辣根素水乳剂、处理方式 3 为 11 L/667 m² 浓度的辣根素水乳剂。处理方式 4 和 5 为对照组,处理方式 4 为 50%多菌灵可湿性粉剂,处理方式 5 为无措施空白对照组,以下同上

图 1 不同处理方式对连作辣椒植株生长状态的影响

Fig.1 Effects of different treatments on the growth status of continuous cropping pepper plants

2.2 不同处理方式对连作辣椒植株发育状态的影响

不同处理方式对连作辣椒发育状态的影响如表1所示,从表1中可以发现,相比对照组方式5,方式1和方式2的辣根素水乳剂处理后,辣椒的总重均高于对照组方式5,差异呈显著状态,同时方式1对辣椒的茎秆的生长有显著促进作用,方式2对辣椒的根部生长有显著促进作用,另外相比对照组方式4,方式1和方式2的辣根素水乳剂处理后,辣椒的总重均高于对照组方式4,但差异并不显著。同时从表1中还可以看到,方式3相比对照组方式4和方式5,对辣椒的发育状态影响并不显著,一定程度上阻碍了辣椒的生长,特别是辣椒茎部发育状态显著低于方式4和方式5,总重也低于方式4和方式5处理的辣椒植株。

表1 不同处理对连作辣椒发育状态的影响
Table 1 Effects of different treatments on the development status of continuous cropping pepper

处理方式	总干重/g	根干重/g	茎干重/g
方式1	0.136*	0.021	0.105*
方式2	0.135*	0.051*	0.075
方式3	0.109	0.043*	0.055*
方式4	0.126	0.032	0.083
方式5	0.117	0.025	0.082

注:* 表示相比对照组方式5的差异显著($p<0.05$),# 表示相比对照组方式4的差异显著($p<0.05$),以下同上

2.3 不同处理方式对连作辣椒植株根系状态的影响

从表3中的数据可以看到,相比对照组方式5,方式1和方式2能有效促进辣椒根系的生长发育,辣椒根系的长度、根系表面积以及体积方面显著增长,与对照组方式5存在显著性差异,其中方式2的处理效果最好,辣椒植株的根系长度、根系表面积、根系体积以及根尖数相比对照组均增加了24.49%、46.62%、76.92%以及31.63%,对辣椒植株根系的整体生长状态发育效果最好,同时从表2中还可以看到,方式1和方式2相比对照组方式4,辣椒

根系的长度、根系表面积以及体积方面有一定程度的增长,但差异并不显著。另外从表2中可以看到,方式3相比对照组方式5,对辣椒根系的影响几乎一致,并且小于方式4对辣椒根系生长的促进作用,从根系长度、根系表面积、根系体积及根尖数等参数都能显示出来。

表2 不同处理方式对连作辣椒根系状态的影响

Table 2 Effects of different treatments on root status of continuous cropping pepper

处理方式	根系长度 /mm	根系表面积 /cm²	根系体积 /cm³	根尖数 /个
方式1	216.77*	38.78*	0.61*	313.36*
方式2	219.37*	39.13*	0.62*	318.57*
方式3	205.62	31.59	0.46	283.19
方式4	209.36	33.12	0.51	296.39
方式5	183.26	30.37	0.43	278.19

2.4 不同处理方式对连作辣椒植株光合能力状态的影响

从表3中的数据可以看到,相比对照组方式4和方式5,方式3对辣椒植株光合作用的影响显著,能够有效促进辣椒植株的光合作用,具体的参数包含PSII最大量子产量Fv/Fm、PSII实际量子产量Y(II)、光化学淬灭系数qP等参数,均有显著程度的提高。同时从表3中可以看到,方式1和2相比对照组方式4和方式5,对辣椒植株光合作用有一定的促进作用,但并不显

表3 不同处理方式对连作辣椒叶片光合作用能力的影响

Table 3 Effects of different treatments on photosynthetic capacity of continuous cropping pepper leaves

处理方式	PSII最大量子产量Fv/Fm	PSII实际量子产量Y(II)	光化学淬灭系数qP	非化学淬灭系数NQP
方式1	0.85	0.38	0.51	1.37
方式2	0.87	0.39	0.52	1.36
方式3	0.93*#	0.46*#	0.67*#	1.61*#
方式4	0.83	0.37	0.49	1.35
方式5	0.79	0.35	0.46	1.33

著。对照组方式 4 相比不作任何处理的方式 5，也均有一定程度的提高，但同样也不显著。

2.5 不同处理方式对连作辣椒植株叶绿素含量的影响

从表 4 中数据可以看到，相比对照组方式 4 和方式 5，方式 3 对辣椒植株能够有效促进辣椒植株的叶绿素含量，差异呈显著状态，这与处理方式 3 促进辣椒植株光合作用有关，具体的参数包含叶绿素 A、叶绿素 B 与叶绿素 A+B 等参数，方式 3 相比对照组方式 5 叶绿素各参数有 17.03%、19.25% 及 22.56% 的增长，提升作用显著。另外从表 4 中可以看到，方式 1 和 2 相比对照组方式 4 和方式 5，对辣椒植株的叶绿素含量有一定的促进作用，但并不显著，这与辣椒植株光合作用的分析结果也是一致的。

表 4 不同处理方式对辣椒叶绿素含量的影响
Table 4 Effect of different treatments on chlorophyll content of pepper

处理方式	叶绿素 A/(mg·g^{-1})	叶绿素 B/(mg·g^{-1})	叶绿素 A+B/(mg·g^{-1})
方式 1	1.98	0.78	2.56
方式 2	1.92	0.75	2.52
方式 3	2.13*#	0.97*#	2.87*#
方式 4	1.88	0.66	2.49
方式 5	1.82	0.62	2.46

2.6 不同处理方式对连作辣椒土壤酶含量的影响

从表 5 中可以看到，方式 1 和方式 2 对土壤酶的活性有显著增强，土壤脲酶、蔗糖酶、酸性磷酸酶及多酚氧化酶相比对照组方式 4 和方式 5 均显著增加，特别是方式 2，对以上几种土壤酶的活性有显著增强，增长率分别为 66.36%、88.71%、120.56% 以及 174.83%，显示了方式 2 对土壤酶的显著改善作用，增强了辣椒植株连作土壤的总体活性。同时从表 5 中可以看到，方式 3 对土壤酶的改善作用一般，跟对照组方式 4 和方式 5 相比，几个土壤酶的活性参数增长较小，并无显著差距。

表 5 不同处理方式对辣椒种植土壤酶活性的影响

Table 5 Effects of different treatments on soil enzyme activities in pepper cultivation

处理方式	脲酶 /(mg·g^{-1}·h^{-1})	蔗糖酶 /(mg·g^{-1}·h^{-1})	酸性磷酸酶 /(mg·g^{-1}·h^{-1})	多酚氧化酶 /(mg·g^{-1}·h^{-1})
方式 1	1.382*#	2.865*#	0.309*#	0.203*#
方式 2	1.563*#	2.989*#	0.363*#	0.228*#
方式 3	1.123	1.903	0.207	0.013
方式 4	1.038	1.893	0.176	0.096
方式 5	0.938	1.589	0.165	0.083

3 讨论与结论

辣椒等经济作物连作对土壤的影响较为显著,容易造成土壤活性降低,土壤菌群失调,导致种植植物的病虫害增加,学者对经济作物连作现象进行了研究,通过采用多种化学肥料及原料来改善连作土壤的活性,对土壤有一定的改进作用,提高了抗病虫害的能力,目前也有学者通过加入各种固态菌混合到连作土壤中,从而起到改善土壤连作效应的作用,提升了土壤的自我修复能力,增强了种植作物的生长活性[10],这些研究与本文的研究结论较为一致。本研究的辣根素为熏蒸型药剂,且有强烈的刺激性,水乳剂、乳油、微乳剂等剂型对操作人员来说并不方便,大胶囊制剂的研制为其应用提供了便利。辣根素虽为生物土壤消毒剂,但与化学土壤消毒剂存在不同的地方,化学土壤消毒剂在杀灭土壤中致病微生物的同时杀灭有益微生物,破坏了土壤中各种微生物群落。而辣根素水乳剂不仅能够实现杀灭土壤细菌,同时能够改善土壤的活性,确保土壤中有益微生物生长,抑制致病微生物繁殖,进而降低致病菌源,保持土壤中微生物种群动态平衡,降低致病菌对蔬菜的危害。为此,本研究通过加入不同浓度的辣根素水乳剂对种植辣椒的生长特性进行了改善,提高了辣椒株高、茎粗等生物量的整体特性,不同浓度的辣根素水乳剂均有提高,但方式 3 的 11 L/667 m² 浓度的辣根素水乳剂对辣椒生长的促进作用更加明显,由此可以看到,对于辣根素水乳剂的应用,浓度需要通过科学调配以达到最好的效果。另外本研究选择的处理方式4,也即50%多菌灵可湿性粉剂,对辣椒的生长状态指标也有一定程度的促进作用,但相比不做处理下的增长效应并不明显。从研究结果也进一步说明,辣根素水乳剂对定植后的农作物不仅可以提高土壤消毒,而且对作物生长也有促进作用。

植物叶绿素等参数是衡量植物光合作用的主要参数,对于改善整体农业作业环境,增强有机物合成有显著作用,植物叶绿素等参数的测量通常采用荧光法,结合本研究的结果,目前土壤中加入不同浓度的辣根素水乳剂均能一定程度上提高辣椒植株的光合作用,相关的光合作用等参数均有一定程度的提高,显示加入辣根素水乳剂能够增加辣椒叶片的光合作用利用率,11 L/667 m² 浓度的辣根素水乳剂对辣椒植株的光合作用影响最为显著,同样的,处理方式4,也即50%多菌灵可湿性粉剂对辣椒植株的光合作用能力提升有限,相比不做处理下的提升效应较小。以上研究成果与文献中给出的辣根素水溶剂对种植蔬菜的叶绿素含量提高类似[11],充分说明了辣根素水乳剂对土壤活性成分的变强,增强了辣椒植株的光合作用影响及连作辣椒的叶绿素含量,增强辣椒的根系活性,提升辣椒作物的生长能力及抗病毒能力。

在农作物种植过程中,适当改善土壤酶活性可以有效提升农作物的生长活力,在本次试验中[12],土壤中添加的辣根素水乳剂能够改善土壤的酶活性,浓度不同对土壤的酶活性改善也有所区别,7 L/667 m² 浓度的辣根素水乳剂以及 9 L/667 m² 浓度的辣根素水乳剂的浓度对土壤的酶活性改善较为显著,显示辣根素水乳剂是土壤有益的催化剂,增强辣椒连作植株生理代谢及养分吸收能力。另外对照组方式4的加入50%多菌灵可湿性粉剂对土壤的改善作用较小,相比没有处理下的对照组方式5提升非常有限。综合本文的研究可以发现,辣根素水乳剂对连作辣椒的生长及光合作用能力增加,同时改善了土壤的酶活性,但不同浓度辣根素水乳剂对土壤及辣椒植株的改善能力不同,后面需要继续通过更多的样本研究,找到辣根素水乳剂的最优浓度配比,通过本研究成果,挖掘了辣根素水乳剂对于农作物土壤活性的改变,对于改善辣椒等经济作物连作的土壤活性、生长能力以及抗病毒能力有重要的理论及实践参考价值,研究成果可结合辣椒等经济作物种植中进行广泛推广。

参考文献

[1] Robin Lacassin, Franck Valli, Nicolas Arnaud. Reply to Comment on "Large-scale geometry, offset and kinematic evolution of the Karakorum fault, Tibet"[J]. Earth and Planetary Science Letters, 2004, 229(1):18-26.

[2] 刘政,李迎宾,孙艳,李慧,曹永松,罗来鑫,李健强. 20%辣根素水乳剂防控棉花黄萎病的研究[J]. 中国科技论文, 2017, 12(24):2817-2821.

[3] 张萌,赵欢,肖厚军,秦松,芶久兰,王正银. 贵州典型黄壤辣椒生长、品质及光合特性对新型肥料的响应[J]. 干旱地区农业研究, 2017, 35(06):187-193.

[4] Nur Alim Bahmid, Laurens Pepping, Matthijs Dekker. Using particle size and fat content to control the release of Allyl isothiocyanate from ground mustard seeds for its application in antimicrobial packaging[J]. Food Chemistry, 2020, 308(3):29-38.

[5] Yuanyuan Zhou, Xiaoya Xu. Allyl isothiocyanate treatment alleviates chronic obstructive pulmonary disease through the Nrf2-Notch1 signaling and upregulation of MRP1[J]. Life Sciences, 2020, 243(7):82-87.

[6] 王彦柠,李迎宾,黄小威,罗来鑫,曹永松,李健强. 辣根素对常见植物病原菌的抑菌活性研究[J]. 中国科技论文, 2018, 13(06):692-697.

[7] 乔岩,董杰,王品舒,杨伍群,张胜菊,杨建国. 不同土壤消毒药剂对甘薯2种土传病害的防治效果[J]. 河南农业科学, 2017, 46(08):92-95.

[8] Nazareth Tiago de Melo, Alonso-Garrido Manuel. Effect of allyl isothiocyanate on transcriptional profile, aflatoxin synthesis, and Aspergillus flavus growth[J]. Food research international(Ottawa, Ont.), 2020, 128(3):28-36.

[9] 钟平安,邵东,黄英金,王强,杨小龙,叶子飘. 不同光环境下辣椒光合特性和瞬时水分利用效率[J]. 生态学杂志, 2019, 38(07):2065-2071.

[10] 宁楚涵,李文彬,张晨,刘润进.丛枝菌根真菌与放线菌对辣椒和茄子的促生防病效应[J].应用生态学报,2019,30(09):3195-3202.

[11] 安东,车永梅,赵方贵,李雅华,杨德翠,刘新.解磷菌3P29促进辣椒生长的生理机制[J].北方园艺,2019(22):8-16.

[12] 李静.低温弱光下辣椒叶片中类胡萝卜素组分的变化及其与品种耐性的关系研究[D].甘肃农业大学,2018.

辣根素水乳剂不同施药量处理对土壤微生物多样性的影响

高晶霞[1]，吴雪梅[2]，高 昱[2]，牛勇琴[2]，王学梅[1]，谢 华[1]

(1. 宁夏农林科学院种质资源研究所，宁夏银川 750002；
2. 宁夏回族自治区彭阳县蔬菜产业发展服务中心，宁夏彭阳 756500)

摘 要：本研究以连作8年土壤为研究对象，设置7 L/667 m^2、9 L/667 m^2、11 L/667 m^2、对照药剂(50%多菌灵可湿性粉剂)、空白对照(不做任何处理)5个处理，采用高通量测序技术，测定16S(细菌)和ITS(真菌)微生物群落结构变化。结果表明：16S分析结果，对不同处理组样本OTUs和Tags数量统计，总Tags数量平均86 000，Unique Tags平均68 000，有物种注释的Tags数量与Unique Tags基本相同；5处理组共有2 253 OTUs，特有OTUs最多是H处理组，其次是I处理组和CK处理组，F处理组和G处理组最低；每个处理样本在门水平注释率在95%以上；Gammaproteobacteria为各处理样品中占比最多细菌，Planctomycetacia在随着处理剂量加重有略微上升，多菌灵处理和对照(CK)中占比大于处理组，Alphaproteobacteria在H处理组有所上升，I处理组和对照(CK)较H处理组减少；Gemmatimonadetes从F处理至H处理表现出先将后升，多菌灵处理后，该菌量最低。Betaproteobacteria在3个处理中不变，多菌灵处理后显著降低，CK组最高。Deltaproteobacteria在各处理组中无明显变化。Thermomicrobia在各个处理中相较对照(CK)都显著上升；Alpha多样性分析，shannon指数中位数都在10.00左右，F处理组最低9.543 6，最高10.055 5，G处理组最低9.698 4，最高10.023 0，H处理组最低9.552 2，最高10.169 1，I处理组是4个处理组中波动最大的，可能和多菌灵处理条件有关；Beta多样性分析，对照(CK)组和F、G、H处理三组聚类存在差异，说明利用辣根素处理，土壤菌群发生变化，I处理组和其他4组处理相聚较远，也反映了多菌灵处理和辣根素处理不同之处，其中H组处理中H2样本离群较远，I组处理中I1样本离群较远；物种差异分析结果，随着药物浓度增加，

Arenimonas 和 Oceanospirillales 菌群降为非优势菌群,推测这 2 种菌群降低对土传病害防治是有利的。多菌灵使用使 I 组 cinetobacter_calcoaceticus、Trueperaceae 和 Deinococcales 菌群增加,推测该几类菌群增加有利于病害防治。ITS 分析结果,5 处理组共有 229 OTUs,特有 OTUs 最多是 F 处理组;物种分类分析,每个处理样本在门水平注释率在 70%左右;各处理样本在科水平上物种分布堆叠图,注释率在 15%~50%,Hypocreaceae 占比最多,在对照(CK)和 I 处理组中含量很高,在辣根素处理组中含量显著降低;Trichocomaceae 和 Chaetomiaceae 在 F 处理组、G 处理组和 I 处理组显著增加,其中 G 处理组最为明显;Botryosphaeriaceae 在对照(CK)和 I 处理组中含量较高,在辣根素处理组中几乎看不到,含量很低;Saccharomycetaceae 在辣根素处理组中含量很多,是优势菌群,在对照(CK)和 I 处理组中很低;Alpha 多样性分析,shannon 指数中位数都在 5.3 左右,G 处理组最低 5.002 2,最高 5.486 8,F 处理组最低 5.343 9,最高 5.758 0,H 处理组最低 4.122 6,最高 5.591 4,是 4 个处理中波动最大的,推测辣根素浓度增加到最高,导致对应真菌群落发生很大变化。I 处理组最低 4.833 1,最高 5.449 3;Beta 多样性分析,H 处理组和 G 处理组真菌方面组内组间差异很难区分,F 处理组和 H 处理、G 处理组差异明显,I 处理组和对照(CK)组组间差异较小,I 处理组和对照(CK)分别与辣根素处理组差异显著;物种差异分析,I 处理组中 Hydropisphaera 为唯一差异真菌,G 处理组中优势差异真菌为 Penicillium。综合分析结果:辣根素浓度应该为 G 处理组浓度最好,同时 50%多菌灵处理同样可以有效进行防治病害。为辣椒土传病害防控提供新防治方法,对宁夏南部山区拱棚辣椒安全、丰产栽培具有重要意义。

关键词:辣根素;施药量;微生物多样性

Effects of Different Dosages of Horseradish Water Emulsion on Soil Microbial Diversity

Gao jing-xia[1], Wu xue-mei[2], Gao yu[2], Niu yong-qin[2], Wang xue-mei[1], Xie hua[1]

(1. *Institute of Germplasm Resources, Ningxia Academy of Agriculture and Forestry Sciences, Yinchuan* 750002, *Ningxia*; 2. *Vegetable Industry Development Service Center of Pengyang County, Ningxia Hui Autonomous Region, Pengyang* 756500, *China*)

Abstract: In this study, the soil of 8 years of continuous cropping was taken as the research object, five treatments, 7 L/667 m², 9 L/667 m², 11L/ L/667 m², control agent (50% carbendazim wetting powder) and blank control (no treatment) were set up. The microbial community structure changes of 16S (bacteria) and ITS (fungi) were determined by high-throughput sequencing technology.The results show that:16S analysis results show that the number of OTUs and Tags in samples of different processing groups is 86 000 on average, the number of Tags in Unique Tags is 68 000 on average, and the number of Tags with species annotation is basically the same as that of Unique Tags; There were 2 253 OTUs in the five treatment groups,the H treatment group had the most unique OTUs, followed by the I treatment group and the CK treatment group, and the F treatment group and the G treatment group had the lowest, the annotation rate of each treatment sample at gate level is above 95%; Gammaproteobacteria was the bacteria with the largest proportion in each treatment sample. Planctomycetacia increased slightly with the aggravation of treatment dose, and the proportion of carbendazim treatment and control (CK) was greater than that of the treatment group. Alphaproteobacteria increased in the H treatment group, while the proportion of I treatment group and control (CK) decreased compared with the H treatment group.From F to H, Gemmatimonadetes showed that it rose first and then rose.Betaproteobacteria remained unchanged in the three treatments, and significantly decreased after carbendazim treatment, with the highest value in CK group.Deltaproteobacteria showed no significant changes in each treatment group.Thermomicrobia in various processing compared with control (CK) increased significantly;According to Alpha diversity analysis, the median shannon index was around 10.00, with the lowest value of 9.543 6 and the highest value of 10.055 5 in the F treatment group, the lowest value of 9.698 4 and the highest value of 10.023 0 in the G treatment group, the lowest value of 9.552 2 and the highest value of 10.169 1 in the H treatment group;Beta diversity analysis, the control group (CK) and F, G, H treatment differences between three groups of clustering, shows that using horseradish grain processing, change of soil bacteria, I in treatment group and the other four groups together far away, also reflects the carbendazim treatment and horseradish differs from that of the group H in processing the H2 samples to far away from the group, group I processing I1 sample stray far away;According to the results of species difference analysis, the flora of Arenimonas and Oceanospirillales decreased to non-dominant flora with the increase of drug concentration, suggesting that the decrease of these two flora was beneficial to the control of soil-borne diseases.The use of carbendazol increased the flora of cinetobacter_calcoaceticus, Trueperaceae and Deinococcales in group I, suggesting that the increase of these flora was beneficial to the disease control.According to ITS analysis, there were 229 OTUs in 5 treatment groups, and F treatment group had the most OTUs;the annotation rate was about 15%~50%, and Hypocreaceae occupied the largest proportion. The content of Hypocreaceae was very high in the control (CK) and I treatment groups, and significantly decreased in the horseradish treatment group; trichocomaceae and Chaetomiaceae were significantly increased in F treatment group, G treatment group and I treatment group, of which G treatment group was the most obvious.The content of Botryosphaeriaceae was high in the CK and I treatment groups, but was almost not seen in the horseradish treatment group; saccharomycetaceae was a dominant flora with a high content in the horseradectin treatment group and was very low in the CK and I treatment groups;In the analysis of Alpha diversity, the median of shannon index was all around 5.3, with the lowest value of 5.002 2 and the highest value of 5.486 8 in the G treatment group, the lowest value of 5.343 9 and the highest value of 5.758 0 in the F treatment group, and the lowest value of 4.122 6 and the highest value of 5.591 4 in the H treatment group, indicating the largest fluctuation among the four treatments. It was speculated that the increase of horseradin concentration to the highest level led to great changes in the corresponding fungal community; The lowest value of treatment group I was 4.833 1, and the highest value was 5.449 3. In the analysis of Beta diversity, it was difficult to distinguish the difference between the treatment group H and the treatment group G in terms of fungi, the difference between the treatment group F and the treatment group H and the treatment group G was significant, the difference between the treatment group I and the control group (CK) was small, and the difference between the treatment group I and the control group (CK) was significant compared with the treatment group horseradish; According to the analysis of species difference, Hydropisphaera was the only differential fungus in the treatment group I, and the dominant differential fungus in the treatment group G was Penicillium. According to the comprehensive analysis, the concentration of horseradish should be the best in the treatment group G, and the treatment with 50% carbendazim could also effectively control diseases.It provides a new method for the prevention and control of soil-borne diseases of pepper and is of great significance to the safe and productive cultivation of capsicum in gongpeng mountain area of southern ningxia.

Key words: horseradish; the medication. microbial diversity

辣椒(Capsicum annuum L.)又名番椒,原产于拉丁美洲热带地区,茄科辣椒属,为一年生草本植物,在我国普遍栽培[1]。辣椒营养丰富,维生素C含量在蔬菜中居第一位[2]。辣椒是宁夏南部山区栽培面积较大蔬菜之一,然而受到耕地的制约,目前蔬菜连作这种现象比较常见,在一定程度上导致土壤层中菌落失衡,病虫害增加,使得土壤病害现象严重[1-4]。与此同时,在辣椒的种植生长过程中,由于使用农药、化肥及外部高频率耕地的影响,使得土壤生产能力下降,农药残留超标,病虫害爆发等诸多问题,进而衍生了区域环境及生态问题,导致辣椒生产环境恶化,质量下滑,俨然已成为影响辣椒栽培可持续发展重要因素[5-9]。土壤消毒处理是预防和控制土传病害最简便、经济、有效技术措施之一[10],辣根素(allyl isothiocyanate)是植物源农药,对环境安全,作为土壤处理剂能有效杀灭多种土壤微生物防控多种土传病害。本研究设置 7 L/667 m²、9 L/667 m²、11 L/667 m²、对照药剂(50%多菌灵可湿性粉剂)、空白对照(CK):不做任何处理 5 个处理,采用高通量测序技术,测定 16S 和 ITS 微生物群落变化,旨在探明 20%辣根素水乳剂不同施药量处理辣椒土壤后对土壤微生物多样性影响,拟为辣椒土传病害防控提供新的防治方法,对宁夏南部山区拱棚辣椒安全、丰产栽培具有重要意义。

1 材料与方法

1.1 试验地点

试验地点位于宁夏彭阳县新集乡拱棚辣椒核心示范基地(连作 8 a 的土壤)。

1.2 供试材料

供试辣椒品种为"巨峰 1 号",购买自宁夏巨丰种苗有限公司;20%辣根素水乳剂,购买自中农齐民(北京)科技发展有限公司;50%多菌灵可湿性粉剂,购买自宁夏尚博农植物保护有限公司;土壤为连作 8 年的土壤。

1.3 试验设计

试验设 5 个处理:

处理 1:7 L/667 m²;处理 2:9 L/667 m²;处理 3:11 L/667 m²;处理 4:对照药剂(50%多菌灵可湿性粉剂);处理 5:空白对照(CK):不做任何处理。

处理 1~3 用量利用滴灌系统施肥器将药剂施入土壤,覆膜密封 3~5 d 后定植;处理 4 按照 2 kg/667 m² 用量拌土撒施。

1.4 指标测定

土壤样品中细菌 16S rRNA 和真菌 ITS2 基因高通量测序由广州 GeneDenovo 生物信息技术有限公司利用 Illumina HiSeq2500 平台进行。利用引物对 16S V3-V4 区域进行扩增:341F(CCTACGGGNGGCWGCAG)和 806R（GGACTACHVGGGTATCTAAT）。使用正向引物 KYO2F(GATGAA GAACGYAGYRAA)和反向引物 ITS4RT(CCTCCGCTTATT GATATGC)扩增 ITS2 区。所有 PCR 反应进行的 15μL Phusion®高保真 PCR 反应混合液(新英格兰生物学实验室);正向和反向引物 0.2 μM,大约 10 ng 模板DNA。热循环包括 98℃初始变性 1 min,98℃变性 10 s,50℃退火 30 s,72℃延伸 30 s,共 30 个循环。72℃ 5 min,将等量的 SYB green 缓冲液(含 SYB green)与 PCR 产物混合,2%琼脂糖凝胶电泳检测。PCR 产物按等密度比混合。然后用 Qiagen 凝胶提取试剂盒(Qiagen,德国)对混合 PCR 产物进行纯化。使用 truseq®DNA PCR-Free Sample Preparation Kit（Illumina,USA)按照制造商的建议添加索引代码,生成测序文库。在 Qubit 2.0 荧光仪(Thermo Scientific)和安捷伦 2100 生物分析仪(Agilent Bioanalyzer 2100)上进行库质量评估。

2 数据分析

2.1 16S 数据分析

2.1.1 不同处理数据处理统计分析

由图 1 可知,reads QC filter 为低质量 reads,Non-overlap 为没有 overlap 的未组装 reads,Tag QC filter 为未通过"tag 过滤"的 tags,以上三者的总和百分比小于 5%,说明测序下机指标数据合格;effective tags 为有效用于后续分析的 tags,有效 Tags 数量占据原始 PE Reads 百分比从 F-1 到 CK-3 处理分别为 91.99%、94.13%、95.16%、93.43%、92.70%、94.31%、95.31%、87.75%、92.81%、91.24%、95.50%、95.55%、94.74%、93.72%和 92.54%,只有 H-2 处理低于 90%,

但也在87.75%,有效Tags数据量是足够进行后续分析,可靠数据。Tags丰度统计显示:检测到的Tags数量每个样品都大于80 000条,Tags总长3 900万以上,最长Tags为474,最短Tags为341~407不等,N50除H-2处理为448,其余都大于460,N90为441,当数据达到饱和后,过多的数据并不会产生更多新的OTU,因此为保证分析时间,测序公司一般对超过8万effetive tag样本进行数据抽取处理,最终随机选取8~9.2万tags进行后续分析。综上所述,数据质量合格。

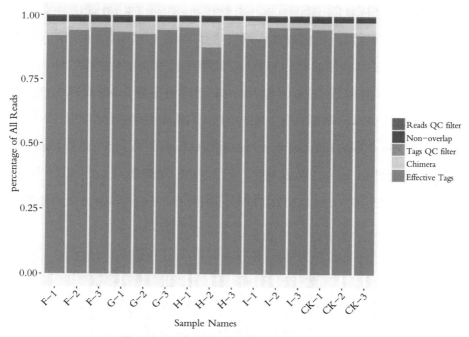

图1 不同处理数据预处理分布图(百分比)

Fig. 1 Distribution diagram of data pretreatment with different processing (percentage)

2.1.2 不同处理组样本OTU分析

OTU(Operational Taxonomic Units)是指为了方便在系统发生学或群体遗传学研究,人为设定的分类单元。利用effective tags之间的序列相似性关系,可以将不同的tags聚类成OTU。获得OTU之后,利用相关软件,根据其丰度和序列信息,能够逐一开展物种注释、群落多样性、组间差异等多种核心分析。由图2可知,对不同处理组样本的OTUs和Tags数量统计,总Tags数量平均86 000,Unique Tags平均68 000,有物种注释的Tags数量与Unique Tags基本相同,没有物种注释的Tags的数量为0,被过滤掉的Tags最低10 306,最高26 383,最高的为H-2处理,可以看出总丰度为1的OTUs所对应的Tags数量,该样本最多,被过滤掉的数据较多,但总测序量较大,数据足够聚类及后续分析;得到OTUs每个样本大于4 500条。

由图3可知,各处理组样本之间共有和特有OTUs的韦恩图和各个集合的数量统计柱状图,对照(CK)vsF处理组:对照(CK)处理组特有1897 OTUs,F处理组特有1 526 OTUs,两组共有3 187 OTUs;CK处理组vsG处理组:CK处理组特有1 837 OTUs,G处理组特有1 634 OTUs,两组共有3 247 OTUs;CK处理组vsH处理组:CK处理组特有2 249 OTUs,H处理组特有1 851 OTUs,两组共有3 233 OTUs;CK处理组vsI处理组:CK处理组特有1 956 OTUs,I处理组特有1 949 OTUs,两组共有3 135 OTUs;F处理组vsG处理组vsH处理组vsI处理组vsCK处理组:F处理组特有611 OTUs,G处理组特有669 OTUs,H处理组特有1 517 OTUs,I处理组特有

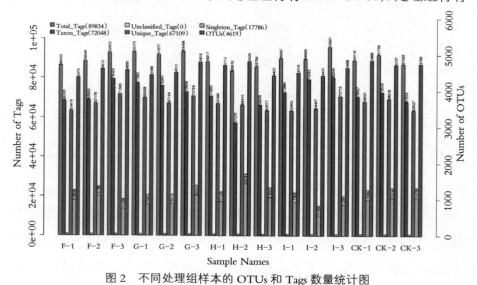

图2 不同处理组样本的OTUs和Tags数量统计图

Figure 2. Quantity statistics of OTUs and Tags for different processing samples

1 005 OTUs,CK 处理组特有 987 OTUs,5 组共有 2 253 OTUs。特有 OTUs 最多的是 H 处理组,其次是 I 处理组和 CK 处理组,F 处理组和 G 处理组最低。

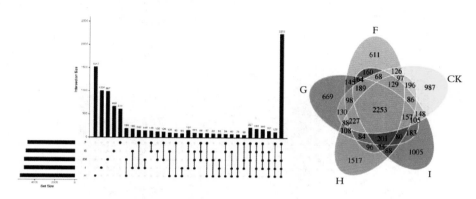

图 3 不同处理组 OTUs 比较韦恩图

Fig.3 Comparison of wynn diagrams between different treatment groups OTUs

2.1.3 不同处理组样本物种分类分析

微生物物种分类一般分为界、门、纲、目、科、属、种 7 个等级,而每个 OTU 代表某类型分类水平集合。因此根据 OTU 的序列信息进行物种注释,能将分析结果与实际的生物学意义进行关联,从而研究群落中物种的变化关系等内容。图 4 是每个处理样本在各分类水平上的序列构成柱形图,可以看出门水平的注释率在 95% 以上,纲水平在 90% 左右,目水平在 80%~90%,科水平在 75% 左右,属水平在 40% 左右,种水平 2%~12%。整体纲水平和目水平注释率正常,各处理样本组内差异较小,H 处理组中 H2 样本纲和目水平注释率比同处理组中的其他样本高出 10%,浮动较为明显,有可能是特异样本,H 处理组另外 2 个样品注释率相近。

优势物种很大程度决定着微生物群落的生态结构以及功能结构,了解群落在各个水平的物种组成情况能有效地对群落结构的形成、改变以及生态影响等进行解读。

我们统计了各个层级分类水平上各样品的物种组成情况,然后用堆叠

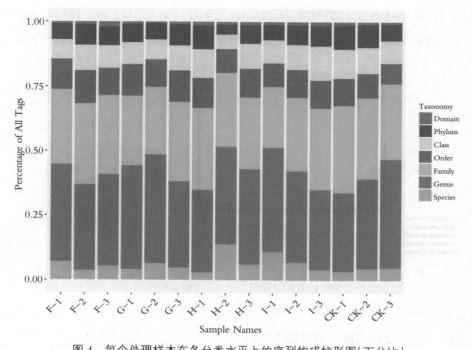

图4 每个处理样本在各分类水平上的序列构成柱形图(百分比)
Fig. 4 Histogram of sequence composition of each treated sample at each classification level (percentage)

图的形式,直观展示不同的样品在各个分类层级水平上的物种丰度变化情况。图5是各处理样本纲水平上的物种分布堆叠图,Gammaproteobacteria为各处理样品中占比最多的细菌,该菌在7 L/667 m² 和9 L/667 m² 的处理下未有明显变化,11 L/667 m² 处理下显著降低,多菌灵处理和对照(CK)基本相同;Planctomycetacia在随着处理剂量的加重有略微上升,多菌灵处理和对照(CK)中占比大于处理组;Alphaproteobacteria在F处理和G处理组无变化,在H处理组有所上升,I处理组和对照(CK)较H处理组减少;Gemmatimonadetes在对照(CK)和F处理组中检测到的量相同,从7 L/667 m² 到9 L/667 m² 明显降低,11 L/667 m² 略微上升,多菌灵处理后,该菌量最低。Betaproteobacteria在3个处理中不变,多菌灵处理后显著降低,CK组最高。Deltaproteobacteria在各处理组中无明显变化。Thermomicrobia在各个处理中相较对照（CK）都显著上升。

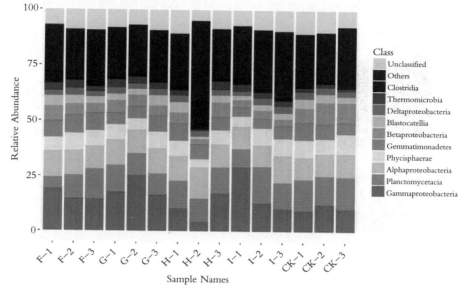

图 5 不同处理样本纲水平上的物种分布堆叠图

Fig. 5 Stacking plots of species distribution at the level of different processing classes

图 6 是各处理样品科水平上物种分布堆叠图，从图中可知，注释率在 75%左右，Planctomycetaceae 在 F、G 和 H 处理组中含量较 I 处理和对照（CK）含量较低；Xanthomonadaceae 在 F 和 G 处理组中较高，处理增加到 11 L/667 m^2 后降低，I 处理组和对照（CK）中较低，处理后出现该菌含量显著增加，但随着处理加强，含量在 11 L/667 m^2 出现降低。Blastocatellaceae 在对照（CK）中最高，F 处理组无明显降低，G 处理组和 I 处理组后显著降低；Pseudomonadaceae 在各处理组中都未有明显变化，多菌灵处理有明显降低。

图 7 是属水平上物种分布堆叠图，从图中可以看出，属水平会出现较为杂乱的细菌分类，和其他土壤情况相同，但从中可以看出，Planctomyces 是主要菌群，在整体样本中未有明显变化，Lysobacter 在 CK 和 F 处理中基本相同，随着处理剂量加强，该类菌明显增加，Pseudomonas 在对照（CK）中很少，在 F、G 和 H 处理注重增加，I 处理组中出现特别明显增加，Pseudoduganella

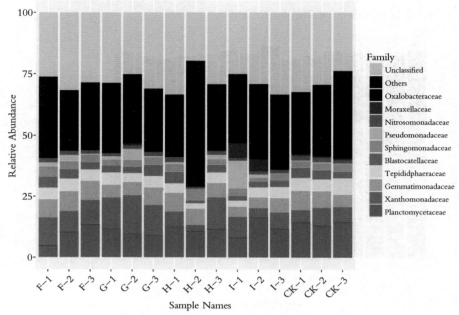

图 6 不同处理样本科水平上的物种分布堆叠图
Fig.6 Species distribution stack diagram at the undergraduate level of different treatment samples

在对照(CK)中含量较多,经过辣根素和多菌灵处理,土壤每个样本中几乎含量降为 0。Bacillus 在经过处理后显著增加,对照(CK)中基本为 0。Acinetobacter 在多菌灵处理组中显著增加,其余各组处理中含量很低。

图 8 是科水平的物种分类热图,结果和图 6 堆叠图一致,只是堆叠图显示前 10 的优势菌群,热图显示很多,并做了均一化处理,以堆叠图为准即可。若进行详细数据挑选,也可用热图进行。

2.1.4 不同处理组样本 Alpha 多样性分析

α 多样性是指特定生境或者生态系统内多样性情况,它可以指示生境被物种隔离的程度,通常利用物种丰富度(种类情况)与物种均匀度(分布情况)2 个重要参数来计算。本项目主要展示 Chao1,ACE,Shannon,observed_species,Simpson 和 Good's Coverage 六大类常用 α 多样性指数以及它们的相关分析结果。总体来说,Chao1 / ACE 指数主要关心样本的物种

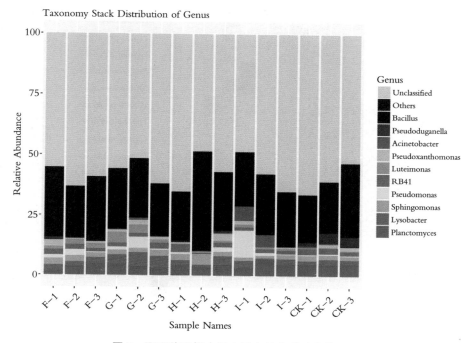

图7 不同处理样本属水平上的物种分布堆叠图
Fig.7 Stack of species distribution at the level of genus in different treatments

丰富度信息;Good's Coverage 反映样本低丰度 OTU 覆盖情况;observed_species 表示能检测到 OTU 种类情况;Simpson / Shannon 主要综合体现物种的丰富度和均匀度。表 1 是上述指标的统计表,从表中可以看出,chao1 指数都在 5 000 以上, 平均值 5 365.059,ACE 指数和 chao1 相近, 平均值 5 425.227, Goods Coverage 反映样本的低丰度 OTU 覆盖情况,平均值在 98.5%,说明绝大多数低丰度 OTU 已经被覆盖到,完整性很好。observed_species 平均值在 4 619.2; Simpson / Shannon 数值平均值分别是 9.882 5 和 0.994 5,丰富度非常高,均匀度很好,说明数据没有问题。

我们通过绘制稀释曲线(rarefaction curve)来评价测序量是否足以覆盖所有类群,并间接反映样品中物种的丰富程度。稀释曲线是利用已测得序列中已知的 OTU 的相对比例,来计算抽取 n 个(n 小于测得 tags 序列总数)tags 时出现 OTU 数量的期望值,然后根据一组 n 值(一般为一组小于总序

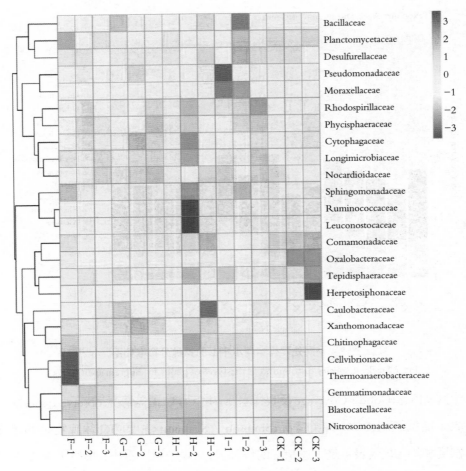

图 8 不同处理样本科水平上的物种分类热图

Fig. 8 Heat maps of species classification at different treatment levels

列数的等差数列)与其相对应的 OTU 数量的期望值做出曲线来。当曲线趋于平缓或者达到平台期时也就可以认为测序深度已经基本覆盖到样品中所有物种。图 9 可以看出抽取的 tags 数量到达 60 000 时,所能得到 OTU 数量的期望值已经到达平台期,而我们数据的 clean tags 平均在 90 000 以上,说明已经覆盖了所有的物种。

图 10 是对 F、G、H 和 I 组的 shannon 指数箱线图,横坐标为各分组情况,各组分别为红绿蓝紫四种颜色区分,纵坐标表示对应 Alpha 多样性指数

表1 不同处理 Alpha 多样性各指标统计表

Table 1 Statistical table of indicators of Alpha diversity with different treatments

样品	chao1	ace	goods_coverage	observed_species	shannon	simpson
F-1	5 178.982 166	5 301.727 003	0.983 566 05	4 375	9.543 648 075	0.994 680 276
F-2	5 517.080 645	5 568.163 757	0.983 139 704	4 612	9.932 021 158	0.996 205 542
F-3	5 177.903 226	5 225.161 793	0.988 413 682	4 585	10.055 555 04	0.997 073 223
G-1	5 303.668 702	5 281.711 834	0.986 324 565	4 450	9.698 494 298	0.994 898 259
G-2	5 231.879 121	5 291.308 974	0.986 506 233	4 511	9.746 224 9	0.995 232 233
G-3	5 599.506 112	5 685.742 759	0.984 410 563	4 810	10.023 039 65	0.996 806 246
H-1	5 435.296 154	5 521.935 4	0.985 019 466	4 713	10.169 125 52	0.997 077 763
H-2	5 323.716 514	5 527.099 61	0.981 846 772	4 829	9.552 294 282	0.990 071 266
H-3	5 257.788 674	5 311.955 463	0.983 481 556	4 432	9.839 555 45	0.996 080 943
I-1	5 325.501 326	5 390.310 483	0.984 852 643	4 518	9.237 130 771	0.982 688 185
I-2	5 054.236 934	5 013.816 711	0.989 355 269	4 433	10.092 810 85	0.995 832 893
I-3	5 301.587 127	5 285.359 234	0.988 708 873	4 658	10.298 046 16	0.997 771 835
CK-1	5 612.320 631	5 633.346 699	0.984 877 475	4 865	10.262 293 26	0.997 301 804
CK-2	5 540.789 894	5 582.687 982	0.984 839 297	4 737	10.045 419 16	0.995 910 033
CK-3	5 615.624 852	5 758.078 441	0.982 265 531	4 760	9.742 201 194	0.991 099 776

大小。图形展示有效数据分布的最大值(直线顶端),最小值(直线底端),中位数(盒子中线),上四分位数(盒子顶边),下四分位数(盒子底边),以及无效数据;shannon 指数中位数都在 10.00 左右,F 处理样本组最低 9.543 6,最高 10.055 5,G 处理组样本最低 9.698 4,最高 10.023 0,H 处理组样本最低 9.552 2,最高 10.169 1,I 处理组样本是 4 个处理组中波动最大的,可能和多菌灵处理条件有关。可以看出 3 个处理组浮动基本相同,G 处理组样本波动最小,H 处理组样本浓度最大,和多菌灵处理最为接近,也说明 G 处理组处理浓度可作为最佳参考浓度。

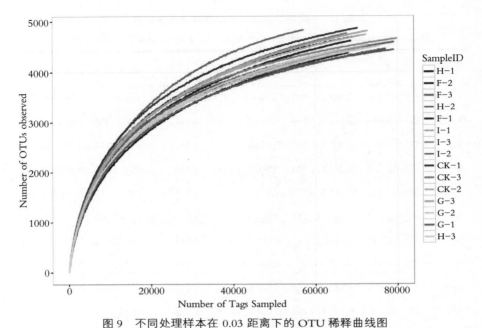

图 9 不同处理样本在 0.03 距离下的 OTU 稀释曲线图
Fig. 9 OTU dilution curves of samples with different treatments at a distance of 0.03

2.1.5 不同处理组样本 Beta 多样性分析

Beta 多样性是不同生态系统之间多样性比较，是物种组成沿环境梯度或者在群落间变化率，用来表示生物种类对环境异质性的反应。一般来说，不同环境梯度下群落 Beta 多样性计算包括物种改变（多少）和物种产生（有无）2 部分。

这种方法借用方差分解可以有效找出数据中最"主要"元素和结构，将复杂的样本组成关系反映到横纵坐标 2 个特征值上，从而达到简化数据复杂度的效果。分析结果中，样品组成越相似，反映在 PCA 图中距离越近，而且不同环境间样品往往可能表现出各自聚集分布情况。从图 11 可以看出，CK 处理组和 F 处理组、G 处理组、H 处理组三组聚类存在差异，说明利用辣根素处理，土壤菌群发生变化，I 处理组和其他 4 组处理相聚较远，反映了多菌灵处理和辣根素处理的不同之处，其中 H 处理组中 H2 离群较远，I 处理组中 I1 样本离群较远。从 PCA 能看出对照 CK 最左，I 处理组最右，微弱的

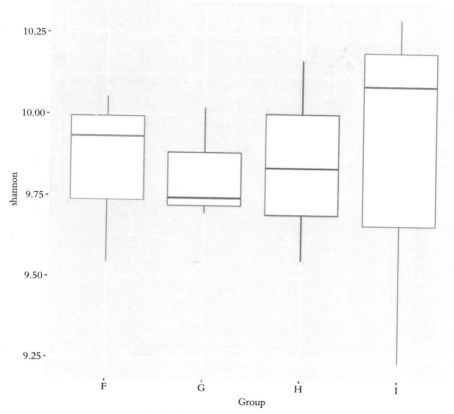

图 10 不同处理组 shannon 指数箱线图比较

Fig.10 Comparison of shannon index boxplot of different treatment groups

看出处理按照 F 处理组、G 处理组、H 处理组和 I 处理组趋势分布。

UPGMA 分类树可以用于研究样本间相似性，解答样本的分类学问题。利用 Mothur 软件，根据 weighted 和 unweighted Unifrac 矩阵信息，可以将样本进行 UPGMA 分类树分类。其中越相似的样本将拥有越短的共同分支。从图 12 中可知，暂不考虑 H2 样本，剩余样本各组间 3 个重复都是相似性最高的，其中 F 处理组和 G 处理组拥有最短的共同分支，处理相近，H 处理组加重，多了一条分支，说明 H 组处理使土壤菌群发生明显变化，I 处理组和 F 处理、G 处理、H 处理条件不同，比这三组多一条分支，菌群变化符合处理条件

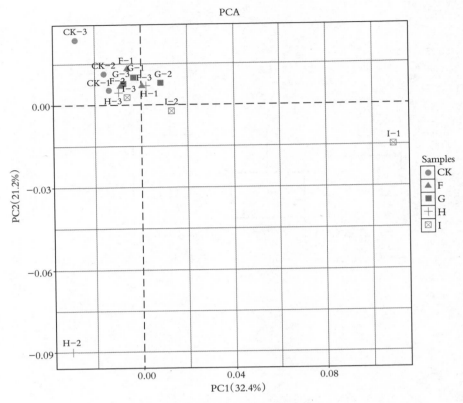

图 11 不同处理组 2D OTU PCA plot
Fig.11 2D OTU PCA plot with different processing groups

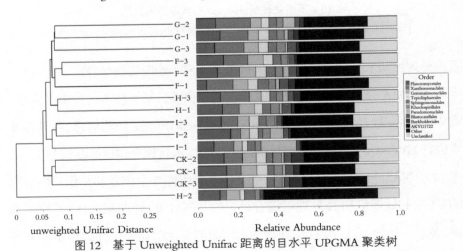

图 12 基于 Unweighted Unifrac 距离的目水平 UPGMA 聚类树
Fig.12 Hierarchical UPGMA clustering tree based on Unweighted Unifrac distance

变化。

表2是Anosim差异分析结果表，其中R值表示差异程度，一般介于(0,1)之间，R>0,说明组间存在差异;R>0.75,大差异;R>0.5,中等差异;R>0.25,小差异。F处理组、G处理组和I处理组R值为1,说明组间差异明显，H处理组相比空白对照(CK),R为0.3704,>0.25是小差异。统计分析可信度用P-value表示,P<0.05表示统计具有显著性,处理组与对照(CK)组相比有差异,多组间比较差异极显著。

表2 基于unweighted unifrac距离的Anosim分析结果表
Table 2 Results of Anosim analysis based on unweighted unifrac distance

Diffs	Rvalue	Pvalue	significant
CK-vs-F	1	0.1	
CK-vs-G	1	0.1	
CK-vs-H	0.370 4	0.1	
CK-vs-I	1	0.1	
F-vs-G-vs-H-vs-I-vs-CK	0.659 3	0.001	★★
F-vs-G-vs-H-vs-I	0.651 2	0.002	★★

2.1.6 不同处理组样本群落功能分析

完整微生物群落研究主要分为物种组成、多样性以及功能研究等几个重要方面。多种证据表明微生物群落功能组成比物种组成与环境关系更为密切,随着分析技术发展,利用多样性测序数据进行群落功能预测已经成为群落研究的重要内容。我们将利用FUNGuild、Tax4fun、FAPROTAX、BugBase等多个预测软件,根据不同数据类型有针对性地完成群落功能预测分析。图13是Tax4fun预测结果,对丰度最大的20个通路和各个样本进行了热图分析,颜色越红说明该样本在对应通路中相关系数越高,就越可能是我们需要关注的通路,从CK空白对照-1-2-3中可以看出,ko00260到ko00500d这几条通路,可以作为候选通路,在处理组中都有不同程度的相关性,其中 ko03010、ko00550、ko00330、ko00230 最为显著。I 处理组中 ko02020 和 ko03070 相比其他处理组相关性低,对应其他处理该通路相关性高,可能是多菌灵和辣根素处理差异导致。在辣根素处理中,F 处理组样本对应 ko00051 和 ko03018 相关性低,G 处理组样本和 H 处理组样本相关性高,推测由于药物浓度增加导致对应通路相关性增强。

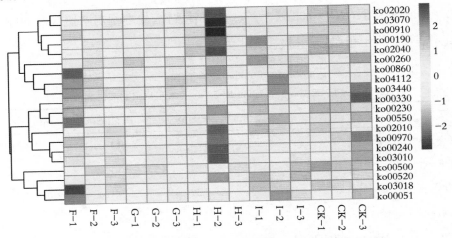

图 13　不同处理组 Pathway 丰度热图(丰度最大的 20 个)

Figure 13　Heat maps of Pathway abundance in different treatment groups (the 20 with the largest abundance)

2.1.7　不同处理样本物种差异分析

通过 LEFse 分析组间菌群差异,可以找出各组间特异的主要菌群,有助于开发 biomaker 等研究。利用 LEFse 软件对差异组间进行分析,LEFse 先对所有组样品间进行 kruskal-Wallis 秩和检验(一种多样本比较时常用的检验方法),将筛选出差异再通过 wilcoxon 秩和检验(一种两样本组比较常用的检验方法)进行两两组间比较,最后筛选出差异使用 LDA(Linear Discriminant Analysis)得出的结果进行排序得到左图,左图展示了不同组中丰度差异显著的物种,柱状图长度代表差异物种的影响大小(即为 LDA Score)。随后通过将差异映射到已知层级结构的分类树上方式得到进化分支图(右图)。在进化分支图中,由内至外辐射圆圈代表了由门至属(或种)的分

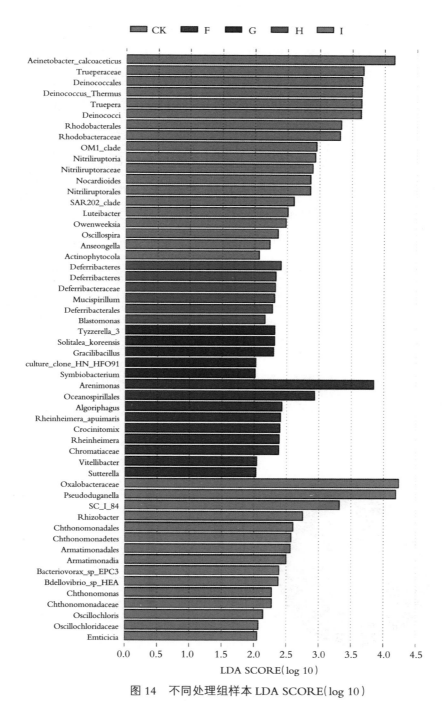

图 14 不同处理组样本 LDA SCORE(log 10)

Fig.14 LDA SCORE(log10) of different processing

类级别。在不同分类级别上每一个小圆圈代表该水平下的一个分类,小圆圈直径大小与相对丰度大小呈正比。着色原则:无显著差异的物种统一着色为黄色。详细见图 14,展示了不同组中丰度差异显著的物种,柱状图的长度代表差异物种的影响大小(即 LDA Score)。从图中可以看到 I 处理组中影响最大的差异物种是 cinetobacter_calcoaceticus,其次是 Trueperaceae、Deinococcales、Deinococcus_Thermus,差异优势菌群也就是排名就前面这几个菌群。F 处理组中影响最大最为显著的是 Arenimonas,其次是 Oceanospirillales,该组中的这两个菌群较 G 处理组和 H 处理组尤为显著,而 G 处理组和 H 处理组中没有最为显著的菌群,但分别有 5 个和 4 个菌群作为候选影响差异物种。空白对照(CK)组中 Oxalobacteraceae 和 Pseudoduganella 两个菌群影响最显著。从该结果中 F 处理组到 G 处理组到 H 处理组,随着药物浓度的增加,Arenimonas 和 Oceanospirillales 菌群降为非优势菌群,推测这两种菌群的降低对土传病害防治是有利的。多菌灵的使用使 I 处理组 cinetobacter_calcoaceticus、Trueperaceae 和 Deinococcales 菌群增加,推测该几类菌群的增加有利于病害防治;CK 为空白对照组中 Oxalobacteraceae 和

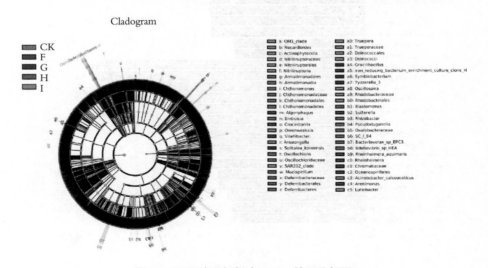

图 15 不同处理组样本 LEFse 差异分析图

Figure 15 LEFse difference analysis of samples in different treatment groups

Pseudoduganella 两个菌群为优势菌群,可能不利于病虫害防治,在其他处理组中该类菌群显著降低,结合病虫害实际防治情况,推测这两类菌需要处理后降为非优势菌群,可能利于防治病害。

2.2 ITS 数据分析

2.2.1 不同处理组样本数据处理统计

测序得到原始数据后,由于 PCR 错误、测序错误等产生大量低质量数据或者无生物学意义数据(例如嵌合体),因此为保证后续分析具有统计可靠性和生物学有效性,我们会在 reads 利用、tags 拼接等多个数据处理过程进行严格质控,最终获得 effective tags 来开展后续 OTU 聚类等多个分析。由图 16 可知,reads QC filter 为低质量 reads,低于 1%;Non-overlap 为没有 overlap 的未组装 reads,低于 5%;tag QC filter 为未通过"tag 过滤"的 tags,低于 1%;effective tags 为有效用于后续分析的 tags,经过过滤后各组得到高质量的 Pair

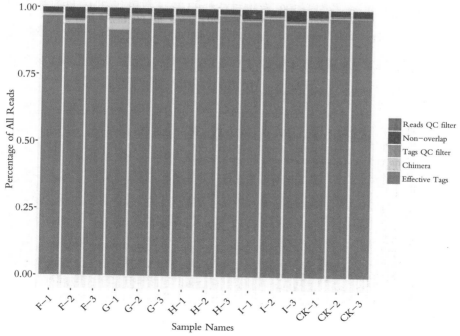

图 16 不同处理组数据预处理分布图(百分比)

Fig.16 Distribution diagram of data preprocessing in different processing groups (percentage)

End Reads 数都大于 90 000，过滤后 Tags 数最低 86 680 条，最高 96 575 条，有效比例最高为 97.33%，最低为 91.46%，都大于 90%以上。Max length 平均 448，N50 最低 370，最高 409，N90 最低 332，最高 349。当数据达到饱和后，过多的数据并不会产生更多新的 OTU，因此为保证分析时间，测序公司一般对超过 8 万 effetive tag 样本进行数据抽取处理，最终随机选取 8 万~9.2 万 tag 进行后续分析。综上所述，数据质量合格。

2.2.2 不同处理组样本 OTU 分析

OTU（Operational Taxonomic Units）是指为了方便在系统发生学或群体遗传学研究，人为设定的分类单元。利用 effective tags 之间序列相似性关系，可以将不同的 tags 聚类成 OTU。获得 OTU 之后，利用相关软件，根据其丰度和序列信息，能够逐一开展物种注释、群落多样性、组间差异等多种核心分析。图 17 是对不同样本 OTUs 和 Tags 数量的统计，总 Tags 数量平均 90000，Unique Tags 平均 16500，有物种注释的 Tags 数量与 Tags 总数基本相同，没有物种注释的 Tags 数量为 0，被过滤掉的 Tags 平均为 599，最高为 CK-2 样品 952，可以看出总丰度为 1 的 OTUs 所对应的 Tags 数量，该样品

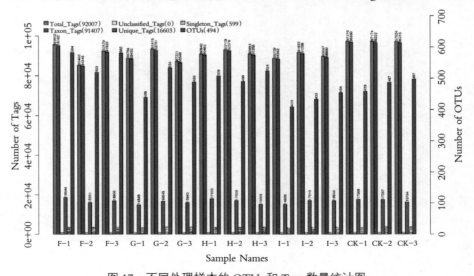

图 17 不同处理样本的 OTUs 和 Tags 数量统计图

Figure 17. The number of OTUs and Tags for different processing samples

最多,被过滤掉的数据较多,但总测序较大,数据足够聚类及后续分析;得到OTUs每个样本平均494条。

图18是各组处理之间共有和特有OTUs韦恩图和各个集合数量统计柱状图,对照(CK)处理组vsF处理组:CK处理组特有180 OTUs,F处理组特有263 OTUs,两组处理共有329 OTUs;CK处理组vsG处理组:对照(CK)处理组特有193 OTUs,G处理组特有208 OTUs,两组处理共有316 OTUs;CK处理组vsH处理组:对照(CK)处理组特有169 OTUs,H处理组特有175 OTUs,两组处理共有340 OTUs;CK处理组vsI处理组:对照(CK)处理组特有169 OTUs,I处理组特有110 OTUs,两组处理共有340 OTUs;F处理组vsG处理组vsH处理组vsI处理组vsCK处理组:F处理组特有163 OTUs,G处理组特有71 OTUs,H处理组特有53 OTUs,I处理组特有48 OTUs,对照(CK)处理组特有75 OTUs,5组处理共有229 OTUs。特有OTUs最多的是F处理组,其次是G处理组和对照(CK)组,H处理组和I处理组最低。

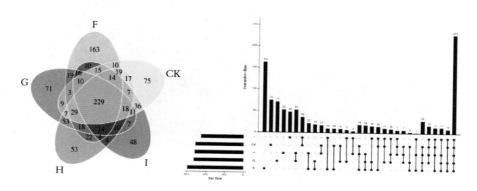

图18 不同处理组 OTUs 比较韦恩图

Fig.18 Comparison of wynn diagrams between different treatment groups OTUs

2.2.3 不同处理组样本物种分类分析

微生物物种分类一般分为界、门、纲、目、科、属、种7个等级,而每个OTU代表某类型分类水平集合。因此根据OTU序列信息进行物种注释,能将分析结果与实际的生物学意义进行关联,从而研究群落中物种变化关系等内容。下图是每个样品在各分类水平上的序列构成柱形图。图19是每个处理样本在各分类水平上序列构成柱形图,从图中可以看出门水平注释率在70%左右,纲水平在20%~65%不等,目水平在20%~65%,科水平在20%~50%,属水平在15%~35%,种水平12%~25%。整体上数据较为正常,真菌和细菌差别较大,物种水平越高,越不稳定,通常真菌要比细菌种水平注释率高,因为真菌相对要研究的多,种水平注释率高,但属和科注释就会出现无规律变化,忽高忽低,实际是经过药物或者外界处理,真菌相对细菌要少得多,变化浮动就会很大。

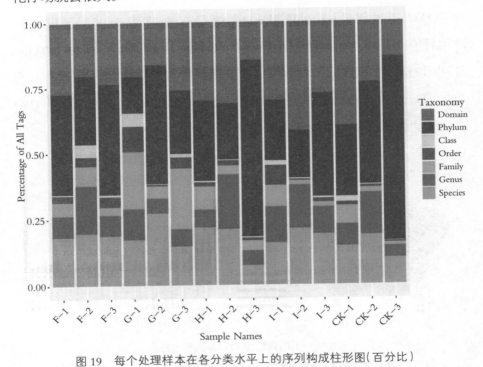

图19 每个处理样本在各分类水平上的序列构成柱形图(百分比)

Fig.19 Histogram of sequence composition of each processing sample at each classification level (percentage)

为了更加清晰研究物种分布分类,对微生物的分类数据进行堆叠图处理,图20为各样本纲水平上物种分布堆叠图,主要包括5大类,占比最多的为

Sordariomycetes,其次为 Eurotiomycetes、Saccharomycetes、Dothideomycetes,最低为 Agaricomycetes。其中 Sordariomycetes 在各个处理中基本无明显变化,Eurotiomycetes 在空白对照(CK)中很少,经过不同处理后都有所增加,尤其是 G 处理组增加最为明显。Saccharomycetes 在 CK 中含量较低,在处理组中显著增加,Dothideomycetes 在 I 处理组和空白对照(CK)组中无明显变化,辣根素处理后出现先降低后上升的变化。Agaricomycetes 总量较低,无明显变化。

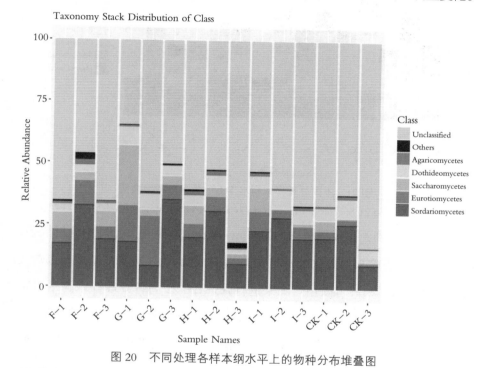

图 20　不同处理各样本纲水平上的物种分布堆叠图

Fig.20 Stacking plots of species distribution at the level of each class of samples under different treatments

图 21 是各处理样本在科水平上的物种分布堆叠图,注释率在 15%~50%,Hypocreaceae 占比最多,在空白对照(CK)和 I 处理组样本中含量很高,在辣根素处理组中含量显著降低;Trichocomaceae 和 Chaetomiaceae 在 F 处理组、G 处理组和 I 处理组显著增加,其中 G 处理组最为明显;Botryosphaeriaceae 在空白对照(CK)和 I 处理组中含量较高,在辣根素处理

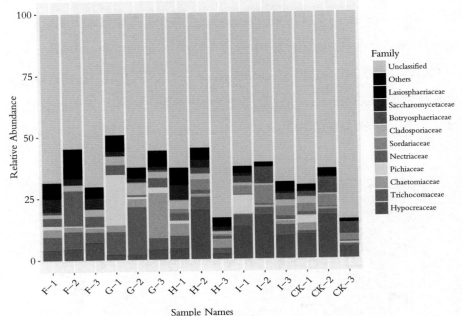

图 21 不同处理样本科水平上的物种分布堆叠图
Fig.21 Species distribution stacks at different treatment levels

组中几乎看不到,含量很低;Saccharomycetaceae 在辣根素处理中含量很多,是优势菌群,在空白对照(CK)和 I 处理组中很低。为了更加清楚地筛选对我们实验相关的菌群,可以用热图来进行筛选,图 22 为各处理样本科水平物种分类热图,结果和上述堆叠图一致,只是堆叠图显示前 10 的优势菌群,热图显示很多,并做了均一化处理,以堆叠图为准即可。若进行详细数据挑选,也可用热图进行。

2.2.4 不同处理组样本 Alpha 多样性分析

α 多样性是指特定生境或者生态系统内的多样性情况,它可以指示生境被物种隔离程度,通常利用物种丰富度(种类情况)与物种均匀度(分布情况)两个重要参数来计。Chao1/ACE 指数主要关心样本物种丰富度信息,从表 3 看出 Chao1 指数平均值为 606.267 7,ACE 指数平均值为 615.448 7,均值接近;Goods Coverage 反映样本低丰度 OTU 覆盖情况,所有样本的有效

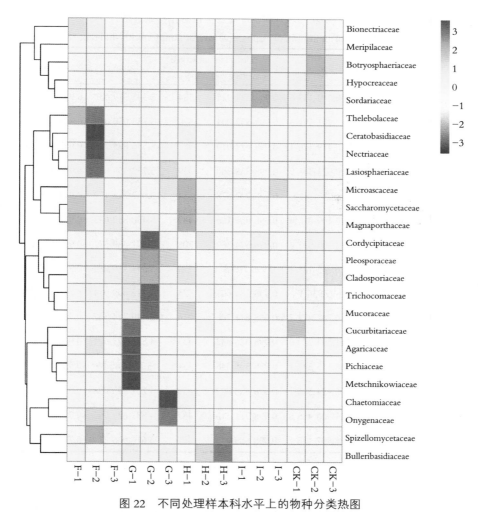

图22 不同处理样本科水平上的物种分类热图

Fig.22 Heat maps of species classification at different treatment levels

覆盖都在99%以上，平均值99.868%；observed_species 表示能检测到 OTU 种类情况，平均值为494，正常范围；Simpson / Shannon 主要综合体现物种丰富度和均匀度，数值居中，数据良好，其中 Simpson 指数平均值为0.919 6，Shannon 指数平均值为5.133 3。

稀释曲线(rarefaction curve)来评价测序量是否足以覆盖所有类群，并间接反映样品中的物种丰富程度。稀释曲线是利用已测得序列中已知的

表3 不同处理样本Alpha多样性各指标统计表

Table 3 Statistical table of indicators of Alpha diversity of samples with different treatments

各处理样本	chao1	ace	goods_coverage	observed_species	shannon	simpson
F-1	692.373 333 3	700.649 375 6	0.998 654 925	584	5.343 979 032	0.932 448 083
F-2	633.681 818 2	652.741 591 3	0.998 552 004	520	5.758 050 044	0.954 882 794
F-3	727.223 880 6	720.449 428 4	0.998 477 019	582	5.536 825 194	0.932 480 705
G-1	528.033 898 3	529.023 348 9	0.998 835 566	439	5.232 554 035	0.940 274 156
G-2	624.48	638.301 819	0.998 737 878	534	5.002 213 172	0.912 575 825
G-3	561.402 985 1	582.109 816 4	0.998 856 126	489	5.486 852 642	0.937 644 807
H-1	641.161 290 3	634.393 933 4	0.998 574 302	508	5.591 480 577	0.951 018 102
H-2	598.181 818 2	620.735 121 4	0.998 703 81	490	5.534 110 088	0.951 588 749
H-3	694.5	699.936 1	0.998 284 601	524	4.122 699 914	0.815 517 587
I-1	517.058 823 5	516.742 872 7	0.998 811 706	410	5.449 314 337	0.961 387 627
I-2	534.267 857 1	541.394 720 6	0.998 826 832	433	4.833 177 621	0.926 108 061
I-3	589.964 912 3	588.113 547 1	0.998 595 348	454	5.372 285 89	0.939 083 539
CK-1	599.368 421 1	606.424 596 7	0.998 688 829	459	5.221 975 096	0.939 216 063
CK-2	583.531 645 6	611.906 485 8	0.998 711 313	487	4.897 639 986	0.927 975 878
CK-3	568.784 810 1	588.807 135 1	0.998 889 073	497	3.616 966 219	0.773 018 117

OTU相对比例，来计算抽取n个（n小于测得tags序列总数)tags时出现OTU数量的期望值,然后根据一组n值(一般为一组小于总序列数的等差数列)与其相对应OTU数量的期望值做出曲线来。当曲线趋于平缓或者达到平台期时也就可以认为测序深度已经基本覆盖到样品中所有的物种。从图23可以看出抽取的tags数量到达60 000时,所能得到OTU数量的期望值到达平台期,而我们数据有物种注释的Tags平均在90 000以上,说明数据可靠,覆盖度足够。

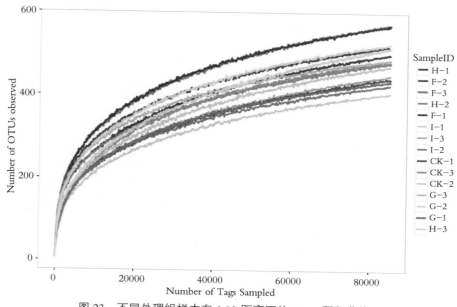

图 23　不同处理组样本在 0.03 距离下的 OTU 稀释曲线图

图 24 是对 F 处理组、G 处理组、H 处理组和 I 处理组 shannon 指数的箱线图,横坐标为各分组情况,各组分别为红绿蓝紫四种颜色区分,纵坐标表示对应 Alpha 多样性指数大小。图形展示有效数据分布最大值(直线顶端),最小值(直线底端),中位数(盒子中线),上四分位数(盒子顶边),下四分位数(盒子底边),以及无效数据;shannon 指数中位数都在 5.3 左右,G 处理组最低 5.002 2,最高 5.486 8,F 处理组最低 5.343 9,最高 5.758 0,H 处理组最低 4.122 6,最高 5.591 4,是 4 个处理中波动最大的,推测辣根素浓度增加到最高,导致对应真菌群落发生很大变化。I 处理组最低 4.833 1,最高 5.449 3。

2.2.5　不同处理组各样本 Beta 多样性分析

Beta 多样性是不同生态系统之间多样性的比较,是物种组成沿环境梯度或者在群落间的变化率,用来表示生物种类对环境异质性的反应。一般来说,不同环境梯度下群落 Beta 多样性计算包括物种改变(多少)和物种产生(有无)2 部分。这种方法借用方差分解可以有效地找出数据中最"主要"的元素和结构,将复杂样本组成关系反映到横纵坐标两个特征值上,从而达到简

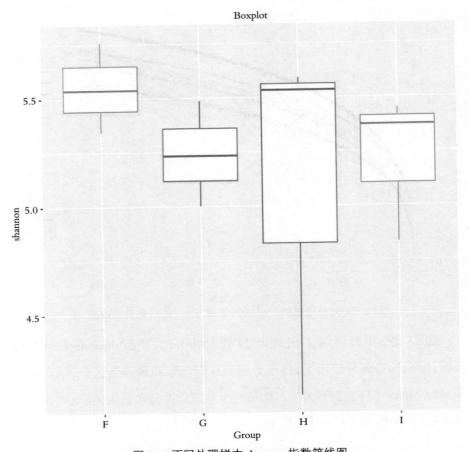

图 24　不同处理样本 shannon 指数箱线图

Fig.24 Shannon index boxplot of samples with different treatments

化数据复杂度效果。分析结果中,样本组成越相似,反映在 PCA 图中距离越近,而且不同环境间样本往往可能表现出各自聚集的分布情况。从图 23 可以看出,各个组间相对较为分散,虽然和各组交叉存在,但各组间相聚不是很远,从 PCA 上来看,结果聚类正常。

图 26 可知,PCoA 主坐标分析是一种展示样本间相似性的分析方式,它的分析思路与 PCA 分析基本一致,都是通过降维方式寻找复杂样本中主要样本差异距离。与 PCA 不同的是,PCoA 主要利用 weighted 和 unweighted Unifrac 等配对信息,因此结果更集中于体现样本间的相异性距离。

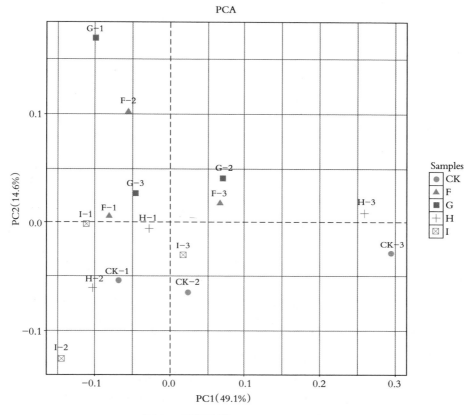

图 25 不同处理 2D OTU PCA plot
Fig.25 2D OTU PCA plot with different processing

基于样本间的 unweighted Unifrac 数据结果，我们可以绘制 PCoA 图形。分析结果中，样本越相似，反映在 PCoA 图中距离越近，而且不同环境间的样本往往可能表现出各自聚集的分布情况。从图 25 可以看出各组处理样本聚类在一起，G 处理组和 H 处理组聚类在一起，F 处理组单独，I 处理组和对照（CK）聚类在一起，这也符合实验的前期处理，F 处理组为低浓度处理，G 处理组和 H 处理组为高浓度处理，I 处理组和空白对照（CK）作为 2 种不同的对照。

图 27 可知，UPGMA 分类树可以用于研究样本间的相似性，解答样本的分类学问题。利用 Mothur 软件，根据 weighted 和 unweighted Unifrac 矩阵信

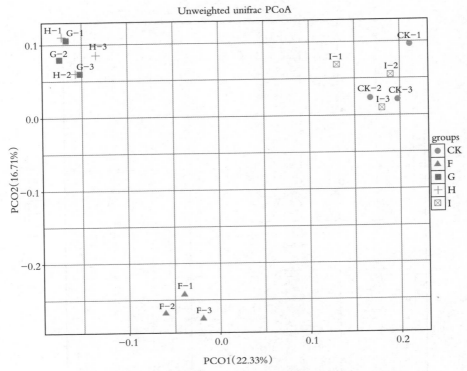

图 26 不同处理样本基于 unweighted Unifrac 距离的 PCoA 分析

Fig.26 PCoA analysis of samples with different treatments based on unweighted Unifrac distance

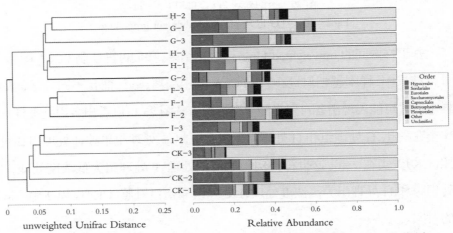

图 27 不同处理样本基于 Unweighted Unifrac 距离的目水平 UPGMA 聚类树

Fig.27 Hierarchical UPGMA clustering tree with Unweighted Unifrac distance based on different processed samples

息,可以将样本进行 UPGMA 分类树分类。其中越相似的样本将拥有越短的共同分支。图 27H 处理组和 G 处理组真菌方面组内组间差异很难区分,F 处理组和 H 处理、G 处理组差异明显,I 处理组和对照(CK)组组间差异较小,I 处理组和对照(CK)分别与辣根素处理组差异显著。

表 4 是 Anosim 差异分析的结果表,其中 R 值表示差异程度,一般介于 (0,1)之间,R>0,说明不同处理组间存在差异;一般 R>0.75,大差异;R>0.5,中等差异;R>0.25,小差异。F 处理组、G 处理组和 H 处理组的 R 值为 1,说明组间差异明显,I 处理组相比空白对照差异很低,也符合上述聚类树分析,I 处理组与对照(CK)组组间差异小。统计分析的可信度用 P-value 表示,P< 0.05 表示统计具有显著性。多组间比较差异极显著。

表 4 不同处理样本基于 unweighted unifrac 距离的 Anosim 分析结果表
Table 4 Results of Anosim analysis of different treatment samples based on unweighted unifrac distance

Diffs	Rvalue	Pvalue	significant
CK-vs-F	1	0.1	
CK-vs-G	1	0.1	
CK-vs-H	1	0.1	
CK-vs-I	0.222 2	0.2	
F-vs-G-vs-H-vs-I-vs-CK	0.814 8	0.001	★★
F-vs-G-vs-H-vs-I	0.780 9	0.001	★★

2.2.6 不同处理组样本群落功能分析

完整的微生物群落研究主要分为物种组成、多样性以及功能研究等几个重要方面。多种证据表明微生物的群落功能组成比物种组成与环境关系更为密切,随着分析技术发展,利用多样性测序数据进行群落功能预测已经成为群落研究的重要内容。我们将利用 FUNGuild、Tax4fun、FAPROTAX、BugBase 等多个预测软件,根据不同的数据类型有针对性地完成群落功能预测分析。基于 OTU 丰度表格信息,利用 FUNGuild 开展真菌功能注释。由于缺乏真菌基因组数据,因此 FUNGuild 并不开展 KEGG 注释,而是通过整合已发表文章数据,进行一种称为 Guild 的功能分类预测。Guild 是一个生态学概念,用于描述物种资源利用吸收所进行的功能分类,FUNGuild 中的 Guild 分类包括动物病原菌、植物病原菌、木质腐生菌等 12 种。开展 Guild 预测,可以从其他生态角度研究真菌功能。图 28 对于 OTU 丰度前 10 的统计:10 类都比对照不同的物种分类水平,其中 3 类 Order 水平,2 类 Genus 水平,其中 4 类腐生营养型,1 类病理营养型,其中 4 类为未定义根内真菌,1 类是植物病原菌,后续研究可以针对这 4 类根内真菌和植物病原菌。

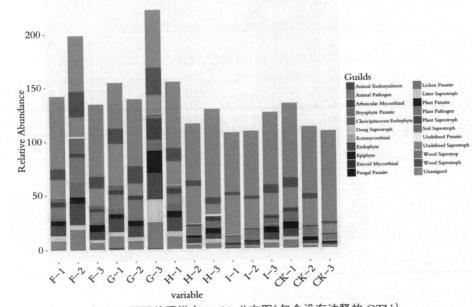

图 28 不同处理样本 Guilds 分布图(包含没有注释的 OTU)

Fig.28 Guilds distribution diagram of different processing samples (including OTU without comments)

2.2.7 不同处理组样本物种差异分析

通过统计学的方法(Metastats 软件)检验两组样品间微生物群落丰度的差异,得到 p 值,然后使用 FDR 对 p 值进行校正得到 q 值,最后通过对 p 值或者 q 值的筛选来评估差异的显著性,找出导致两组样品组成差异的物种。

通过 LEFse 分析组间菌群差异,可以找出各组间特异的主要菌群,有助于开发 biomaker 等研究。利用 LEFse 软件对差异组间进行分析,LEFse 先对所有组样品间进行 kruskal-Wallis 秩和检验（一种多样本比较时常用的检验方法）,将筛选出的差异再通过 wilcoxon 秩和检验(一种两样本成组比较常用的检验方法）进行两两组间比较,最后筛选出的差异使用 LDA(Linear Discriminant Analysis)得出的结果进行排序得到左图,左图展示了不同组中丰度差异显著的物种,柱状图的长度代表差异物种的影响大小（即 LDA Score)。随后通过将差异映射到已知层级结构的分类树上方式得到进化分支图(图 7.1)。在进化分支图中,由内至外辐射的圆圈代表了由门至属(或种)的分类级别。在不同分类级别上的每一个小圆圈代表该水平下的一个分类,小圆圈直径大小与相对丰度大小呈正比。着色原则:无显著差异的物种统一着色为黄色。图 29、图 30 为各个处理组样本之间的物种差异,4 个处理组和 1 个空白对照(CK)一起比较,最终结果里只出现 F 组处理、G 处理组和 I 处理组的差异物种,证明多组间比较,H 处理组和对照(CK)组没有特异的差

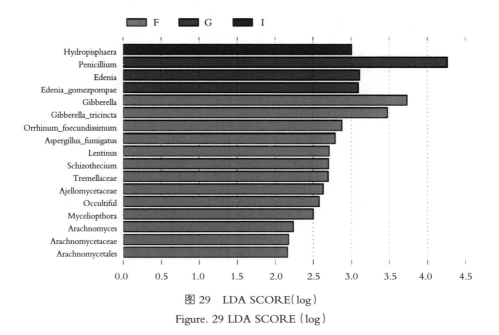

图 29　LDA SCORE(log)

Figure. 29 LDA SCORE (log)

异物种。I 处理组中 Hydropisphaera 为唯一差异真菌,经过多菌灵处理,只有这一种真菌和其他组存在差异,证明多菌灵处理后该真菌成为差异优势真菌,推测可能是该类真菌的增加导致病害的减轻。G 处理组中优势差异真菌为 Penicillium,得分显著高于另外两类真菌,该真菌可能是辣根素浓度到达一定值后有效防治病害的优势真菌。F 处理组中的差异真菌较多,推测因为辣根素的浓度偏低,大多真菌存活,未能有效防治病害,而处理组浓度过高,导致基本所有的差异真菌不能存活,或者没有显著优势,导致结果显示无差异菌群。综上所述,辣根素浓度应该为 G 处理组的浓度最好,同时多菌灵处理同样可以有效地进行防治病害。

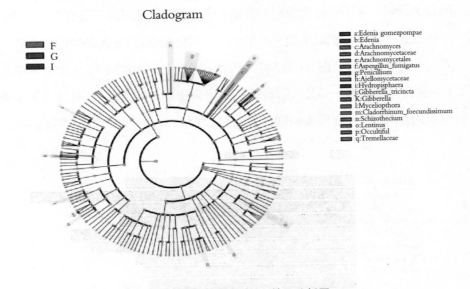

图 30　不同处理样本 LEFse 差异分析图

Fig. 30 LEFse difference analysis diagram of samples with different treatments

3　讨论

辣根素来源于天然产物,其主要成分烯丙基异硫氰酸酯本身也是一种食品添加剂(吴华等,2013)对人较为安全,从作用特性上来看,具有速效、高效和低毒等特点,且在环境中易于分解,对环境造成污染的程度极小,是绿色、无公害、有机蔬菜生产中防治病虫害的重要手段,符合精品、安全"菜篮子"产品供给需求和现代都市型农业发展的要求(毛科军等,2015)。

目前已经在多地开展了番茄、草莓、甜瓜等多种蔬菜,蚜虫、白粉病、马唐等多种病虫草害的实际防治效果和示范应用(杨龙等,2018;肖长坤等,2010;包敏辉,2020)。辣根素用于土壤熏蒸处理可有效杀灭土壤中镰刀菌、腐霉、曲霉、青霉等多种微生物等病害,具有防控多种土传病害的综合效果(Angus et al,1994;Dunne et al,2003)。王彦柠等(2018)研究表明辣根素对供试的 26 种植物病原真菌、卵菌及细菌的菌丝生长,孢子萌发和菌落生长均呈现出不同程度的熏蒸抑制活性,证明辣根素具有广谱高效的抑菌活性,在农业病害防治中具有光明的应用前景。

鉴于辣根素的高效杀菌作用,本研究分别设置 7 L/667 m^2、9 L/667 m^2、11 L/667 m^2、对照药剂(50%多菌灵可湿性粉剂)、空白对照(不做任何处理)5个处理,采用高通量测序技术,测定 16S(细菌)和 ITS(真菌)的微生物群落结构变化,研究不同浓度辣根素对连作 8 年辣椒地土壤微生物的变化。结果表明,科水平 Planctomycetaceae 在不同浓度辣根素处理组中含量较 50%多菌灵处理和对照(CK)含量较低;Xanthomonadaceae 在 7 L/667 m^2 和 9 L/667 m^2 处理组中较高,处理增加到 11 L/667 m^2 后降低,50%多菌灵处理组和对照(CK)中较低,处理后出现该菌含量显著增加,但随着处理的加强,含量在 11 L/667 m^2 出现降低。Blastocatellaceae 在对照(CK)中最高,F 处理组无明显降低,G 处理组和 I 处理组后显著降低;Pseudomonadaceae 在各处理组中都未有明显变化,多菌灵处理有明显降低。属水平 Planctomyces 是主要菌群,在整体样本中未有明显变化,Lysobacter 在 CK 和 7 L/667 m^2 处理中基本相同,随着处理剂量的加强,该类菌明显增加,Pseudomonas 在对照(CK)中很少,在不同浓度辣根素处理中逐渐增加,50%多菌灵处理组中出现特别明显的增加,Pseudoduganella 在对照 CK 中含量较多,经过辣根素和多菌灵处理,土壤每个样本中几乎含量降为 0。Bacillus 在经过处理后显著增加,对照(CK)

中基本为0。Acinetobacter 在多菌灵处理组中显著增加,其余各组处理中含量很低。经过不同浓度辣根素和50%多菌灵处理,细菌和真菌的菌群都发生明显的变化。

不同样本处理的物种差异分析表明50%多菌灵处理组中影响最大的差异物种是 cinetobacter_calcoaceticus,其次是 Trueperaceae、Deinococcales、Deinococcus_Thermus。7 L/667 m² 处理组中影响最大最为显著的是 Arenimonas,其次是 Oceanospirillales,该组中的这两个菌群较 9 L/667 m² 处理组和 11 L/667 m² 处理组尤为显著,而 9 L/667 m² 处理组和 11 L/667 m² 处理组中没有最为显著的菌群,但分别有 5 个和 4 个菌群作为候选影响差异物种。空白对照(CK)组中 Oxalobacteraceae 和 Pseudoduganella 两个菌群影响最显著。从低浓度辣根素到高浓度辣根素处理,随着药物浓度的增加,Arenimonas 和 Oceanospirillales 菌群降为非优势菌群,推测这两种菌群的降低对土传病害的防治是有利的。多菌灵的使用使 I 处理组 cinetobacter_calcoaceticus、Trueperaceae 和 Deinococcales 菌群增加,推测该几类菌群的增加有利于病害的防治;CK 为空白对照组中 Oxalobacteraceae 和 Pseudoduganella 两个菌群为优势菌群,推测可能不利于病虫害的防治,在其他处理组中该类菌群显著降低,结合病虫害实际防治情况,推测这两类菌需要处理后降为非优势菌群,可能利于防治病害。

在不同辣根素处理下真菌的多样性结果显示,处理组和空白对照(CK)一起比较,差异分析结果里只出现 7 L/667 m² 组处理、9 L/667 m² 处理组和 50%多菌灵处理组的差异物种,证明多组间比较,11 L/667 m² 处理组和对照(CK)组没有特异的差异物种。50%多菌灵处理组中 Hydropisphaera 为唯一差异真菌,经过多菌灵处理,只有这一种真菌和其他组存在差异,证明多菌灵处理后该真菌成为差异优势真菌,推测可能是该类真菌的增加导致病害的减轻。9 L/667 m² 处理组中 Penicillium 为优势差异真菌,得分显著高于另外两类真菌,该真菌可能是辣根素浓度到达一定值后有效防治病害的优势真菌。7 L/667 m² 处理组中的差异真菌较多，推测因为辣根素的浓度偏低，大多真菌存活，未能有效防治病害，而处理组浓度过高，导致基本所有的差异真菌不能存活，或者没有显著优势，导致结果显示无差异菌群。综上所述，辣根素浓度应该为 9 L/667 m² 处理组的浓度最好，同时多菌灵处理同样可以有效地进行防治病害。

4　结论

本试验通过3个浓度梯度 7 L/667 m²、9 L/667 m² 和 11 L/667 m² 的辣根素以及两个对照 50%多菌灵可湿性粉剂和空白的土壤处理比较，证明 9 L/667 m² 处理组的浓度是最佳的辣根素土壤处理浓度，相比 7 L/667 m² 和 11 L/667 m² 可以有效地防治病虫害的发生。空白对照组中 Oxalobacteraceae 和 Pseudoduganella 两个细菌菌群为优势菌群，推测可能不利于病虫害的防治，在不同浓度的辣根素和 50%多菌灵处理组中该类菌群显著降低，结合病虫害实际防治情况，推测降低 Oxalobacteraceae 和 Pseudoduganella 菌群为非优势菌群，可能利于防治病害。真菌 Hydropisphaera 和 Penicillium 的增加可能有利于病虫害的防治，为土壤改良工作打下良好的基础。

参考文献

[1]　Dunne C P, Dell B, Hardy G E S.The effect of biofumigants on the vegetative growth of five Phytophthoraspecies in vitro[J]. Acta Horticulturae,2003,60(2):45-51.

[2]　J. F. Angus,P. A. Gardner,J. A. Kirkegaard,J. M. Desmarchelier.Biofumigation：Isothiocyanates released from brassica roots inhibit growth of the take-all fungus[J]. Plant and Soil, 1994,16(2):107-112.

[3]　包敏辉.北海地区设施栽培甜瓜根结线虫病绿色防控综合技术[J].长江蔬菜,2020(03):50-52.

[4]　毛科军,樊敏,陈曦丹.新常态下天津现代都市型农业发展的新思路、新目标与新举措[J].天津农业科学,2015(7):1-6.

［5］ 王彦柠,李迎宾,黄小威,罗来鑫,曹永松,李健强.辣根素对常见植物病原菌的抑菌活性研究[J].中国科技论文,2018,13(06):692-697.

［6］ 吴华,冯俊涛,何军,等.辣根素的生物活性研究进展[J].中国生物防治学报,2013,29(2):301-306.

［7］ 肖长坤,张涛,陈海明,等.20%辣根素水乳剂对设施草莓土壤的效果[J].中国蔬菜,2010(21):29-31.

［8］ 杨龙,李柏润,杨爱宾等.20%辣根素水乳剂防治番茄真菌病害的效果[J].天津农业科学,2018(6):63-66.

伴生燕麦对连作辣椒叶片保护酶活性及根际土壤微环境的影响

高晶霞[1]，吴雪梅[2]，高　昱[2]，谢　华[1]

（1.宁夏农林科学院种质资源研究所，宁夏银川　750002；
2.宁夏回族自治区彭阳县蔬菜产业发展服务中心，宁夏彭阳　756500）

摘　要：以辣椒亨椒新冠龙品种为试验材料，设伴生燕麦和单作辣椒2个处理，采用盆栽试验，研究伴生燕麦对连作辣椒叶片保护酶活性及根际土壤微生物环境的影响。结果表明：伴生燕麦处理可以提高辣椒叶片可溶性糖含量、过氧化氢酶活性和丙二醛含量，分别比CK增加了62.2%、83.3%和43.75%，伴生燕麦处理降低了辣椒叶片的可溶性蛋白含量，但辣椒果实维生素C含量和可溶性糖含量明显高于CK，分别比CK增加了54.36%和19.72%，伴生燕麦处理的辣椒果实可溶性蛋白和干物质含量低于CK,分别减少了44.44%和7.6%；伴生燕麦处理的蒸腾速率、气孔导度、净光合速率和细胞间CO_2浓度均高于CK，分别比CK减少了60.7%、91.77%、10.28%和53.77%，但叶绿素含量高于CK,比对照增加了3.3%；伴生燕麦处理的荧光参数F_0、F_m和F_v均高于CK，分别降低了21.65%、22.12%和20.1%，荧光参数F_0/F_m和F_v/F_m与CK并无差异，伴生处理的F_v/F_0大于CK,比对照增加了2.18%；伴生燕麦处理的过氧化氢酶、纤维素酶和蔗糖酶的活性均高于CK,分别增加了4.35%、127.66%和3.95%，土壤脲酶活性低于CK，比对照降低了10.72%。综上所述，伴生燕麦可以提高辣椒叶片保护酶的活性，改善连作辣椒根际土壤微环境，为拱棚辣椒丰产栽培提供理论依据。

关键词：辣椒；伴生；连作障碍；保护酶活性；根际土壤

Effects of Associated Oat on Protective Enzyme Activity and Rhizosphere Soil Microenvironment of Continuously Cultivated Pepper Leaves

Gao jing-xia[1], Wu xue-mei[2], Gao yu[2], Xie hua[1]

(1. *Institute of Germplasm Resources, Ningxia Academy of Agriculture and Forestry Sciences, Yinchuan* 750002, *Ningxia*; 2. *Vegetable Industry Development Service Center of Pengyang County, Ningxia Hui Autonomous Region, Pengyang* 756500, *China*)

Abstract: The effects of associated oats on the protective enzyme activity of pepper leaves and the microbial environment of rhizosphere soil were studied by pot experiment. The results show that: The soluble sugar content, catalase activity and malondialdehyde content of pepper leaves were increased by 62.2%, 83.3% and 43.75%, respectively, compared with CK, the soluble sugar content, catalase activity and malondialdehyde content of pepper leaves were increased by 62.2%, 83.3% and 43.75%, respectively, compared with CK, The soluble protein and dry matter contents of pepper fruits treated with oat were lower than CK, which decreased by 44.44% and 7.6%, respectively; Transpiration rate, stomatal conductance, net photosynthetic rate and intercellular CO_2 concentration of associated oat were all lower than CK, with decreases of 60.7%, 91.77%, 10.28% and 53.77%, respectively, and higher chlorophyll content than CK, with an increase of 3.3%; Fluorescence parameters F0, Fm and Fv of associated oat treatment were all lower than CK, with decreases of 21.65%, 22.12% and 20.1%, respectively. Fluorescence parameters F0/Fm and Fv/Fm had no difference with CK, while Fv/F0 of associated oat treatment was greater than CK, with an increase of 2.18%; The activities of catalase, cellulase and sucrase treated with oat were all higher than those of CK, increasing by 4.35%, 127.66% and 3.95%, respectively. The soil urease activity was lower than that of CK, decreasing by 10.72%. In summary, the associated oat could improve the activity of protective enzymes in pepper leaves, improve the soil microenvironment of rhizosphere of pepper in continuous cropping, and provide theoretical basis for the cultivation of high yield of pepper in arch shed.

Key words: Pepper; Associated; Continuouscroppingobstacle; Protectiveenzyme activity; Rhizosphere

辣椒为茄科辣椒属一年生或有限多年生植物,在我国南北各地广泛栽培。据农业部大宗蔬菜体系统计,近年来我国辣椒种植面积为150万~200万 hm², 占全国蔬菜总种植面积的8%~10%[1]。然而,随着辣椒种植面积的扩大和连作年限的增加,连作障碍日益严重,轻者减产30%左右,严重的减产60%以上,甚至绝产[2]。

目前制约我国设施农业健康可持续发展的瓶颈就是连作障碍[3],连作障碍引起土壤养分不均衡和次生盐渍化。而且长期连作造成辣椒植株生育迟缓、长势下降、品质变劣,产量下降等,连作病害频发已成为辣椒生产中较为严重的问题。大量研究表明,利用植物间的化感作用原理合理安排伴生或间套作不仅可以提高蔬菜产量和品质,还可以有效降低病虫害改善土壤生态环境,从而减轻连作障碍的发生[4,5]。伴生栽培可以合理地利用时间、空间和作物间的相互影响,达到抑制杂草滋生和病虫害蔓延的效果,减少化学药剂使用量,保持生态平衡[6,7]。目前常有的伴生作物为禾本科作物,禾本科作物具有较强的化感作用,其根系分泌物能够提高土壤微生物多样性。如小麦伴生西瓜处理的根际土壤微生物总数及放线菌、细菌数量均比单作显著增加,而土壤真菌的比例明显降低;同时提高了西瓜根际土壤微生物生物量碳、氮、磷含量[8];伴生禾本科作物,特别是大麦能够提高番茄根际土壤酶活性,增加土壤微生物数量,降低根结线虫的发生[9];花生与麦类作物(大麦、燕麦、小麦)混作可以提高花生铁的含量,改善植株铁营养[10];与单作辣椒相比,玉米与辣椒间作缓解了Cd、Pb等重金属对辣椒的毒害作用,提高了土壤酶活性以及菌落数量[11]。Tringovska[12]等研究发现,万寿菊、紫苏、莴苣和白芥伴生番茄可以有效提高对根结线虫的抗性,其中万寿菊和白芥的伴生防效较好,分别达到 53.45%和 46.38%。但伴生栽培也存在与主栽作物的竞争关系,可能影响到主栽作物的产量和生长发育,这种不利影响可以通过筛选合适的伴生植物、调节养分和水分管理来克服[13-14]。

本试验以辣椒为主栽作物,以燕麦为伴生作物,通过盆栽试验探讨了伴生燕麦对连作辣椒叶片保护酶活性及根际土壤环境的影响,旨在为提高克服辣椒连作障碍能力以及生产实践提供理论指导。

1 材料与方法

1.1 供试材料

供试辣椒品种:"亨椒新冠龙"(宁夏嘉禾源种苗有限公司提供),伴生品种:带皮燕麦(购买自宁夏银川市西北农资城)。所用肥料为果蔬专用有机肥(购买自宁夏尚博农植物保护有限公司)。供试土壤:彭阳县新集乡辣椒核心试验基地辣椒连作 7 a 的土壤(全氮 0.82 g/kg,全磷 1.05 g/kg,全钾 22.8 g/kg,有机质 14.5 g/kg,碱解氮 78 mg/kg,有效磷 45.9 mg/kg,速效钾 122 mg/kg,pH=8.46,全盐 0.27 g/kg)。

1.2 试验方法

试验于 2019 年 4 月 20 日在彭阳县新集乡拱棚辣椒核心试验示范基地进行,将辣椒连作土壤与有机肥混匀(10V:1V)并均匀装入盆中(40 cm×29 cm)。设置 2 个处理:CK 为单作辣椒处理;T1 为伴生燕麦处理。辣椒催芽后播种于穴盘中,待 4 叶期定植到盆中,每盆 1 株;定植 5 d 后进行伴生处理,在距辣椒植株 5 cm 处分别撒播 30 粒大麦种子,当大麦长至 25 cm 左右时留茬 5 cm。每个处理 5 盆,3 次重复,随机排列,放置温室棚内进行常规管理。试验期间不使用任何药剂,人工除草,定量浇水。伴生 30 d 结束,采集辣椒叶片测定酶活性和生长指标;利用剥落分离法采集辣椒根际土壤,4℃下保存土样用于测定土壤酶活性。

1.3 指标测定

1.3.1 辣椒叶片指标的测定

可溶性糖采用蒽酮比色法,可溶性蛋白采用考马斯亮蓝 G-250 染色法,过氧化氢酶活性采用紫外吸收法,丙二醛含量采用改进的硫代巴比妥酸法。每个样品重复测定 3 次。

1.3.2 辣椒果实品质的测定

维生素 C 采用 2,6-二氯靛酚滴定法,可溶性糖采用蒽酮比色法,可溶性蛋白采用考马斯亮蓝 G-250 染色法,干物质用分析天平称量。每个样品重复测定 3 次。

1.3.3 辣椒叶片叶绿素含量的测定

定植 30 d 后,采集生长点下第 2 片展开叶进行叶绿素含量的测定,采用 SPAD502 叶绿素仪进行测定,每个叶片重复测定 5 次,取平均值。

1.3.4 辣椒叶片光合参数和荧光参数的测定

定植 30d 后,采用便携式光合测定系统(Li-6400,美国)于 10:00~11:00 进行光合参数的测定,净光合速率(Pn)、气孔导度(Gs)、细胞间 CO_2 浓度(Ci)和蒸腾速率(Tr)由光合测定系统直接读出,每个叶片重复测定 3 次,取平均值。荧光采用 PAM2100 调制荧光仪(德国 Walz 公司生产)进行测定。每个处理重复测定 3 次。

1.3.5 土壤酶活性的测定

脲酶活性用靛酚比色法;过氧化氢酶活性用高锰酸钾滴定法;蔗糖酶活性用比色法;纤维素酶用 DNS 法。

1.4 数据分析

用 Excel 2019 和 SPSS 软件进行数据分析处理,不同处理间的多重比较采用 Duncan's 新复极差法($p \leqslant 0.05$)。

2 结果分析

2.1 伴生燕麦对连作辣椒叶片保护酶活性的影响

由表1可以看出,伴生燕麦辣椒叶片可溶性糖含量明显高于CK,为0.73%,比CK增加了62.2%;不同处理的辣椒叶片可溶性蛋白含量低于CK,但差异不明显;辣椒叶片的过氧化氢酶活性明显高于CK,为0.11 μg·g^{-1}·min^{-1},比CK增加了83.3%。伴生燕麦辣椒叶片丙二醛含量与CK无差异,均为0.16 mmol/g。由此说明,伴生燕麦可以促进叶片保护酶活性的升高。

表1 不同处理对连作辣椒叶片保护酶活性的影响
Table 1 Effects of different treatments on protective enzyme activities in continuously cultivated pepper leaves

处理 treatment	可溶性糖 soluble sugar/%	可溶性蛋白 soluble protein/%	过氧化氢酶 catalase(μg·g^{-1}·min^{-1})	丙二醛 MDA(mmol/g)
处理1	0.73	0.27	0.11	0.16
CK	0.45	0.29	0.06	0.16

2.2 伴生燕麦对连作辣椒叶片光合特性的影响

由表2可以看出,不同处理辣椒叶片光合特性有差异,伴生燕麦辣椒叶片蒸腾速率高于CK,为5.98 mmol·m^{-2}·s^{-1},比CK增加了60.7%;伴生燕麦处理的辣椒叶片气孔导度高于CK,为1 467.5 mmol·m^{-2}·s^{-1},比CK增加了91.77%;伴生燕麦处理的净光合速率高于CK,为19.74 μmmol·m^{-2}·s^{-1},比CK增加了10.28%;伴生处理的辣椒叶片细胞间CO$_2$浓度高于CK,为

表2 不同处理对连作辣椒光合特性的影响
Table 2 Effects of different treatments on photosynthetic characteristics of continuous cropping pepper

处理 treatment	蒸腾速率 transpiration rate /mmol·m^{-2}·s^{-1}	气孔导度 stomatal conductance /mmol·m^{-2}·s^{-1}	净光合速率 net photosynthetic rate /μmmol·m^{-2}·s^{-1}	细胞间CO$_2$浓度 intercellular thetic rate /μmol·mol^{-1}
处理1	5.98	1467.50	19.74	345.03
CK	2.35	120.70	17.71	159.50

345.03 μmol·mol⁻¹,比 CK 增加了 53.77%。

2.3 伴生燕麦对连作辣椒叶片叶绿素含量的影响

由图 1 可知，伴生燕麦处理的辣椒叶片叶绿素含量显著高于 CK,为 65.04 SPAD,比 CK 增加了 3.3%。由此说明,伴生燕麦可以促进辣椒叶片中光合色素含量的升高,从而增强了叶片捕捉和利用光能的能力,同时也影响了光能在叶绿体中的分配[6]。

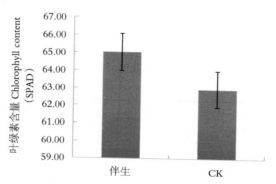

图 1 伴生燕麦对连作辣椒叶绿素含量的影响

Fig.1 Effect of associated oats on chlorophyll content of continuous cropping pepper

2.4 伴生燕麦对连作辣椒荧光特性的影响

由表 3 可以看出,不同处理的辣椒叶片荧光参数有差异,伴生燕麦辣椒叶片荧光参数 F0 低于 CK,比 CK 降低了 21.65%;伴生燕麦处理的辣椒叶片荧光参数 Fm 低于 CK,比 CK 降低了 22.12%;伴生燕麦处理的辣椒叶片荧光参数 Fv 低于 CK,比 CK 降低了 20.1%;伴生燕麦处理的辣椒叶片荧光参数 F0/Fm 和 Fv/Fm 与 CK 并与差异,均为 0.76;伴生燕麦处理的辣椒叶片

表 3 不同处理对连作辣椒叶片荧光参数的影响

Table 3 Effects of different treatments on the fluorescence parameters of continuously cultivated pepper leaves

处理 treatment	F0	Fm	Fv	F0/Fm	Fv/Fm	Fv/F0
处理 1	512.80	2187.00	1674.20	0.24	0.76	3.28
CK	654.50	2749.97	2095.47	0.24	0.76	3.21

荧光参数 Fv/F0 大于 CK,为 3.28,比 CK 增加了 2.18%。

2.5 伴生燕麦对连作辣椒根际土壤酶活性的影响

由表 4 可以看出,不同处理辣椒根际土壤酶活性有差异,伴生燕麦的土壤过氧化氢酶活性高于 CK,为 0.24 mg·g^{-1}·30 min^{-1},比 CK 增加了 4.35%;伴生燕麦处理的土壤脲酶活性低于 CK,比 CK 降低了 10.72%;伴生燕麦处理的土壤纤维素酶活性明显高于 CK,为 1.07 mg·g^{-1}·h^{-1},比 CK 增加了 127.66%;伴生燕麦土壤纤维素酶活性高于 CK,为 24.19 mg·g^{-1}·h^{-1},比 CK 增加了 3.95%。

表 4 不同处理对连作辣椒根际土壤酶活性的影响
Table 4 Effects of different treatments on enzyme activity in rhizosphere soil of continuous cropping pepper

处理 treatment	过氧化氢酶 catalase /ml·g^{-1}·30 min^{-1}	脲酶 urease /mg·g^{-1}·h^{-1}	纤维素酶 cellulase /mg·g^{-1}·h^{-1}	蔗糖酶 sucrase /mg·g^{-1}·h^{-1}
处理	0.24	64.12	1.07	24.19
CK	0.23	71.82	0.47	23.27

2.6 伴生燕麦对连作辣椒果实品质的影响

由表 5 可以看出,不同处理的辣椒果实维生素 C 含量明显高于 CK,为 193.8 mg/100 g,比 CK 增加了 54.36%;伴生燕麦处理的辣椒果实可溶性糖含量明显高于 CK,为 0.85 g/100 g,比 CK 增加了 19.72%;伴生燕麦辣椒果实可溶性蛋白含量明显低于 CK,比 CK 减少了 44.44%;伴生燕麦辣椒干物质含量低于 CK,比 CK 减少了 7.6%。说明伴生燕麦栽培可以提高辣椒果实的品质。

表 5 不同处理对连作辣椒果实品质的影响
Table 5 Effects of different treatments on the quality of pepper in continuous cropping

处理 treatment	维生素 C vitaminC (mg/100 g)	可溶性糖 soluble sugar (g/100 g)	可溶性蛋白 soluble protein (g/100 g)	干物质含量 soluble protein (g/100 g)
处理	193.80	0.85	0.15	10.10
CK	125.03	0.71	0.27	10.93

3 讨论与结论

蔬菜连作障碍是农业可持续发展的一个重大问题,作物的生长发育、光合作用、酶活性等生理过程都受其影响,解决连作障碍已成为蔬菜可持续发展的棘手问题。研究表明,自毒作用是导致设施蔬菜连作障碍重要原因之一,由于连续种植,根系分泌物释放的自毒物质积累到一定浓度,就会抑制下茬蔬菜的生长[15]。伴生可以合理地利用作物间相生相克的原理,调控植物生长发育和改善土壤理化性质,达到生态平衡[16]。

本试验通过测定伴生燕麦对辣椒叶片酶活性、果实品质、光合特性、叶绿素含量、荧光参数和土壤酶活性等指标的影响,结果显示:伴生燕麦提高了辣椒叶片中可溶性糖含量、过氧化氢酶活性、辣椒果实维生素C含量以及可溶性糖含量,伴生处理降低了辣椒叶片的光合参数和荧光参数。

土壤酶活性是评价生态环境质量优劣的重要指标,土壤酶活性提高能够改善土壤环境和养分的转化,从而促进作物对养分的吸收[17]。土壤酶参与有机质的分解和腐殖质的形成,是土壤生物活性的综合表现,其催化土壤中的生物化学反应,影响土壤养分的形成和积累[18,19],其中,过氧化氢酶主要来源于细菌、真菌以及植物根系的分泌物,它能破坏土壤中生化反应生成的过氧化氢,将过氧化氢分解为水和氧,减轻活性氧对植物的危害[20]。脲酶是一种对土壤有机态氮分解转化起非常重要作用的酶,它直接参与尿素形态转化能够催化尿素水解生成氨、水和二氧化碳,可以作为土壤生态系统变化的敏感指标,其活性反应土壤有机态氮向有效态氮的转化能力和土壤无机态氮的供应能力[21]。

本试验结果显示,伴生燕麦处理提高了土壤中过氧化氢酶、纤维素酶和蔗糖酶的活性,说明伴生燕麦具有刺激土壤酶活性的提高,促进土壤有机成分的转化,改善辣椒连作土壤生物学环境的作用。类似的结果在其他作物的研究中得到了证实,伴生小麦提高了西瓜根区土壤多酚氧化酶和蔗糖酶的活性[22];麦类与棉花套作能够增加土壤中脲酶和蔗糖酶活性[23]。本试验中,伴生燕麦处理降低了连作辣椒根际脲酶活性,可能与植株生理代谢和养分吸收有关,有待进一步研究。

参考文献

[1] 王立浩,张正海,曹亚从,张宝玺."十二五"我国辣椒遗传育种研究进展及其展望[J].中国蔬菜,2016,(1):1-7.

[2] 胡亮,陈宾.辣椒常见几种病害的发生及防[J].现代园艺,2015,(23):112-113.

[3] 喻景权,周杰."十二五"我国设施蔬菜生产和科技进展及其展望[J].中国蔬菜,2016,(9):18-30.

[4] Fu X,Li C,Zhou X,Liu S,Wu F. Physiological response and sulfur metabolism of the V. dahlia infected tomato plants in tomato/potato onion companion cropping.Scientific Reports,2016,6:36445.

[5] Gao D,Zhou X,Duan Y,Fu X,Wu F. Wheat cover crop promoted cucumber seedling growth through regulating soil nutrient resources or soil microbial communities? Plant & Soil,2017,418(1-2):459-475.

[6] 韩哲.伴生小麦提高黄瓜霜霉病抗性的生理生化机制[D].哈尔滨:东北农业大学.2012.

[7] 吴凤芝,潘凯,刘守伟.设施土壤修复及连作障碍克服技术[J].中国蔬菜,2013,(13):39.

[8] 徐伟慧,吴凤芝.西瓜根际土壤酶及微生物对小麦伴生的响应[J].浙江农业学报,2016,28(9):1588-1594.

[9] 杨瑞娟,王腾飞,周希,刘爱荣,陈双臣,杨英军.禾本科作物伴生对番茄根区土壤酶活性、微生物及根结线虫的影响[J].中国蔬菜,2017,(3):38-42.

[10] 左元梅,张福锁.不同禾本科作物与花生混作对花生根系质外体铁的累积和还原力的影响[J].应用生态学报,2004,15(2):221-225.

[11] 杨晶.辣椒与玉米间作对重金属吸收的影响及其机理的研究[D].沈阳:沈阳大学.2016.

[12] Tringovska I,Yankova V,Markova D,Mihov M. Effect of companion plants on tomato greenhouse production.Scientia Horticulturae,2015,186:31-37.

[13] Borowy A.2012.Growth and yield of stake tomato under no-tillage cultivation using hairy vetch as a living mulch.Acta Sci Polonorum-Hortorum,11(2):229-252

[14] Kolota E,Adamczewska-Sowinska K. Living mulches in vegetable crops production: perspectives and limitations.Acta Scientiarum Polonorum-Hortorum Cultus,2013,12(6):127-142.

[15] 喻景权,杜尧舜.蔬菜设施栽培可持续发展中的连作障碍问题[J].沈阳农业大学学报,2000(1):124-126.

[16] 张福建,陈昱,杨有新,等.2018.伴生大麦和芥菜对连作辣椒叶片保护酶活性及根际土壤环境的影响[J].中国蔬菜,2018(2):42-46.

[17] 官欢欢,尤一泓,林勇明,吴承祯,李键.不同林龄木麻黄纯林土壤酶活性与土壤养分研究[J].江西农业大学学报,2017,39(3):516-524.

[18] GARCI/A-GIL J C,PLAZA C,SOLER-ROVIRA P,et al.Long-term effects of municipal solid waste compost application on soil enzyme activities and microbial biomass[J].Soil Biology and Biochemistry,2000,32(13):1907-1913.

[19] 邱莉萍,刘军,王益权,等.土壤酶活性与土壤肥力的关系研究[J].植物营养与肥料学报,2004,10(3):277-280.

[20] 唐艳领.微生物肥在设施辣椒连作障碍克服中的应用研究[D].郑州:河南农业大学,2014.

[21] 张为政,祝廷成,张镇媛,等.作物茬口对土壤酶活性和微生物的影响[J].土壤肥料,1993(5):12-14.

[22] 徐伟慧.伴生小麦对西瓜生长及枯萎病抗性调控的机理研究[D].哈尔滨:东北农业大学.2016.

[23] 孟亚利,王立国,周治国,陈兵琳,王瑛,张立桢,卞海云,张思平.麦棉两熟复合根群体对棉花根际非根际土壤酶活性和土壤养分的影响[J].中国农业科学,2005,38(5):904-910.

嫁接砧木对拱棚连作辣椒生长发育、产量及品质的影响

高晶霞[1],吴雪梅[2],陈凯张[3],牛勇琴[2],王学梅[1]

(1.宁夏农林科学院种质资源研究所,宁夏银川 750002;
2.宁夏回族自治区彭阳县蔬菜产业发展服务中心,宁夏彭阳 756500;
3.宁夏回族自治区彭阳县科技服务中心,宁夏彭阳 756500)

摘 要:以引选的6份辣椒砧木("神根二号"、"青园砧木"、"新峰四号"、"砧强"、"神根F1"、"天骄F1")为研究对象,采用随机区组排列,对拱棚连作辣椒生长指标、产量及品质进行了研究。结果表明:拱棚连作辣椒株高、茎粗、开展度、单株结果数、净光合速率,"神根二号"辣椒砧木材料均高于其他砧木及"亨椒1号"。"神根二号"小区总产量、折合667 m^2产量、折合667 m^2产值均最高,分别为171.5 kg,4 888.5 kg,6 746.1元,分别比"青园砧木"、"新峰四号"、"砧强"、"神根F1"、"天骄F1"、"亨椒1号"(CK)667 m^2产量增幅17.9%、2.7%、1.8%、2.2%、9.7%、3.4%;"神根二号"、"神根F1"维生素C含量最高,分别为96.2 mg·g^{-1}、95.4 mg·g^{-1},其他4份辣椒砧木维生素C含量不显著,"亨椒1号"维生素C含量最低,为38.2 mg·g^{-1}。综合分析"神根二号"辣椒砧木材料是适合彭阳地区辣椒嫁接的优良砧木。

关键词:砧木;连作;辣椒;生长发育;产量;品质

Effects of Grafting Stock on Growth, Yield and Quality of Continuous Cropping Pepper

GAOJing xia[1], WU Xuemei[2], CHEN Kaizhang[3], NIU Yongqin[2], WANG Xuemei[1]

(1. *Ningxia Academy of Agriculture and Forestry Plant Resources, Ningxia, Yinchuan 750002; 2. Ningxia Hui Autonomous Region, Pengyang County Agricultural Technology Extension and Service Center, Pengyang, Ningxia, Pengyang 756500; 3. Ningxia Hui Autonomous Region Pengyang county science and Technology Service Center, Ningxia, Pengyang 756500*)

Abstract: Six selected pepper rootstocks ('Shengeng No.2' 'Qing Yuan rootstock' 'Xinfeng No.4' 'Zhenqiang' 'Shengeng F1' and 'Tianjiao F1') were selected as the research objects. Using random group arrange, growth index, yield and quality ofPepper continuous cropping were studied. The results showed that plant height, stem diameter, development degree, plant number per plant and net photosynthetic rate, the rootstocks of 'Shengeng No.2' were all higher than other rootstocks and 'Henjiao No.1', the total output of 'Shengeng No.2', the yield equivalent to 667 m², and the value of 667 m² were highest, 171.5 kg, 4 888.5 kg, 6746.1 RMB, respectively. Compared with 'Qingyuan' 'Xinfeng No.4' 'Zhenqiang' 'Shengeng F1' 'Tianjiao F1' (CK), yield increased 17.9%, 2.7%, 1.8%, 2.2%, 9.7%, 3.4%, respecty, the highest content of vitamin C in 'Shengeng F1' and 'Shengeng No.2', 96.2、95.4 mg·g^{-1}, respectively, there was no significant difference in vitamin C content between the other 4 pepper rootstocks. The content of vitamin C of 'Henjiao No.1' was the lowest, 38.2 mg·g^{-1} respectly, comprehensive analysis, 'Shengeng No.2' rootstock was a good rootstock suitable for grafting pepper in Pengyang area.

Key words: rootstock; continuous cropping; pepper; growth and development; yield; quality

辣椒(*Capsicum annuum*)属茄科一年生草本植物,又叫番椒、海椒、辣子、辣角、秦椒等。辣椒因其营养丰富、味道鲜美而在世界各地广泛种植,其产量在茄科蔬菜中仅次于番茄,成为重要的蔬菜作物之一[1,2]。我国是辣椒种植大国,年种植面积超过 130 万 hm^2,是世界最大的生产国和消费国。宁夏地区光照资源充足、热量丰富、昼夜温差大,自然环境条件适宜辣椒栽培。近年来栽培面积逐步增大,年播种面积已达 1.33 hm^2。但随着栽培面积的逐年加大,连作、重茬等导致土传病害逐年加重,尤其是辣椒疫病、根基腐病严重,造成大面积死秧,对辣椒生产构成了严重威胁[3,4]。虽然通过采取引进抗病新品种、高垄地膜覆盖、合理密植等措施,在一定程度上缓解了病害的发生,但由于辣椒疫霉病初次浸染的部位是根茎部,一般施药很难达到防治效果。在全区主要辣椒产区调查,辣椒疫病几乎在所有辣椒种植区都有发生,流行年份,会造成大面积死秧。一般造成的产量损失约 35%,农民每年 667 m^2 投入的防治费用为 700~1 000 元。这不仅加重了农民负担,还严重污染了产品和环境[5]。

宁夏南部山区,以彭阳县为代表的拱棚辣椒栽培近 1.3 hm^2,其中彭阳县近 0.67 hm^2[6,7]。栽培技术逐步成熟,辣椒产量及质量较高。但是多年来彭阳辣椒连作现象较为普遍,导致病虫害发生严重。嫁接技术可防止土传病害、克服连作障碍、提高抗病性等,已经在瓜类、茄子、番茄、辣椒等蔬菜作物上得到广泛应用[8-14]。该研究以引进的不同嫁接砧木("神根二号"、"青园砧木"、"新峰四号"、"砧强"、"神根 F1"、"天骄 F1")进行辣椒嫁接处理,观察比较嫁接砧木亲和性和对产量、品质的影响,以期筛选出适合彭阳地区辣椒嫁接的优良砧木,为宁南山区嫁接辣椒生产栽培提供依据。

1 材料与方法

1.1 试验材料

接穗品种为"亨椒 1 号";砧木品种为"神根二号"、"青园砧木"、"新峰四号"、"砧强"、"神根 F1"、"天骄 F1"。

1.2 试验方法

试验于2017年4月在宁夏彭阳县新集乡拱棚辣椒示范基地（连作辣椒3年以上）进行。1月中旬在宁夏嘉禾源种业有限公司进行育苗,采用劈接法进行嫁接。以5个砧木嫁接苗为处理,以"亨椒1号"（自根苗,CK）为对照,每处理嫁接600株。嫁接苗3叶1心时定植到彭阳新集乡拱棚辣椒示范基地,单垄双行种植,株距35 cm,垄宽70 cm,垄高25 cm。每小区46株,3次重复,其他管理同常规生产。

在辣椒盛花期、盛果期、采收中后期分别调查1次株高、茎粗、开展度、叶绿素含量、净光合速率（每小区测10株取平均值）；采果盛期采收坐果天数相同的果实,测定果实单果重,可溶性糖含量、可溶性蛋白质含量、维生素C含量、可溶性固形物含量（每小区测5个果取平均值）；统计小区产量。

1.3 项目测定

采用铝蓝比色法测定辣椒果实中维生素C含量；采用苯酚硫酸比色法测定果实中可溶性糖含量；采用紫外分光光度法测定果实中可溶性蛋白质含量；采用PRO-101型糖度计测定果实中可溶性固形物含量；采用叶绿素仪（CSPAD502）测定相对叶绿素含量（SPAD）,每片叶重复3次,取平均值；采用Li-6400型光合仪测定净光合速率（Pn）,每小区选有代表性的5片叶,10:00~11:00测定,每叶重复测定3次,取平均值。

1.4 数据分析

采用Excel 2003软件对试验数据进行整理；利用SPSS统计分析软件进行显著性分析。

2 结果与分析

2.1 不同嫁接砧木对连作辣椒株高的影响

由图1可知,引选的6份辣椒砧木材料对拱棚连作辣椒株高有差异。"神根二号"、"砧强"、"亨椒1号"（自根苗）株高在辣椒生长的3个时期,株高均高于其他处理,"青园砧木"株高最低。

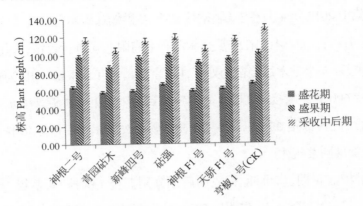

图 1 嫁接砧木对连作辣椒株高的影响

Fig. 1 Effect of grafted rootstock on plant height of continuous cropping pepper

2.2 不同嫁接砧木对连作辣椒茎粗的影响

由图2可知,引选的6份辣椒砧木材料对拱棚连作辣椒茎粗有差异。在辣椒生长的3个时期,"神根二号"、"砧强"、"亨椒1号"(自根苗)茎粗均高于其他处理,采收中后期,"青园砧木"、"新峰四号"、"神根F1"、"天骄F1"茎粗差异不显著,范围在14.05~14.74 mm。

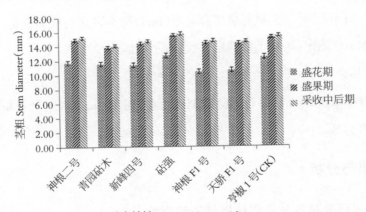

图 2 嫁接砧木对连作辣椒茎粗的影响

Fig. 2 Effect of grafted rootstock on the stem diameter of continuous cropping pepper

2.3 不同嫁接砧木对连作辣椒开展度的影响

由图 3 可知,引选的 6 份辣椒砧木材料对拱棚连作辣椒开展度有差异。在辣椒生长的 3 个时期,"新峰四号"、"砧强"、"亨椒 1 号"(自根苗)开展度均高于其他处理,采收中后期,"青园砧木"、"神根 F1"、"神根二号"、"天骄 F1"开展度差异不显著,范围在 71.60~79.80 cm。

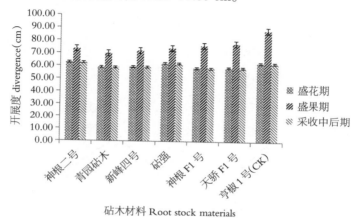

图 3　嫁接砧木对连作辣椒开展度的影响

Fig. 3 Effect of grafted rootstock on the development of continuous cropping pepper

2.4 不同嫁接砧木对连作辣椒叶绿素的影响

由图 4 可知,引选的 6 份辣椒砧木材料对拱棚连作辣椒叶绿素含量有

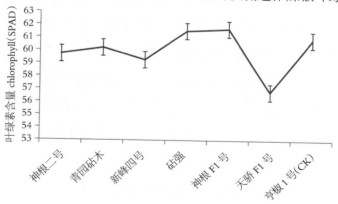

图 4　嫁接砧木对连作辣椒叶绿素含量的影响

Fig. 4 Effect of grafted rootstock on chlorophyll content in continuous cropping pepper

差异。在辣椒生长的 3 个时期,"神根二号"叶绿素含量均高于其他处理,"青园砧木"、"新峰四号"、"神根 F1"、"砧强"叶绿素含量差异不显著,"天骄 F1"叶绿素含量最低为 56.77 SPAD。

2.5 不同嫁接砧木对连作辣椒净光合速率的影响

由图 5 可知,引选的 6 份辣椒砧木对拱棚连作辣椒净光合速率有差异。在辣椒采收盛期,"神根二号"净光合速率均高于其他 5 份砧木材料,"新峰四号"、"神根 F1"净光合速率次之,"青园砧木"、"砧强"、"天骄 F1"、"亨椒 1 号"(自根苗)净光合速率最低。

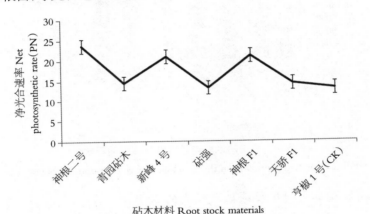

图 5 嫁接砧木对连作辣椒净光合速率的影响

Fig. 5 Effect of grafted rootstock on net photosynthetic rate of continuous cropping pepper

2.6 不同嫁接砧木对连作辣椒单株结果数的影响

由图 6 可知,引选的 6 份辣椒砧木材料对拱棚连作辣椒单株结果数有差异。在辣椒采收盛期,"神根二号"、"砧强"单株结果数高于其他 4 份砧木材料,"新峰四号"、"神根 F1"、"天骄 F1"单株结果数次之,"青园砧木"、"亨椒 1 号"单株结果数最少。

2.7 不同嫁接砧木对连作辣椒单果质量的影响

由图 7 可知,引选的 6 份辣椒砧木对拱棚连作辣椒单果质量有差异。在辣椒采收盛期,"亨椒 1 号"(自根苗)单果质量最高,为 87.18 g,"神根

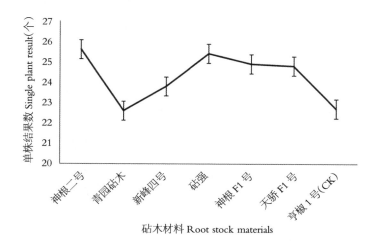

图 6　嫁接砧木对连作辣椒单株结果数的影响

Fig. 6 Effect of grafted rootstock on single plant results of continuous cropping pepper

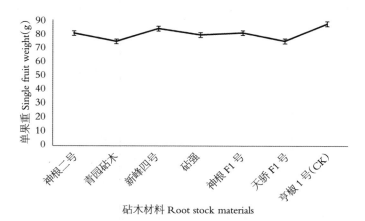

图 7　嫁接砧木对连作辣椒单果重的影响

Fig. 7 Effect of grafted rootstock on single fruit weight of continuous cropping pepper

二号"、"新峰四号"单果质量次之,"青园砧木"、"天骄 F1"单果质量最低,"砧强"、"神根 F1"单果质量无差异。

2.8　不同嫁接砧木对连作辣椒产量的影响

从表 1 可以看出,引选的 6 份辣椒砧木材料对连作辣椒产量及品质存在显著性差异。"神根二号"小区总产量、折合 667 m^2 产量、折合 667 m^2 产值均最高,分别为 171.5 kg,4 888.5 kg,6 746.1 元,"神根二号"667 m^2 产量分别

表 1 嫁接砧木对拱棚连作辣椒产量的影响

Table 1 Effect of grafted rootstock on the yield of pepper in continuous cropping of arch shed

处理 Treatment	小区产量 Plot yield(kg·(23.4 m²)⁻¹)				折合 667 m² 产量/kg Amount to 667 m² yield	折合 667 m² 产值/元 Amount to 667 m² output value	折合 667 m² 产量比对照±/kg Amountto 667m² comparative yield comparative control	折合 667 m² 产量比对照±/% Amountto 667m² comparative yield comparative control	折合 667 m² 产量显著性 Amountto 667 m² yieldsignificance		位次 Precedence
	I	II	III	平均					5%	1%	
神根二号	169.5	173.2	171.8	171.5	4 888.5	6 746.1	165.3	3.5	a	A	1
青园砧木	142.3	140.7	139.4	140.8	4 013.4	5 538.5	-709.8	-15.0	ab	AB	7
新峰四号	168.0	164.8	167.3	166.7	4 751.7	6 557.3	28.5	0.6	b	AB	4
砧强	170.2	169.6	165.5	168.4	4 800.1	6 624.1	76.9	1.6	b	B	2
神根 F1 号	169.2	167.4	166.7	167.8	4 783.0	6 600.5	82.5	1.8	b	B	3
天骄 F1 号	152.8	155.6	156.2	154.9	4 415.3	6 093.1	-307.9	-6.5	c	C	6
亨椒 1 号(CK)	167.4	165.5	164.1	165.7	4 723.2	6 518.0	—	—	d	D	5

注:6 月中旬开始采收,每 12 d 彭阳县北环市场调查辣椒价格,平均单价 1.38 元·kg⁻¹

Note:Starting in mid June, the average price of pepper was 1.38 yuan. kg⁻¹ per 12D Pengyang County North Ring market

比"青园砧木"、"新峰四号"、"砧强"、"神根 F1 号"、"天骄 F1 号"、"亨椒 1 号"(CK)增幅 17.9%、2.7%、1.8%、2.2%、9.7%、3.4%,"神根 F1"、"新峰四号"、"砧强"小区总产量、折合 667 m² 产量、折合 667 m² 产值差异不显著($P<0.05$),"青园砧木"小区总产量、折合 667 m² 产量、折合 667 m² 产值最低。

2.9 不同嫁接砧木对连作辣椒品质的影响

由图 8~10 可知,引选的 6 份辣椒砧木材料辣椒果实品质存在显著性差异,"神根二号"、"神根 F1"维生素 C 含量最高,分别为 96.2 mg·g^{-1}、95.4 mg·g^{-1},其他 4 份辣椒砧木材料维生素 C 含量不显著,"亨椒 1 号"(自根苗)维生素 C

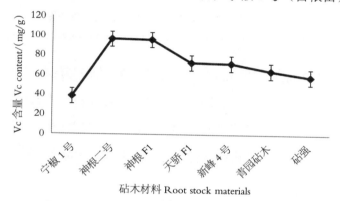

图 8　嫁接砧木对连作辣椒维生素 C 含量的影响

Fig. 8 Effect of grafted rootstock on vitamin C content in continuous cropping pepper

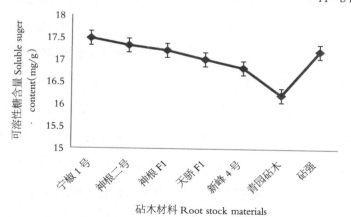

图 9　嫁接砧木对连作辣椒可溶性糖含量的影响

Fig. 9 Effect of grafted rootstock on soluble sugar content in continuous cropping pepper

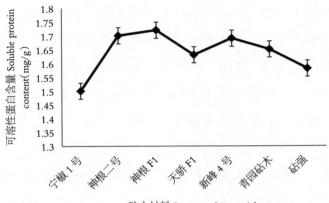

图 10　嫁接砧木对连作辣椒可溶性蛋白质含量的影响

Fig. 10 Effect of grafted rootstock on soluble protein content in continuous cropping pepper

含量最低,为 38.2 mg·g⁻¹。"亨椒 1 号"(自根苗)可溶性糖含量最高,为 17.48 mg·g⁻¹,"新峰四号"、"青园砧木"可溶性糖含量最低,其他 4 份砧木材料可溶性糖含量差异不明显。"神根二号"、"神根 F1"辣椒可溶性蛋白质含量最高,"亨椒 1 号"(自根苗)最低,其他 4 份砧木材料可溶性蛋白质含量不显著。

3　讨论与结论

不同砧木嫁接对辣椒生长、产量有不同的影响,利用抗性强的砧木嫁接辣椒,可使植株生长势增强,促进植株的生长发育,延长果实的收获期,提高产量。前人等通过对黄瓜和茄子的试验,证明利用抗性较强的砧木嫁接能提高接穗的抗性,促进植株生长发育。本试验结果表明:"神根二号"辣椒砧木材料,连作辣椒株高、茎粗、开展度均高于其他砧木及对照;单果重、单株结果数、小区总产量、折合亩产量、折合亩产值均最高;"神根 F1"维生素 C 含量最高,分别为 96.2 mg/g、95.4 mg/g,其他 4 份辣椒砧木材料维生素 C 含量不显著,"亨椒 1 号"(自根苗)维生素 C 含量最低,为 38.2 mg/g。

由于拱棚栽培辣椒品种比较单一,连作障碍严重,根系自毒产物增加,使植株的抵抗力逐渐下降,为各种土传疾病提供了发病条件。利用抗性强品种和抗性强砧木嫁接已经成为目前替代化学药品控制土传病害的有效的方法之一。本研究结果表明,"神根二号"辣椒砧木材料适合彭阳地区辣椒嫁接的优良砧木,为宁南山区嫁接辣椒生产栽培提供理论依据,提升宁夏蔬菜产业的竞争力和广大菜农的经济收入。

参考文献

[1] 戴雄泽,刘志敏.我国辣椒产业现状及发展趋势[J].饮食文化研究,2006(4).

[2] 王永平,张绍刚,张蜻,等.我国辣椒产业发展现状及趋势[J].河北农业科学,2009(6):141-144.

[3] 赵鑫,周宝利,林桂荣,等.辣椒嫁接效果试验研究[J].北方园艺,2000(4):8-10.

[4] 冯义.大棚辣椒病虫害绿色综合防控技术[J].长江蔬菜,2014(13):48-49.

[5] 孙红霞,武琴,郑国祥,等.EM对茄子、黄瓜抗连作障碍和增强土壤生物活性的效果[J].土壤,2001(5):264-267.

[6] 邹琦.植物生理学实验指导[M].北京:中国农业出版社,1998.

[7] 马瑞,惠浩剑,马守才,等.大棚辣椒品种引进观察试验[J].现代农业科技,2012(7):149-150.

[8] 冯春梅,莫云彬,潘晓飚,等.不同砧木嫁接对西瓜抗病性及主要经济性状的影响[J].中国农学通报,2006,22(2):289-291.

[9] 徐敬华,黄丹枫,支月娥.PAL活性与嫁接西瓜枯萎病抗性传递的相关性[J].上海交通大学学报(农业科学版),2002,22(1):12-16.

[10] 何莉莉,侯丽霞,葛晓光,须晖,李天来.嫁接番茄抗叶霉病效果及其与体内几种抗性物质的关系[J].沈阳农业大学学报,2001,32(2):99-101.

[11] 周宝利,林桂荣.不同茄子砧木防病增产效果与POD同工酶关系[J].北方园艺,1998,3:14-15.

[12] 张凤丽,周宝利,王茹华,何雨.嫁接茄子根系分泌物的化感效应[J].应用生态学报,2005,16(4):750-753.

[13] 陈绍莉,周宝利,王茹华,等.嫁接对茄子根系分泌物中肉桂酸和香草醛的调节效应[J].应用生态学报,2008,19(11):2394-2399.

[14] 高永利.辣椒嫁接栽培技术及其应用前景[J].北方园艺,2009,(7):175-176.

不同嫁接砧木对连作辣椒生长发育及 N、P、K 分配特性的影响

高晶霞[1],吴雪梅[2],赵云霞[1],牛勇琴[2],颜秀娟[1],裴红霞[1],谢　华[1]

(1.宁夏农林科学院种质资源研究所,宁夏银川　750002;
2.宁夏回族自治区彭阳县蔬菜产业发展服务中心,宁夏彭阳　756500)

摘　要:以引选的 12 份辣椒砧木为研究对象,采用随机区组排列,对拱棚连作辣椒亲和性、生长指标及植株 N、P、K 分配特性进行了研究。结果表明:"CT09"、"CT10"砧木嫁接辣椒成活率最高,分别为 98.15%、98.61%;连作辣椒株高、茎粗、单株结果数、净光合速率,"CT09"辣椒砧木均高于其他砧木及"巨峰 1 号"(CK),"CT05"根系活力最强,为 17.32 μg/FW·h,"CT09"砧木辣椒单株产量、小区产量、折合亩产量、产值均高于其他砧木及自根苗,分别为 1.97 kg、299.44 kg、5 690.2 kg、11 949.4 元,比对照增幅 16.6%,产值增加 1 698.2 元;"CT09"、"CT10"砧木嫁接辣椒果实维生素 C 含量最高,分别为 65.14 mg/100 g、63.26 mg/100 g,"CT03"、"CT07"砧木辣椒根部全氮含量最大,分别为 2.180 g/100 g、2.002 g/100 g,8817、CT02、CT03、CT07 砧木根全磷含量较高,"CT01"砧木辣椒茎全氮含量最高,为 0.379 g/100g,综合分析:"CT09"、"CT03"辣椒砧木材料适合彭阳地区辣椒嫁接生产。

关键词:砧木;连作;辣椒;生长发育;N、P、K 分配

Effects of Different Grafted Rootstocks on Growth and Distribution Characteristics of N, P and K in Continuous Cropping Pepper

Gao Jing-xia[1], Wu Xue-mei[2], Zhao Yun-xia[1], Niu Yong-qin[2],
Yan Xiu-juan[1], Pei Hong-xia[1], Xie Hua[1]

(1. *Ningxia Academy of Agriculture and Forestry Plant Resources*, Ningxia, Yinchuan, 750002; 2. *The Ningxia Hui Autonomous Region*, Pengyang County *AgriculturalTechnology Extension and Service Center*, Pengyang, Ningxia 756500)

Abstract: 12 selected capsicum rootstocks were selected as research objects, Random block arrangement, The affinity, growth index and distribution characteristics of N, P and K in Capsicum under continuous cropping in arch shed were studied. The results show that: "CT09" and "CT10" rootstocks had the highest survival rates, 98.15% and 98.61% respectively. the plant height, stem diameter, fruit number per plant and net photosynthetic rate of continuous cropping pepper, "CT09" pepper rootstock was higher than other rootstocks and "Jufeng 1" (CK), "CT05" has the strongest root activity, 17.32 ug/FW.h, "CT09" rootstock pepper yield per plant, plot yield, yield per mu, output value were higher than other rootstocks and self-rooted seedlings, 1.97 kg, 299.44 kg, 5 690.2 kg and 11 949.4 yuan respectively, the increase was 16.6% compared with the control, "CT09" and "CT10" rootstocks had the highest vitamin C content in the grafted pepper fruit, 65.14 mg/100g and 63.26 mg/100g respectively, "CT03" and "CT07" rootstocks had the highest total nitrogen content in Capsicum root, 2.180 g/100 g and 2.002 g/100 g, respectively, The total phosphorus content of Rootstocks 8817, CT02, CT03 and CT07 was higher, "CT01" rootstock has the highest total nitrogen content in pepper stem, 0.379 g/100 g, Comprehensive analysis: "CT09" and "CT03" pepper rootstock materials suitable for pepper grafting production in Pengyang area.

Key words: rootstock; continuous cropping; pepper; growth and development; N, P, K distribution

辣椒(*Capsicum annuum L.*)又名番椒,原产于拉丁美洲热带地区,茄科辣椒属,为一年生草本植物,在我国普遍栽培[1]。辣椒营养丰富,维生素 C 含量在蔬菜中居第一位[2]。辣椒是宁夏地区栽培面积较大的蔬菜之一。但由于连作种植,土壤连作障碍、病虫害问题日趋严重,尤其是辣椒疫病蔓延速度快,造成产量和经济效益下降[3]。嫁接技术是克服辣椒连作障碍行之有效的措施之一,已经在瓜类蔬菜、茄子、番茄等蔬菜作物上得到广泛应用[4],近几年在辣椒栽培上的应用也越来越多[5]。本试验引进 12 份砧木材料嫁接辣椒,观察比较其亲和性、抗病性,以及对连作辣椒生长发育、产量、品质及植株 N、P、K 分配的影响,以期能筛选出适合宁夏彭阳地区辣椒嫁接的优良砧木,为宁夏冷凉区嫁接辣椒生产栽培提供理论依据。

1 材料与方法

1.1 试验地点

试验地点位于宁夏彭阳县新集乡拱棚辣椒核心示范基地(连作辣椒 3 a 以上)。

1.2 供试材料

接穗品种为"巨峰 1 号"(宁夏巨丰种苗有限公司提供),砧木材料为"CT09"、"CT05"、"CT08"、"CT06"、"CT10"、"CT07"、"CT01"、"CT02"、"CT03"、"CT04"、"CT11"、"8817"(均由宁夏天缘种苗有限公司提供)。

1.3 试验设计

试验苗期在宁夏嘉禾源种苗有限公司进行。采用劈接法进行嫁接。设 12 个处理,分别为"CT01"、"CT02"、"CT03"、"CT04"、"CT11"、"8817"、"CT05"、"CT06"、"CT07"、"CT08"、"CT09"、"CT10"。每个处理嫁接 288

株,以"巨峰1号"自根苗(CK)作对照。嫁接苗3叶1心时定植到彭阳县新集乡拱棚辣椒示范基地,单垄双行种植,株距35 cm,垄宽70 cm,垄高25 cm。每小区46株,3次重复,其他管理同常规生产。

1.4 调查项目

砧木嫁接成活后统计嫁接成活率,定植后第三十天、六十天调查1次株高、茎粗、叶片叶绿素含量、净光合速率(每个小区测5株取平均值);整个生育期观察记载发病情况;采果盛期,采收坐果天数相同的果实,测定果实单果质量,可溶性糖含量、蛋白质含量、维生素C含量、可溶性固形物(每个小区测定5个果取平均值);统计小区产量。

用铝蓝比色法测定辣椒果实中维生素C含量,苯酚硫酸比色法测定可溶性糖含量,紫外分光光度法测定可溶性蛋白质含量,PRO-101型糖度计测定可溶性固形物含量。

用叶绿素仪(CSPAD502)测定相对叶绿素含量,并以SPAD值表示。每片叶重复3次,取平均值。测定时间与品质指标测定时间一致。

用美国产Li-6400型光合测定系统测定净光合速率(P_n)每小区选有代表性的5片叶上午10:00~11:00进行测定,每叶重复测定3次,取平均值。

N、P、K测定方法:参照NY/T-2017-2011方法测定。

白粉病分级标准:未发病0级;小于5%为1级;6%~10%为3级;11%~20%为5级;21%~40%为7级;大于40%为9级。

病叶率%=有病叶数/调查总叶数×100%,病情指数={(被害级植株×代表值)÷(调查植株总数×受害总重级代表值)}×100。

用Excel2003对数据进行整理,采用SPSS软件进行显著性差异分析[6]。

2 结果分析

2.1 不同砧木辣椒成活率的比较

从表1中可以看出,不同辣椒砧木材料与接穗的成活率有差异,

"CT01"、"CT09"、"CT10"、"CT04"砧木材料与接穗亲和性较好，分别为98.26%、98.26%、98.61%、98.61%，"CT03"、"CT06"材料与辣椒亲和性次之，均为96.30%、96.18 %，"CT01"、"CT02"砧木材料与辣椒亲和性最差，为84.03%。

表1 不同砧木嫁接辣椒成活率比较
Table 1 Affinity of Grafted Pepper with Different Rootstocks

材料 Material Science	成活株数 Number of surviving plants /株	总株数 Total plant number /株	成活率 Survival rate /%
CT01	283	288	98.26%
CT02	242	288	84.03%
CT03	277	288	96.18%
CT04	284	288	98.61%
CT11	276	288	95.83%
8817	273	288	94.91%
CT05	276	288	95.83%
CT06	277	288	96.18%
CT07	269	288	93.52%
CT08	270	288	93.75%
CT09	283	288	98.26%
CT10	284	288	98.61%
巨峰1号（CK）	—	—	—

2.2 不同砧木对嫁接辣椒植物学性状的影响

从表2可以看出，不同砧木辣椒植物学性状有差异，始花期，"CT09"株高、茎粗均最大，分别为42.6 cm、6.97 mm，"CT05"、"巨峰1号"自根苗（CK）次之，8817株高、茎粗最小，分别为27.6 cm，5.25 mm，其他砧木辣椒株高、茎粗分别在33.2~37.8 cm，5.25~6.53 mm；盛花期，"CT05"辣椒株高最高，为72.4 cm，"CT09"辣椒茎粗最粗，为11.91 mm，"CT07"、"巨峰1号"（CK）辣椒株高次之,"8817"砧木株高最低,为53.6 cm,其他砧木株高、茎粗分别在55.2~67.2 cm,9.32~11.90 mm;采收中后期,"CT10"、"巨峰1号"自根苗(CK)辣椒株高最高,分别为116.8 cm、117.2 cm,"8817"砧木辣椒株高最低,为94.8 cm,其他砧木辣椒株高分别在99.8~116.0 cm,茎粗分别在14.16~15.59 mm;不同砧木辣椒叶色均为绿色,"CT05"、"CT06"、"CT01"、"8817"、"CT11"、"CT03"、"巨峰1号"(CK)植株长势较强,其他砧木植株长势均为强。

表2 不同砧木对嫁接辣椒植物学性状的影响

Table 2 Effects of Different Rootstocks on Botanical Characters of Grafted Pepper

项目 Subject 处理 Treatment	始花期 Initial flowering stage		盛花期 Flowering stage		中后期 Middle late stage		叶色 Leaf color	植株长势 Plant growth potential
	株高 Plant height (cm)	茎粗 Stem diameter (mm)	株高 Plant height (cm)	茎粗 Stem diameter (mm)	株高 Plant height (cm)	茎粗 Stem diameter (mm)		
CT10	37.8	5.96	64.2	9.88	116.8	15.72	绿	强
CT09	42.6	6.97	62.4	11.91	99.8	15.59	绿	强
CT08	34.4	6.16	62.4	11.90	111.8	15.20	绿	强
CT05	40.0	6.88	72.4	11.29	113.8	14.59	绿	较强
CT06	38.0	6.53	67.2	11.09	103.8	13.85	绿	较强
CT07	37.0	5.45	67.8	11.59	116.0	15.57	绿	强
CT01	35.8	5.91	57.0	9.35	102.6	14.78	绿	较强
CT04	34.5	6.00	55.2	9.17	114.8	15.36	绿	强
8817	27.6	5.25	53.6	9.62	94.80	14.16	绿	较强
CT02	37.4	6.23	65.4	11.49	102.6	15.22	绿	强
CT11	33.2	5.82	59.8	10.93	111.0	14.62	绿	较强
CT03	36.6	6.18	66.6	11.2	106.2	14.71	绿	较强
巨峰1号(CK)	38.0	6.72	70.0	10.1	117.2	14.68	绿	较强

2.3 不同砧木对嫁接辣椒果实性状的影响

从表3可以看出,不同砧木果实性状有差异,"CT09"辣椒果长最长,为25.94 cm,"CT06"果长次之,"CT11"果长最短,为20.86 cm,其他砧木辣椒果

表 3 不同砧木对嫁接辣椒果实性状的影响
Table 3 Effects of Different Rootstocks on Fruit Characters of Grafted Pepper

处理 Treatment	果长 Fruit length (cm)	果粗 Rough fruit (mm)	肉厚 Thickness (mm)	单果质量 Single fruit quality(g)	单株结果数 Number of fruit per plant(个)	果柄长 Fruit stalk length(cm)	萼片宽 Sepals width (mm)	果色 Fruit color	辣味 Pungent taste
CT10	23.14	47.38	4.00	85.40	20.6	4.14	24.71	绿	微辣
CT09	25.94	51.32	4.49	90.33	21.8	4.62	23.98	绿	微辣
CT08	23.58	43.58	3.86	84.62	19.6	4.60	26.68	绿	微辣
CT05	22.75	42.12	3.19	82.30	19.2	4.68	24.45	绿	微辣
CT06	25.40	47.49	4.24	89.62	19.4	4.62	24.83	绿	微辣
CT07	23.96	45.15	3.46	82.35	20.2	4.52	25.92	绿	微辣
CT01	22.20	46.93	3.35	85.19	17.6	4.12	24.45	绿	微辣
CT04	23.44	43.23	3.56	80.81	19.0	4.04	24.30	绿	微辣
8817	22.46	43.04	3.58	78.53	16.4	4.46	23.34	绿	微辣
CT02	23.56	46.89	3.32	85.25	18.2	4.50	23.67	绿	微辣
CT11	20.86	44.70	3.64	80.47	18.8	3.98	25.20	绿	微辣
CT03	24.40	46.53	3.78	87.42	20.8	4.60	25.10	绿	微辣
巨峰1号(CK)	24.82	47.56	3.84	88.76	19.0	4.64	25.93	绿	微辣

长分别在22.20~24.82 cm；"CT09"辣椒果粗值最大，为51.32 mm，"CT05"果粗值最小，为42.12 cm，其他砧木辣椒果粗值在43.03~47.49 mm，"CT09"辣椒肉厚值最大，为4.49 mm，"CT05"辣椒肉厚值最小，为3.19 mm，其他砧木辣椒肉厚值分别在3.32~4.24 mm；"CT09"辣椒单果质量最大，为90.33 g，"CT06"辣椒单果质量次之，"8817"辣椒单果质量最小，为78.53 g，其他砧木辣椒单果质量分别在80.47~88.76 g；CT09辣椒单株结果数最多，为21.8个，"8817"单株结果数最少，为16.4个，其他砧木辣椒单株结果数分别在17.6~20.8个；"CT09"、"CT08"、"CT05"、"CT03"、"巨峰1号"(CK)辣椒果柄长较长，分别为4.62 cm、4.60 cm、4.68 cm、4.60 cm、4.64 cm，"CT04"、"CT11"果柄长较短，分别为4.04 cm、3.98 cm，其他砧木辣椒果柄长分别在4.12~4.52 cm；"CT08"辣椒萼片宽值最大，为26.68 mm，"CT09"、"8817"、"CT02"辣椒萼片宽值最小，为23.98 mm、23.34 mm、23.67 mm，其他砧木辣椒萼宽值分别24.40~25.92 mm；不同砧木辣椒果面颜色均为绿色，辣味均微辣。

2.4 不同砧木对辣椒发病率及病情指数的影响

从图1~2可知，不同砧木辣椒白粉病发病有差异，"巨峰1号"(CK)辣椒发病率最高，为100%，"CT11"、"CT08"、"CT09"辣椒发病率次之，"CT10"辣

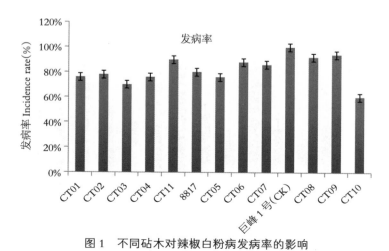

图1 不同砧木对辣椒白粉病发病率的影响

Fig. 1 Effects of Different Rootstocks on the incidence of powdery mildew in Pepper

椒发病率最小，为60%，其他砧木辣椒发病率分别在76%~88%，"CT06"、"CT09"辣椒病情指数最高，分别为65.1、65.5，"巨峰1号"(CK)辣椒病情指数次之，为64.9，"CT10"辣椒病情指数最小，为33.3，其他砧木辣椒病情指数分别在37.4~54.7。

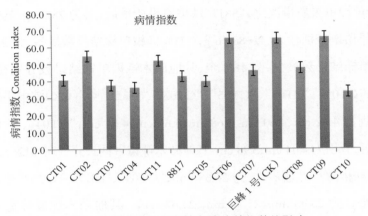

图2 不同砧木对辣椒白粉病病情指数的影响

Fig. 2 Effects of Different Rootstocks on disease index of pepper powdery mildew

2.5 不同砧木对辣椒根系活力的影响

从图3中可知,不同辣椒砧木嫁接辣椒的根系活力有差异,"CT05"根系活力最强,为17.32 μg/FW·h,"CT04"、"CT06"、"CT08"根系活力次之,分

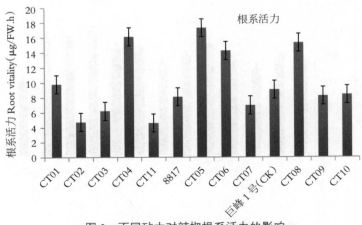

图3 不同砧木对辣椒根系活力的影响

Fig. 3 Effects of Different Rootstocks on root activity of pepper

别为 16.12 μg/FW·h、14.27 μg/FW·h、15.31 μg/FW·h,"8817"、"巨峰 1 号"（自根苗）、"CT09"、"CT10"根系活力差异不显著,"CT02"、"CT11"根系活力弱,分别为 4.61 μg/FW·h、4.73 μg/FW·h。

2.6 不同砧木对辣椒果实品质的影响

从图 4~6 可知,不同砧木嫁接辣椒,其果实品质有差异,"CT09"、"CT10"砧木嫁接辣椒果实维生素 C 含量最高,分别为 65.14 mg/100 g、63.26 mg/100 g,"CT02"维生素 C 含量次之,"CT03"、"CT04"砧木材料、"巨峰 1 号"（CK）果实维生素 C 含量差异不显著,分别为 45.03 mg/100 g、44.84 mg/100 g、44.82 mg/100 g,"CT08"、"CT07"、"8817"、"CT11"果实维生素 C 含量低,分别为 41.86 mg/100 g、39.8 mg/100 g、39.62 mg/100 g、33.25 mg/100 g;"CT09"、"CT10"、"CT02"砧木品种嫁接辣椒,果实可溶性糖含量最高,分别为 34.63 mg/g、32.58 mg/g、32.98 mg/g,"CT01"、"CT02"、"CT04"、"CT11"、"8817"砧木果实可溶性糖含量次之,且差异不显著,"CT05"、"CT06"砧木辣椒果实可溶性糖含量最低,分别为 25.21 mg/g、24.12 mg/g;"CT08"砧木果实可溶性蛋白含量最高,为 2.22 mg/g,"CT09"、"CT06"、"CT07"可溶性蛋白含量次之,且差异不显著,"巨峰 1 号"（CK）果实可溶性蛋白含量最低,为 1.47 mg/g。

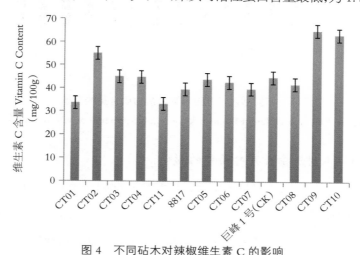

图 4 不同砧木对辣椒维生素 C 的影响

Fig. 4 Effects of Different Rootstocks on vitamin C in Pepper

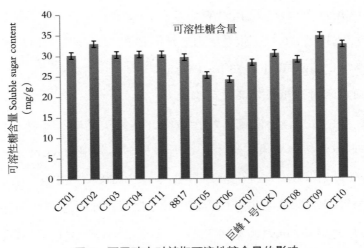

图 5 不同砧木对辣椒可溶性糖含量的影响

Fig. 5 Effects of Different Rootstocks on soluble sugar content in Pepper

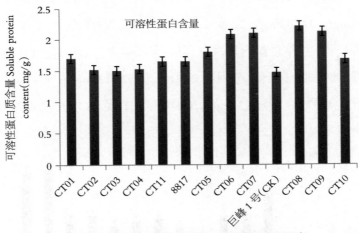

图 6 不同砧木对辣椒可溶性蛋白质含量的影响

Fig. 6 Effects of Different Rootstocks on soluble protein content in Pepper

2.7 不同砧木对辣椒叶片叶面积的影响

从图7可以看出,不同砧木嫁接辣椒叶片叶面积有差异,"CT06"材料辣椒叶片叶面积最大,为76.12 cm²,其次是"巨峰1号"(CK),为65.45 cm²,"CT10"砧木辣椒叶片叶面积最小,为45.68 cm²,其他砧木辣椒叶片叶面积差异不显著,分别在45.68~57.45 cm²。

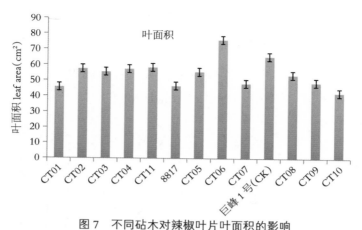

图 7 不同砧木对辣椒叶片叶面积的影响

Fig. 7 Effects of Different Rootstocks on leaf area of pepper

2.8 不同砧木对辣椒叶片叶绿素含量的影响

从图 8 可以看出,不同砧木辣椒叶绿素含量有差异,"CT09"叶片叶绿素含量最高,为 61.33 SPAD,"CT10"叶片叶绿素含量次之,为 57.97 SPAD,其他砧木叶绿素含量在 45.0 SPAD~57.06 SPAD,"CT04" 叶片叶绿素含量最低,为 44.53 SPAD。

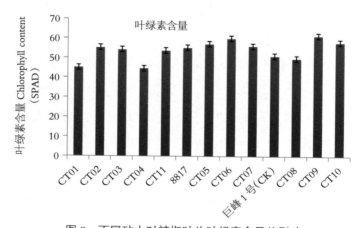

图 8 不同砧木对辣椒叶片叶绿素含量的影响

Fig. 8 Effects of Different Rootstocks on chlorophyll content in Pepper Leaves

2.9 不同砧木对辣椒净光合速率的影响

从图 9 可以看出,不同砧木辣椒净光合速率有差异,"CT09"、"CT02"净光合速率最大,分别为 23.6 μmol·m²·s⁻¹、22.8 μmol·m²·s⁻¹,"CT08"、"CT10"净光合速率次之,分别为 20.9 μmol·m²·s⁻¹、21.15 μmol·m²·s⁻¹,"CT01"、"巨峰 1 号"(自根苗)最小,分别为 12.9 μmol·m²·s⁻¹、13.06 μmol·m²·s⁻¹,其他砧木净光合速率无差异,分别在 13.25~16.3 μmol·m²·s⁻¹。

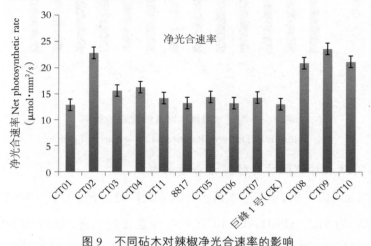

图 9　不同砧木对辣椒净光合速率的影响

Fig. 9 Effects of Different Rootstocks on net photosynthetic rate of pepper

2.10　不同砧木对辣椒植株 N、P、K 分配的影响

从表 4 可以看出,不同砧木对辣椒根部 N、P、K 分配有差异,"CT03"、"CT07"砧木辣椒根部全氮含量最大,分别为 2.180 g/100 g、2.002 g/100 g,"8817"砧木、"巨峰 1 号"(CK)次之,分别为 1.769 g/100 g、1.810 g/100 g,其他砧木全氮含量分别在 1.030~1.735 g/100 g,"CT04"砧木根全氮含量最小,为 0.996 g/100 g;"8817"、"CT02"、"CT03"、"CT07"砧木根全磷含量较高,分别为 0.163 g/100 g、0.167 g/100 g、0.168 g/100 g、0.197 g/100 g,其他砧木根全磷含量次之,在 0.113~0.147 g/100 g,"CT04"、"CT05"根全磷含量最小,为 0.099 g/100 g、0.090 g/100 g;"CT07"、"CT03"、"CT11"根全钾含量最高,分别为 2.606 g/100 g、2.522 g/100 g、2.536 g/100 g,"CT01"根全钾含量最小,为 1.358 g/100 g,其他砧木根全钾含量分别在 1.555~2.301 g/100 g。不同砧木对辣椒茎部 N、P、K 分配有差异,"CT01"砧木辣椒茎全氮含量最高,为

表4 不同砧木对辣椒植株 N、P、K 分配的影响
Table 4 Effects of Different Rootstocks on Distribution of N, P and K in Pepper Plants

植株部位 Plant location	样品名称 Sample name	全氮 Total nitrogen g/100 g	5%显著水平 5%Significant level	全磷 Total phosphorus g/100g	5%显著水平 Significant level	全钾 Total potassium g/100g	5%显著水平 Significant level
根 root	CT07	2.002	b	0.197	a	2.606	a
	巨峰1号(CK)	1.769	d	0.142	e	2.110	c
	8817	1.810	c	0.163	c	2.093	cd
	CT01	1.030	l	0.083	k	1.358	h
	CT02	1.705	f	0.167	b	2.188	bc
	CT03	2.180	a	0.168	b	2.522	a
	CT04	0.996	m	0.090	j	1.555	g
	CT11	1.557	j	0.113	h	2.125	c
	CT11	1.659	h	0.131	g	2.536	a
	CT05	1.200	k	0.099	i	1.747	f
	CT08	1.690	g	0.143	de	2.301	b
	CT06	1.735	e	0.147	d	1.969	b
	CT09	1.584	i	0.136	f	1.941	e
茎 stem	CT07	0.273	a	0.031	b	0.624	h
	巨峰1号(CK)	0.264	c	0.039	b	0.777	f
	8817	0.247	cd	0.032	b	0.694	g
	CT01	0.379	h	0.039	b	0.781	e
	CT02	0.252	bc	0.037	b	1.231	c
	CT03	0.271	a	0.044	b	1.212	d
	CT04	0.250	g	0.028	b	1.373	b
	CT11	0.232	c	0.127	a	1.513	a
	CT11	0.259	a	0.047	b	0.548	i
	CT05	0.241	f	0.040	b	0.357	m
	CT08	0.263	b	0.033	b	0.439	j
	CT06	0.231	de	0.026	b	0.395	l
	CT09	0.270	e	0.043	b	0.408	k

续表

植株部位 Plant location	样品名称 Sample name	全氮 Total nitrogen g/100 g	5%显著水平 5%Significant level	全磷 Total phosphorus g/100g	5%显著水平 Significant level	全钾 Total potassium g/100g	5%显著水平 Significant level
叶 leaf	CT07	0.579	j	0.059	g	0.534	m
	巨峰1号(CK)	0.698	c	0.058	h	0.555	l
	8817	0.677	e	0.054	j	0.663	k
	CT01	0.610	h	0.075	c	0.725	j
	CT02	0.668	f	0.064	e	0.772	i
	CT03	0.701	b	0.054	j	0.961	e
	CT04	0.690	d	0.057	i	1.081	b
	CT11	0.652	g	0.054	j	0.825	g
	CT11	0.692	d	0.077	b	0.811	h
	CT05	0.591	i	0.078	a	1.024	c
	CT08	0.730	a	0.065	d	0.982	d
	CT06	0.693	d	0.060	f	1.101	a
	CT09	0.702	b	0.027	k	0.865	f
果实 fruit	CT07	0.162	c	0.019	g	0.227	c
	巨峰1号(CK)	0.163	c	0.024	e	0.263	b
	8817	0.132	e	0.024	e	0.179	g
	CT01	0.189	a	0.036	b	0.269	a
	CT02	0.114	g	0.018	h	0.172	h
	CT03	0.121	f	0.026	d	0.199	e
	CT04	0.111	g	0.019	g	0.170	h
	CT11	0.149	d	0.021	f	0.232	c
	CT11	0.168	b	0.039	a	0.270	a
	CT05	0.169	b	0.012	j	0.224	d
	CT08	0.164	c	0.013	i	0.174	h
	CT06	0.152	d	0.021	f	0.221	d
	CT09	0.122	f	0.028	c	0.190	f

0.379 g/100 g，"CT03"、"CT07"、"CT11"砧木辣椒及"巨峰1号"（CK）茎全氮含量次之，分别为0.271 g/100 g、0.273 g/100 g、0.259 g/100 g、0.264 g/100 g，CT06砧木辣椒茎全氮含量最小为0.231 g/100 g，其他砧木辣椒茎全氮含量分别0.232~0.252 g/100 g；"CT11"砧木辣椒茎全磷、全钾含量最高，为0.127 g/100 g、1.513 g/100 g 其他砧木及自根苗辣椒茎部全磷含量无差异，分别在0.026~0.047 g/100 g，"CT05"砧木辣椒茎部全钾含量最低，为0.357 g/100 g，其他砧木及自根苗辣椒茎部全钾含量分别在0.395~1.373 g/100 g。不同砧木对辣椒叶部N、P、K分配有差异，"CT08"砧木辣椒叶部全氮含量最高，为0.730 g/100 g，"CT07"砧木辣椒叶部全氮含量最低，为0.579 g/100 g，其他砧木及自根苗辣椒叶部全氮含量分别在0.591~0.702 g/100 g；"CT05"砧木辣椒叶部全磷含量最高，为0.078 g/100 g，"CT01"、"CT03"砧木辣椒叶部全磷含量最低，为0.054 g/100 g，其他砧木及自根苗辣椒全磷含量分别在0.057~0.077 g/100 g，"CT04"、"CT06"砧木辣椒叶部全钾含量最高，分别为1.081 g/100 g、1.101 g/100 g，"CT07"砧木辣椒叶部全钾含量最低，为0.534 g/100 g，其他砧木及自根苗辣椒叶部全钾含量分别在0.555~1.024 g/100 g。不同砧木对辣椒果实N、P、K分配有差异，"CT01"砧木辣椒果实全氮含量最高，为0.189 g/100 g，"CT02"、"CT04"砧木辣椒果实全氮含量最低，为0.114 g/100 g，其他砧木及自根苗辣椒果实全氮含量分别在0.122~0.169 g/100 g；"CT11"砧木辣椒果实全磷含量最高，为0.039 g/100 g，"CT05"砧木辣椒果实全磷含量最低，为0.012 g/100 g，其他砧木及自根苗辣椒果实全磷含量分别在0.013~0.036 g/100 g；"CT01"砧木辣椒果实全钾含量最高，为0.269 g/100 g，"CT02"、"CT04"、"CT08"砧木辣椒果实全钾含量最低，分别为0.172 g/100 g、0.170 g/100 g、0.174 g/100 g，其他砧木及自根苗辣椒果实全钾含量分别在0.179~0.0263 g/100 g。

2.11 不同砧木对嫁接辣椒产量、产值的影响

从表5可以看出，不同砧木辣椒产量、产值有差异，"CT09"砧木辣椒单

株产量、小区产量、折合亩产量、产值均高于其他砧木及自根苗,分别为1.97 kg、299.44 kg、5690.2 kg、11 949.4元,比对照增幅16.6%,产值增加1698.2元,"CT03"砧木辣椒单株产量、小区产量、折合亩产量、产值次之,分别为1.82 kg、276.64 kg、5256.9 kg、11 039.5元,比对照增幅7.7%,产值增加788.3元,"8817"砧木辣椒单株产量、小区产量、折合亩产量、产值最低,分别为1.29 kg、196.08 kg、3726.1 kg、7 824.8元,比对照增幅-23.7%,产值增加-2 426.4元,其他砧木及自根苗辣椒单株产量、小区产量、折合亩产量、产值分别在1.05~1.76 kg、228.0~267.52 kg、4 332.6~5 083.6 kg、9 098.5~10 675.6元。

表5 不同砧木对嫁接辣椒产量、产值的影响

Table 5 Effects of Different Rootstocks on Yield and Value of Grafted Pepper

处理 Treatment	单株产量 Grain weight per plant(kg)	小区产量 Plot yield (kg/35.1 m^2)	折合总产量 Equivalent total output (kg/667 m^2)	折合总产值 Equivalent gross output (元/667 m^2)	比对照 ±(kg)	比对照 ±(%)	位次 Rank
CT10	1.76	267.52	5 083.6	10 675.6	202.1	4.1	3
CT09	1.97	299.44	5 690.2	11 949.4	808.7	16.6	1
CT08	1.68	255.36	4 852.6	10 190.5	-28.9	-0.6	6
CT05	1.58	240.16	4 563.7	9 583.8	-317.8	-6.5	8
CT06	1.74	264.48	5 025.9	10 554.4	144.4	3	4
CT07	1.66	252.32	4 797.8	10 075.4	-83.7	-1.7	7
CT01	1.50	228.00	4 332.6	9 098.5	-548.9	-11.2	12
CT04	1.54	234.08	4 448.2	9 341.2	-433.3	-8.9	10
8817	1.29	196.08	3 726.1	7 824.8	-1155.4	-23.7	13
CT02	1.55	235.60	4 477.1	9 401.9	-393.4	-8.1	9
CT11	1.51	229.52	4 361.5	9 159.2	-520	-10.7	11
CT03	1.82	276.64	5 256.9	11 039.5	375.4	7.7	2
巨峰1号(CK)	1.69	256.88	4 881.5	10 251.2	—	—	5

3 讨论与结论

3.1 砧木与辣椒接穗的亲和性比较

良好的砧木首先应具备与接穗有较高的嫁接亲和力和共生亲和力。嫁接成活率是判定砧木与接穗亲和性的一个重要指标[7],不同砧木对嫁接成活率影响甚为明显[8],本试验选用的 12 份砧木与"巨峰 1 号"接穗嫁接,嫁接成活率有所差异,"CT01"、"CT09"、"CT10"、"CT04"砧木材料与接穗亲和性最好,分别为 98.26%、98.26%、98.61%、98.61%,其他砧木成活率均在 84%以上,具有良好的嫁接亲和性。

3.2 不同砧木对嫁接辣椒生长发育的影响

嫁接植株的生育状况决定于接穗,对养分吸收能力取决于砧木。与接穗相对较弱的根系相比,砧木具有发达的根系,具有更强的吸收能力。通过嫁接可以提高其对矿质元素的吸收[9-11]。本试验结果表明,嫁接增加了辣椒对矿质营养元素的吸收。"CT03"、"CT07"砧木辣椒根部全氮含量最大,分别为 2.180 g/100 g、2.002 g/100 g,"8817"、"CT02"、"CT03"、"CT07"砧木根全磷含量较高,"CT07"、"CT03"、"CT11"根全钾含量最高,分别为 2.606 g/100 g、2.522 g/100 g、2.536 g/100 g,CT01"砧木辣椒茎全氮含量最高,为 0.379 g/100 g,"CT11"砧木辣椒茎全磷含量最高,为 0.127 g/100 g,"CT08"砧木辣椒叶部全氮含量最高,为 0.730 g/100 g,"CT05"砧木辣椒叶部全磷含量最高,为 0.078 g/100 g,"CT04"、"CT06"砧木辣椒叶部全钾含量最高,分别为 1.081 g/100 g、1.101 g/100 g,"CT01"砧木辣椒果实全氮含量最高,为 0.189 g/100 g,"CT11"砧木辣椒果实全磷含量最高,为 0.039 g/100 g,"CT01"砧木辣椒果实全钾含量最高,为 0.269 g/100 g,因此嫁接有利于辣椒的生长发育,提高产量。

3.3 不同砧木对嫁接辣椒产量和品质的影响

嫁接辣椒比自根辣椒具有更强的吸水吸肥能力,生长旺盛,从而有利于产量的提高[12-13]。本试验中"CT09"砧木辣椒单株产量、小区产量、折合亩产量、产值均高于其他砧木及自根苗，分别为 1.97 kg、299.44 kg、5 690.2 kg、11 949.4 元，比对照增幅 16.6%，产值增加 1 698.2 元，"CT03"砧木辣椒单株产量、小区产量、折合亩产量、产值次之，分别为 1.82 kg、276.64 kg、5 256.9 kg、11 039.5 元，比对照增幅 7.7%，产值增加 788.3 元。前人已经证明，嫁接可能对果实品质产生一定影响。多数研究认为，嫁接对蔬菜果实品质的影响通常是负面的，但这种影响程度取决于砧木类型与品种，通过筛选合适的砧木，可以将这种影响降低到最低限度[14,15]。本试验中"CT09"、"CT10"砧木嫁接辣椒果实维生素 C 含量最高，分别为 65.14 mg/100 g、63.26 mg/100 g，"CT02"维生素 C 含量次之，"CT09"、"CT10"、"CT02"砧木品种嫁接辣椒，果实可溶性糖含量最高，分别为 34.638 mg/g、32.588 mg/g、32.98 mg/g，"CT08"砧木果实可溶性蛋白含量最高，为 2.22 mg/g，"巨峰 1 号"（CK）果实可溶性蛋白含量最低，为 1.47 mg/g。

综上所述，以"CT09"和"CT03"砧木嫁接辣椒可明显提高产量和品质，因此具有较高的应用和推广价值。

参考文献

[1] 喻景权,杜尧舜.蔬菜设施栽培可持续发展中的连作障碍问题[J].沈阳农业大学学报,2000,31(1):124-126.

[2] 陈晓红,邹志荣.温室蔬菜栽培连作障碍研究现状及防治措施[J].陕西农业科学,2002,(12):16-20.

[3] 吴凤芝,赵凤艳,刘元英.设施蔬菜连作障碍原因综合分析与防治措施[J].东北农业大学学报,2000,31(3):241-247.

[4] 吕卫光,张春兰,袁飞.嫁接减轻设施黄瓜连作障碍机制初探[J].华北农学报,2000,15(增刊):153-156.

[5] 张继梅,葛志东.嫁接技术在薄皮甜瓜上的应用效果.中国西瓜甜瓜,2002,(1):26-27.

[6] 高梅秀,李树和.不同砧木对茄子抗病性、生理活性及产量的影响[J].园艺学报,

2001,28(5):463-465.

[7] 张斌祥.辣椒嫁接砧木对产量和抗病性的影响[J].甘肃农业科技,2009,(6):44-45.

[8] 郁继华,秦舒浩.黄瓜品种间嫁接苗和自根苗光合特性研究[J].兰州大学学报.2001,(6):63-68.

[9] Santos,H.S.,& Goto,R.（2004）.Sweet pepper grafting tocontrol phytophthora blight under protected cultivation.Horticul-tura Brasileira,22(1):45-49.

[10] 满为群,杜维广,张桂茹,等.高光效大豆几项光合生理指标的研究[J].作物学报,2003,29(5):697-700.

[11] 凌云昕,王凤春.茄子辣椒新技术[M].北京:中国农业出版社,2000.

[12] 樊东隆.靖远县日光温室辣椒嫁接栽培技术[J].甘肃农业科技,2006(10):41.

[13] 朱贤聪.辣椒嫁接栽培效果初探[J].上海农业科技,2007(4):88-90.

[14] 张红梅,金海军,余纪柱,解静.不同南瓜砧木对嫁接黄瓜生长和果实品质的影响.内蒙古农业大学学报,2007,28(3):177-181.

[15] 周俊国,扈惠灵.NaCl胁迫对不同砧木的嫁接黄瓜产量和品质的影响.核农学报,2010,24(4):851-855.

不同穗砧组合对辣椒生理特性及对产量和品质的影响

高晶霞[1],吴雪梅[2],高　昱[2],牛勇琴[2],裴红霞[1],谢　华[1]

(1. 宁夏农林科学院种质资源研究所,宁夏 银川 750002;
2. 宁夏回族自治区彭阳县蔬菜产业发展服务中心,宁夏 彭阳 756500)

摘　要:以"神根"和"神威4F1"为砧木,6个牛角椒品种为接穗,共12个穗砧组合,研究各砧穗组合的辣椒生长势、光合特性、果实品质、产量以及生理生化特性。结果表明:2个砧木,6种接穗共12个组合,嫁接辣椒成活率均在94%~100%,处理1中,神根/卓越组合辣椒开展度最宽,叶片荧光参数F0最大,Fm最大,且辣椒单株产量、小区产量、折合总产量、产值均最高,神根/金惠13-E辣椒叶片净光合速率最大、叶片丙二醛含量最高、辣椒果实维生素C含量、可溶性蛋白含量最高,神根/超越土壤蔗糖酶活性最强,神根/犇腾2号辣椒果实干物质含量最高。处理2中,神威4F1/金惠13-E株高、茎粗均最高,神威4F1/金惠TN-1叶片荧光参数Fm、Fv最大,叶片可溶性蛋白含量最高为0.31%,神威4F1/犇腾2号叶片过氧化物酶、可溶性糖含量最高,神威4F1/超越过氧化氢酶含量最高,叶片丙二醛含量最高,土壤脲酶活性最强、果实维生素C含量最高,神威4F1/超越土壤脲酶活性最强,辣椒单株产量、小区产量、折合总产量、产值均最高,分别为2.48 kg/株、327.4 kg/35.1 m²、6 221.5 kg/667 m²、6 843.7 元/667 m²。综合分析结果:神根/卓越、神威4F1/卓越,2个组合表现好,为拱棚辣椒嫁接高产、优质栽培提供科学依据。

关键词:穗砧组合;辣椒;生理特性;光合特性;产量;品质

Effects of Different Rootstock-scionCombinations on Photosynthetic Characteristics and Yield and Quality of Capsicum

Gao jingxia[1], Wu xuemei[2], Gao yu[2], Niu yongqin[2], Xie hua[2]

Abstract: The growth potential, photosynth etic characteristics, fruit quality, yield and physiological and biochemical characteristics of each anvil were studied with "shenwei 4F1" and "shenwei 4F1" as rootstocks. The results show that: There were 12 combinations of 6 scions and 2 rootstocks. The survival rate of grafted pepper was between 94% and 100%, In treatment 1, the combination of root and zhuyue had the widest degree of development, the maximum fluorescence parameters of leaves were F0 and Fm, and the highest yield per plant, plot yield, total output value and output value, the highest net photosynthetic rate in the leaves of shengan/jinhui 13−e pepper was the highest malondialdehyde content in the leaves, the highest vitamin C content in the fruits, the highest soluble protein content, the highest suprase activity in shengan/beyond the soil, and the highest dry matter content in the fruits of shengan/beneng 2 pepper. In treatment 2, the plant height and stem thickness of shenwei 4F1 / jinhui 13−e were the highest, the fluorescence parameters Fm and the maximum soluble protein content of shenwei 4F1 / jinhui tn−1 leaves were the highest, and the maximum soluble protein content of shenwei 4F1 / jinhui tn−1 leaves was 0.31%, The highest content of catalase, the highest content of malondialdehyde in the leaves, the highest activity of urease in the soil, the highest content of vitamin C in the fruit, the highest activity of urease in the soil, the highest yield of pepper per plant, plot yield, equivalent total output value and output value were 2.48 kg/plant, 327.4 kg / 35.1 m², 6 221.5 kg / 667 m² and 6 843.7 yuan/667 m², respectively. Comprehensive analysis results: shengen/zhuoyue and shenwei 4F1 / zhuoyue performed well, which provided scientific basis for the cultivation of gongpeng pepper grafting with high yield and high quality.

Key words: anvil ear combination; pepper; physiological characteristics; photosynthetic characteristics; yield; quality

辣椒(*Capsicum annuum L.*)又名番椒,原产于拉丁美洲热带地区,茄科辣椒属,为一年生草本植物,在我国普遍栽培[1]。辣椒营养丰富,维生素 C 含量在蔬菜中居第一位[2]。近年来,拱棚辣椒是宁夏设施农业的重要组成部分,目前,在宁夏南部山区,以彭阳县为代表的拱棚辣椒栽培,近 20 万亩,其中彭阳县近 10 万亩。每年 4 月上旬前后定植,6 月中旬前后开始上市,到 9 月底前后拉秧,是宁夏辣椒主要栽培模式,产品销往北京、西安、兰州、青海、呼和浩特、银川及武汉、重庆、郑州等区内外市场。栽培技术逐步成熟,产量质量较高,市场链完善,平均亩效益 5 000 元以上,高的达 12 000 元以上,平均纯收入 2 500 元以上,已成为南部山区的一个知名品牌和当地农民增收致富的重要产业[2]。但是,由于当地气候和地理条件的限制,种植品种单一,很难轮作倒茬。自 2015 年以来,连作障碍所导致的危害愈加明显,最突出的问题是土传病害辣椒疫病、枯萎病在整个拱棚辣椒种植区发病率较高,2018 年辣椒病毒病又相继发生,造成减产减收。目前,因轮作困难、药剂防治效果差且破坏生态环境,而选育抗病品种存在抗病种质资源匮乏、周期长等方面的局限,嫁接换根技术已成为目前解决上述问题的有效途径之一[3]。但是前人大部分是筛选不同的优良辣椒砧木提高辣椒的产量,很少通过研究穗砧组合的搭配不同,砧木与接穗彼此之间作用,会使砧木和接穗的一些性状发生改变,进而影响到辣椒的生长、结果、产量、品质以及生理生化特性等各个方面,最终影响经济效益[4]。本研究以 2018 年筛选出的 2 个抗性强的"神根"和"神威 4F1"为砧木,接穗品种分别为"超越"、"卓越"、"金惠 TN-1"、"金惠 13-E"、"泰达"、"犇腾 2 号"6 个牛角椒品种,比较各砧穗组合的辣椒生长势、光合特性、果实品质、产量以及生理生化特性,明确不同穗砧组合间存在的差异,比较穗砧组合之间的优劣,为选择适宜的穗砧组合提供理论依据,以期为拱棚辣椒嫁接高产、优质栽培提供理论依据。

1 材料和方法

1.1 试验地点

试验地点位于宁夏彭阳县新集乡拱棚辣椒示范基地（连作辣椒 3 年以上）。

1.2 供试材料

砧木品种："神根"和"神威 4F1",接穗品种："超越"、"卓越"、"金惠 TN-1"、"金惠 13-E"、"泰达"、"犇腾 2 号"均为宁夏嘉禾源种苗有限公司提供。

1.3 试验设计

试验苗期在宁夏嘉禾源种苗有限公司进行。采用劈接法进行嫁接。试验设 2 个处理,共 12 个组合,即处理 1(A、B、C、D、E、F),处理 2(A、B、C、D、E、F)。每个处理嫁接 200 株,嫁接苗 3 叶 1 心时定植到彭阳县新集乡拱棚辣椒示范基地,单垄双行种植,株距 35 cm,垄宽 70 cm,垄高 25 cm。每小区 46 株,3 次重复,其他管理同常规生产。

1.4 调查项目

1.4.1 成活率

嫁接 10 d 后,调查辣椒嫁接苗的成活株数,计算成活率。

1.4.2 生长指标

辣椒生长盛期调查株高、茎粗、开展度（每个小区测 5 株取平均值）。

1.4.3 叶片酶活性

过氧化氢酶采用紫外吸收法,可溶性糖采用蒽酮比色法,可溶性蛋白采用考马斯亮蓝 G-250 染色法,过氧化物酶采用愈创木酚法,丙二醛采用硫代巴比妥酸法。

1.4.4 病害调查

整个生育期观察记载辣椒病毒病发病情况,计算发病率。

1.4.5 产量

采果盛期,采收坐果天数相同的果实,测定辣椒单果质量,统计小区产量,计算 667 m² 产量。

1.4.6 品质

采果盛期,采收大小均匀一致的果实,送宁夏农林科学院种质资源研究所实验室测定品质。辣椒果实维生素 C 含量采用铝蓝比色法测定,辣椒果实可溶性糖含量采用苯酚硫酸比色法测定,辣椒果实可溶性蛋白质含量采用紫外分光光度法测定,辣椒果实可溶性固形物含量采用 PRO-101 型糖度计测定。

1.4.7 光合和荧光测定

用美国产 Li-6400 型光合测定系统测定光合,每小区选有代表性的 5 片叶上午 10:00-11:00 进行测定,每叶重复测定 3 次,取平均值。荧光用 Handy PEA 便携式植物荧光仪。

1.4.8 土壤酶活性测定

过氧化氢酶采用高锰酸钾滴定法,脲酶采用苯酚钠-次氯酸钠比色法,土壤蔗糖酶、纤维素酶采用 3,5 二硝基水杨酸比色法。

1.5 相关分析和系统聚类分析

采用唐启义[5]等的 DPS 数据处理系统。显著性差异分析采用 SPSS 软件进行分析。

2 结果分析

2.1 不同穗砧组合嫁接成活率调查

与接穗的亲和度是衡量一个砧木是否能够使用的首要指标,亲和度越高,嫁接成活率越高[6,7]。从表 1 可以看出,不同穗砧组合嫁接辣椒成活率无差异,处理 1 与 6 个不同接穗组合,除与 B 接穗组合,嫁接成活率为 94%以外,其他嫁接成活率均为 100%;处理 2 与 6 个不同接穗组合,其嫁接成活率均为 100%。说明 2 个砧木分别与 6 个接穗组合,嫁接成活率均较好。

表 1 不同穗砧组合嫁接辣椒成活率比较
Table 1 Survival rate comparison of grafting pepper with different combination of panicle and stock

处理 treatment	穗砧组合 rootstock-scion combination	接穗名称 the name of the scion	数量 quantity /株	成活率 rate of survival /%
处理 1/神根	A	超越	200	100
	B	卓越	200	94
	C	金惠 TN-1	200	100
	D	金惠 13-E	200	100
	E	泰达	200	100
	F	犇腾 2 号	200	100
处理 2/神威 4F1	A	超越	200	100
	B	卓越	200	100
	C	金惠 TN-1	200	100
	D	金惠 13-E	200	100
	E	泰达	200	100
	F	犇腾 2 号	200	100

2.2 不同穗砧组合对辣椒生长的影响

从表 2 可以看出,不同穗砧组合嫁接辣椒,辣椒株高有差异,处理 1 中,C 组合辣椒株高最高,为 95.6 cm,E 组合辣椒株高最低,为 72.8 cm,其他组合辣椒株高分别在 78.0~84.8 cm,处理 2 中,D 组合辣椒株高最高,为 92.2 cm,E 组合辣椒株高最低,为 71.4 cm,其他穗砧组合辣椒株高分别在 78.4~86.6 cm;不同穗砧组合嫁接辣椒,辣椒茎粗差异较小,处理 1 中,F 组合辣椒茎粗最粗,为 14.12 mm,其他砧穗组合辣椒茎粗分别在 12.69~13.49 mm,处理 2 中,D 组合辣椒茎粗最粗,为 13.98 mm,其他穗砧组合辣椒茎粗分别在 11.63~

12.64 mm；不同穗砧组合嫁接辣椒，辣椒开展度有差异，处理1中，B组合嫁接辣椒开展度最宽，为72.8 cm，其他组合辣椒的开展度分别在61.6~69.0 cm，处理2中，F组合辣椒开展度最大，为74.0 cm，E组合辣椒开展度最小，为58.8 cm，其他组合辣椒开展度分别在68.4~72.4 cm。

表2 不同穗砧组合辣椒生长势比较
Table 2 Comparison of growth potential of pepper with different combination of panicle and anvil

处理 treatment	株高 plant height/cm	茎粗 stem diameter/mm	开展度 implement activities/cm
处理1			
A	84.0	12.81	61.6
B	87.0	12.83	72.8
C	95.6	13.31	64.6
D	84.8	12.69	62.4
E	72.8	13.49	67.0
F	78.0	14.12	69.0
处理2			
A	78.4	11.66	68.4
B	86.6	12.44	72.4
C	79.8	12.18	71.6
D	92.2	13.98	65.8
E	71.4	11.63	58.8
F	86.0	12.53	74.0

2.3 不同穗砧组合对辣椒病毒病发病率的影响

从图1可以看出，不同穗砧组合辣椒病毒病发病率有差异，处理1中，A组合、C组合辣椒病毒病发病率最高为2%，B组合、D组合、F组合辣椒病毒病发病率次之，为0.67%，E组合辣椒病毒病发病率最低为0.0%；处理2中，A组合、B组合辣椒病毒病发病率最高为0.67%，其他组合辣椒病毒病发病率最低为0.0%。

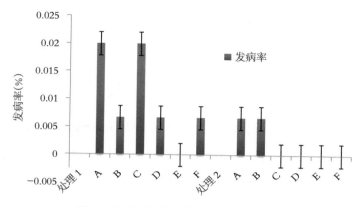

图 1 不同砧穗组合辣椒病毒病发病率比较

Fig. 1 Comparison of incidence of pepper plants with different anvil combinations

2.4 不同穗砧组合对辣椒叶片光合特性的影响

从表3可以看出,不同穗砧组合辣椒叶片蒸腾速率有差异,处理1中,D组合辣椒叶片蒸腾速率最高,为4.56 mmol·m^{-2}·s^{-1},A组合、C组合辣椒叶片蒸腾速率最低,分别为2.57 mmol·m^{-2}·s^{-1}、2.69 mmol·m^{-2}·s^{-1},其他组合辣椒叶片蒸腾速率分别在2.99~3.86 mmol·m^{-2}·s^{-1},处理2中,D组合、F组合辣椒叶片蒸腾速率最高,分别为6.51 mmol·m^{-2}·s^{-1}、5.68 mmol·m^{-2}·s^{-1},E组合辣椒叶片蒸腾速率最低,为3.43 mmol·m^{-2}·s^{-1},其他组合辣椒叶片蒸腾速率分别在4.16~5.08 mmol·m^{-2}·s^{-1};不同穗砧组合辣椒叶片气孔导度有差异,处理1中,E组合辣椒叶片气孔导度最大,为1 529.31 mmol·m^{-2}·s^{-1},D组合、F组合辣椒叶片气孔导度次之,A组合、B组合、C组合辣椒叶片气孔导度最低,分别为152.88 mmol·m^{-2}·s^{-1}、200.53 mmol·m^{-2}·s^{-1}、129.07 mmol·m^{-2}·s^{-1},处理2中,D组合、辣椒叶片气孔导度最大,为1 348.6,B组合、F组合辣椒叶片气孔导度次之,A组合、C组合、E组合辣椒叶片气孔导度最低,分别为592.38 mmol·m^{-2}·s^{-1}、596.53 mmol·m^{-2}·s^{-1}、610.31 mmol·m^{-2}·s^{-1};不同穗砧组合辣椒叶片净光合速率有差异,处理1中,D组合、F组合辣椒叶片净光合速率最大,分别为18.01 μmmol·m^{-2}·s^{-1}、19.07 μmmol·m^{-2}·s^{-1},A组合、E组合辣椒叶片净光合速率次之,B组合、C组合辣椒叶片净光合速率最低,分别为

12.30 $\mu mmol \cdot m^{-2} \cdot s^{-1}$、12.32 $\mu mmol \cdot m^{-2} \cdot s^{-1}$,处理 2 中,各个穗砧组合净光合速率无差异,分别在 20.75~22.11 $\mu mmol \cdot m^{-2} \cdot s^{-1}$;不同穗砧组合辣椒叶片 CO_2 胞间浓度有差异,处理 1 中,D 组合辣椒叶片 CO_2 胞间浓度最高,为 336.27 ppm,B 组合、F 组合辣椒叶片 CO_2 胞间浓度次之,A 组合、E 组合辣椒叶片 CO_2 胞间浓度最低,分别为 234.1 ppm、253.88 ppm,处理 2 中,B 组合、D 组合、E 组合、F 组合辣椒叶片 CO_2 胞间浓度最高,分别为 3 126 ppm、346.27 ppm、327.44 ppm、337.20 ppm,A 组合、C 组合辣椒叶片 CO_2 胞间浓度最低,分别为 299.81 ppm、279.80 ppm。

表3 不同穗砧组合辣椒叶片光合特性比较
Table 3 Comparison of photosynthetic characteristics of pepper leaves with different anvil combinations

处理 treatment		蒸腾速率 transpiration rate ($mmol \cdot m^{-2} \cdot s^{-1}$)	气孔导度 stomatal conductance ($mmol \cdot m^{-2} \cdot s^{-1}$)	净光合速率 net photosynthetic rate ($\mu mmol \cdot m^{-2} \cdot s^{-1}$)	CO_2 胞间浓度 intercellular CO_2 concentration /ppm
处理 1	A	2.57	152.88	16.29	234.13
	B	3.07	200.53	12.30	304.87
	C	2.69	129.07	12.32	239.27
	D	4.56	1 040.20	18.01	336.27
	E	2.99	1 529.31	16.75	253.88
	F	3.86	990.87	19.07	329.47
处理 2	A	4.16	592.38	20.87	299.81
	B	5.08	990.40	20.75	312.60
	C	4.85	596.53	21.63	279.80
	D	6.51	1 348.60	22.11	346.27
	E	3.43	610.31	21.19	327.44
	F	5.68	1 069.33	21.45	337.20

2.5 不同穗砧组合对辣椒叶片荧光的影响

从表 4 可以看出,不同穗砧组合辣椒叶片荧光有差异,处理 1 中,B 组合荧光参数 F0 最大为 569.11，A 组合荧光参数 F0 最小，为 514.89，其他组合荧光参数 F0 分别在 526.88~542.56，处理 2 中，各个组合叶片荧光参数无差异，分别在 531.13~554.56；处理 1 中，B 组合叶片荧光参数 Fm 最大为 2 735.78，E 组合叶片荧光参数 Fm 最小为 2 140.78，其他组合叶片荧光参数 Fm，分别为 2 404.5~2 645.25，处理 2 中，C 组合叶片荧光参数 Fm 最大为 2 401.75，F 组合叶片荧光参数 Fm 最小为 2 079.75，其他组合叶片荧光参数 Fm 分别在 2 201.11~2 398.67；处理 1 中，A 组合、B 组合、C 组合叶片荧光参数 Fv 最大，分别为 2 079.44、2 166.67、2107.25，D 组合、F 组合叶片荧光参数 Fv 次之，E 组合辣椒荧光参数 Fv 最小为 1598.22，处理 2 中，C 组合、E 组合叶片荧光参数 Fv 最大为 1 865.50、1 844.11，A 组合、B 组合、D 组合叶片荧光参数 Fv 次之，F 组合叶片荧光参数 Fv 最小为 1 548.63；处理 1 中，E 组合叶片荧光参数 F0/Fm 最大，为 0.26，其他组合叶片荧光参数 F0/Fm 分别在 0.20~0.22，处理

表 4 不同穗砧组合辣椒叶片荧光参数比较

Table 4 Comparison of fluorescence parameters of pepper leaves with different spike combinations

处理 treatment		F0	Fm	Fv	F0/Fm	Fv/Fm	Fv/F0
处理 1	A	514.89	2 594.33	2 079.44	0.20	0.80	4.04
	B	569.11	2 735.78	2 166.67	0.21	0.79	3.82
	C	538.00	2 645.25	2 107.25	0.20	0.80	3.91
	D	539.56	2 504.22	1 964.67	0.22	0.78	3.67
	E	542.56	2 140.78	1 598.22	0.26	0.74	2.97
	F	526.88	2 404.50	1 877.63	0.22	0.78	3.56
处理 2	A	533.44	2 323.89	1 790.44	0.23	0.77	3.37
	B	547.22	2 201.11	1 653.89	0.25	0.75	3.03
	C	536.25	2 401.75	1 865.50	0.23	0.77	3.50
	D	528.44	2 267.78	1 739.33	0.23	0.77	3.29
	E	554.56	2 398.67	1 844.11	0.24	0.76	3.34
	F	531.13	2 079.75	1 548.63	0.26	0.74	2.91

2中,各个组合叶片荧光参数F0/Fm差异较小,分别在0.23~0.26;不同穗砧组合辣椒荧光参数Fv/Fm无差异,处理1中,各个组合叶片荧光参数分别在0.74~0.80,处理2中,各个组合叶片荧光参数分别在0.74~0.77;不同穗砧组合叶片荧光参数Fv/F0有差异,处理1中,A组合叶片荧光参数Fv/F0最大为4.04,E组合叶片荧光参数Fv/F0最小为2.97,其他组合叶片荧光参数Fv/F0分别在3.56~3.91,处理2中,F组合叶片荧光参数Fv/F0最小为2.91,其他组合叶片荧光参数Fv/F0分别在3.03~3.50。

2.6 不同穗砧组合对辣椒叶片保护酶活性的影响

从图2~6可知,不同穗砧组合辣椒叶片过氧化物酶活性差异显著,处理1中,E组合叶片过氧化物酶活性最高,为1 999.15 μg/g·min,D组合、F组合次之,A组合、B组合、C组合叶片过氧化物酶活性最低,分别为379.1 μg/g·min、252.1 μg/g·min、152.2 μg/g·min,处理2中,F组合叶片过氧化物酶活性最高为1 807.33 μg/g·min,A组合、C组合辣椒叶片过氧化物酶活性最低,分别为168.7 μg/g·min、57.7 μg/g·min,B组合、D组合、E组合叶片过氧化物酶活性次之;不同穗砧组合叶片可溶性糖含量差异明显,处理1中,F组合叶片可溶性糖含量最高为1.65%,D组合叶片可溶性糖含量次之,A组合、B组合、C组合E组合辣椒叶片可溶性糖含量分别在0.7%~0.92%,处理2中,F

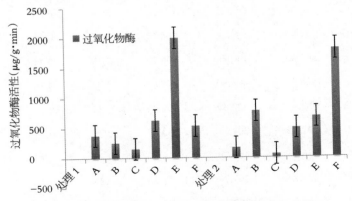

图2 不同砧穗组合辣椒叶片过氧化物酶活性比较

Fig. 2 Comparison of catalase activity in pepper with different spike combinations

组合叶片可溶性糖含量最高为1.7%,B、E组合叶片可溶性糖含量次之,其他组合叶片可溶性糖含量分别在0.91%~0.94%;不同穗砧组合叶片可溶性蛋白含量有差异,处理1中,C组合叶片可溶性蛋白含量最高为0.35%,E组合叶片可溶性蛋白含量最低为0.07%,其他组合叶片可溶性蛋白含量分别在0.13%~0.25%,处理2中,C组合叶片可溶性蛋白含量最高为0.31%,D组合、E组合叶片可溶性蛋白含量最低为0.1%、0.09%,其他组合叶片可溶性蛋白

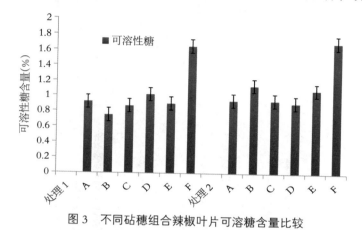

图3　不同砧穗组合辣椒叶片可溶糖含量比较

Fig. 3 Comparison of soluble sugar content in leaves of pepper with different anvil combinations

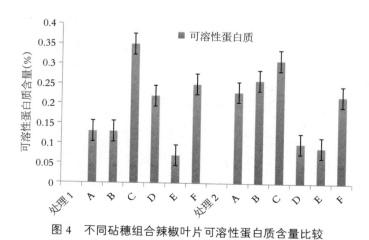

图4　不同砧穗组合辣椒叶片可溶性蛋白质含量比较

Fig. 4 Comparison of soluble protein content in leaves of pepper with different anvil combinations

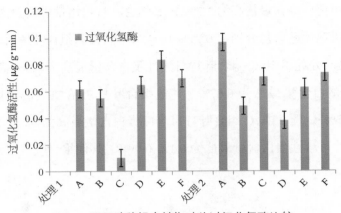

图 5 不同砧穗组合辣椒叶片过氧化氢酶比较

Fig. 5 Comparison of catalase activity in pepper leaves with different anvil combinations

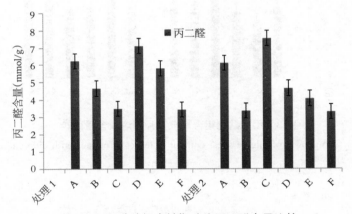

图 6 不同砧穗组合辣椒叶片丙二醛含量比较

Fig. 6 Comparison of malondialdehyde content in pepper leaves with different anvil combinations

含量分别在 0.22%~0.26%；不同砧穗组合叶片过氧化氢酶活性有差异，处理 1 中，E 组合叶片过氧化氢酶活性最高为 0.084 μg/g·min，C 组合叶片过氧化氢酶含量最低，为 0.01 μg/g·min，其他组合叶片过氧化氢酶活性分别在 0.055~0.07 μg/g·min，处理 2 中，A 组合叶片过氧化氢酶活性最高，为 0.097 μg/g·min，D 组合叶片过氧化氢酶活性最低，为 0.038 μg/g·min，其他组合叶片过氧化氢酶活性分别在 0.049~0.074 μg/g·min；不同穗砧组合叶片丙二醛含量差异显著，D 组合叶片丙二醛含量最高为 7.11 mmol/g，A 组合、

E组合叶片丙二醛含量次之，其他组合叶片丙二醛含量分别在3.43~4.65 mmol/g，处理2中，A组合、C组合叶片丙二醛含量最高，分别为6.09 mmol/g、7.54 mmol/g，其他组合叶片丙二醛含量分别在3.3~4.64 mmol/g。

2.7 不同穗砧组合对连作辣椒根系活力的影响

从图7可以看出，不同穗砧组合辣椒根系活力差异显著，处理1中，B组合辣椒根系活力最强为4.76 μg/g·h，D组合辣椒根系活力次之，为4.25 μg/g·h，A、C、E组合辣椒根系活力差异不显著，F组合根系活力最弱为2.27 μg/g·h；处理2中，A、B、D组合辣椒根系活力表现较强，分别为3.70 μg/g·h、3.97 μg/g·h、3.79 μg/g·h，C组合根系活力次之，为3.20 μg/g·h，E、F组合根系活力较弱，分别为2.32 μg/g·h和1.77 μg/g·h。

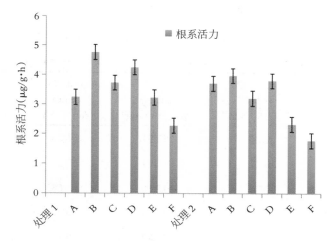

图7 不同砧穗组合辣椒根系活力比较

Fig. 7 Comparison of root activity of pepper with different anvil combinations

2.8 不同穗砧组合对连作辣椒土壤酶活性的影响

从表5可以看出，不同穗砧组合土壤过氧化氢酶活性差异不显著，处理1中，A~F组合土壤过氧化氢酶活性分别在0.26~0.27 ml·g^{-1}·30 min^{-1}，处理2中，A~F组合土壤过氧化氢酶活性分别在0.26~0.28 ml·g^{-1}·30 min^{-1}；不同穗砧组合土壤蔗糖酶活性差异显著，处理1中，A组合土壤蔗糖酶活性最强为21.84 ml·g⁻¹·30 min⁻¹,E组合、F组合土壤蔗糖酶活性最弱为12.22 ml·g⁻¹·30 min⁻¹、12.05 ml·g⁻¹·30 min⁻¹,其他组合土壤蔗糖酶活性分别在17.16~19.94 ml·g⁻¹·30 min⁻¹,处理2中,E组合、F组合土壤蔗糖酶活性最强,分别为20.58 ml·g⁻¹·30 min⁻¹、17.66 ml·g⁻¹·30 min⁻¹,C组合土壤蔗糖酶活性最弱为6.09 ml·g⁻¹·30 min⁻¹,其他组合土壤蔗糖酶活性分别在9.13~13.74 ml·g⁻¹·30 min⁻¹;不同穗砧组合土壤纤维素酶活性有差异,处理1中,E组合土壤纤维素酶活性最强为9.33 mg·g⁻¹·30 min⁻¹,A组合土壤酶活性最低为1.53 mg·g⁻¹·30 min⁻¹,其他组合土壤纤维素酶活性分别在2.87~3.68 mg·g⁻¹·h⁻¹,处理2中,C组合土壤纤维素酶活性最强为2.01 mg·g⁻¹·h⁻¹,E组合、F组合土壤纤维素酶活性最弱为1.19 mg·g⁻¹·h⁻¹、1.06 mg·g⁻¹·h⁻¹;不同穗砧组合土壤脲酶活性有差异,处理1中,B组合土壤脲酶活性最强为63.72 mg·g⁻¹·h⁻¹,A组合土壤酶活性最弱为

表5 不同穗砧组合连作辣椒土壤酶活性比较

Table 5 Comparison of soil enzyme activity of pepper with different anvil combinations

处理 treatment		过氧化氢酶 catalase ($ml·g^{-1}·30\ min^{-1}$)	蔗糖酶 sucrase ($ml·g^{-1}·30\ min^{-1}$)	纤维素酶 cellulase ($mg·g^{-1}·h^{-1}$)	脲酶 urease ($mg·g^{-1}·h^{-1}$)
处理1	A	0.27	21.84	1.53	14.32
	B	0.27	19.94	3.24	63.72
	C	0.26	17.16	3.68	22.97
	D	0.27	19.93	3.21	36.12
	E	0.27	12.22	9.33	46.25
	F	0.27	12.05	2.87	49.37
处理2	A	0.26	10.50	1.62	66.89
	B	0.27	9.13	1.33	75.71
	C	0.27	6.09	2.01	27.40
	D	0.28	13.74	1.62	48.41
	E	0.27	20.58	1.19	35.19
	F	0.26	17.66	1.06	27.97

14.32 mg·g^{-1}·h^{-1},其他组合土壤脲酶活性分别在 22.97~49.37 mg·g^{-1}·h^{-1},处理2中,A 组合、B 组合土壤脲酶活性最强,分别为 66.89 mg·g^{-1}·h^{-1}、75.71 mg·g^{-1}·h^{-1},其他组合土壤脲酶活性分别在 27.40~48.41 mg·g^{-1}·h^{-1}。

2.9 不同穗砧组合对连作辣椒产量的影响

从表 6 可以看出,不同穗砧组合嫁接辣椒产量有差异,处理 1 中,B 组合辣椒单株产量、小区产量、折合总产量、产值均最高,分别为 2.57 kg/株、339.2 kg/35.1 m^2、6 445.8 kg/667 m^2、7 090.4 元/667 m^2,D 组合辣椒单株产量、小区产量、折合总产量、产值次之,F 组合辣椒单株产量、小区产量、折合总产量、产值均最低,分别为 2.01 kg/株、265.3 kg/35.1 m^2、5041.5 kg/667 m^2、5 545.7 元/667 m^2,处理 2 中,B 组合辣椒单株产量、小区产量、折合总产量、产值均最高,分别为 2.48 kg/株、327.4 kg/35.1 m^2、6 221.5 kg/667 m^2、

表6 不同穗砧组合嫁接辣椒产量对比
Table 6 Yield comparison of grafting pepper with different combination of panicle and stock

处理 treatmnet		单株产量 per plant yield (kg)	小区产量 cell production (kg/35.1 m^2)	折合总产量 equivalent output (kg/667 m^2)	产值 the output value (元/667 m^2)
处理 1	A	2.24	295.7	5 619.1	6 181.0
	B	2.57	339.2	6 445.8	7 090.4
	C	2.36	311.5	5 919.4	6 511.3
	D	2.55	336.6	6 396.4	7 036.0
	E	2.05	270.6	5 142.2	5 656.4
	F	2.01	265.3	5 041.5	5 545.7
处理 2	A	2.26	298.3	5 668.5	6 235.4
	B	2.48	327.4	6 221.5	6 843.7
	C	2.23	294.4	5 594.4	6 153.8
	D	2.41	318.1	6 044.8	6 649.3
	E	2.03	268.0	5 092.8	5 602.1
	F	2.00	264.0	5 016.8	5 518.5

6 843.7 元/667 m²,D 组合辣椒单株产量、小区产量、折合总产量、产值次之,F 组合辣椒单株产量、小区产量、折合总产量、产值最低,分别为 2.0 kg/株、264.0 kg/35.1 m²、5 016.8 kg/667 m²、5 518.5 元/667 m²。

2.10 不同穗砧组合对连作辣椒品质的影响

从表 7 可以看出,不同穗砧组合辣椒果实维生素 C 含量差异极显著($P<0.05$),处理 1 中,D 组合辣椒果实维生素 C 含量最高 151.0 g/100 g,F 组合辣椒果实维生素 C 含量次之,为 127.9 g/100 g,其他组合辣椒果实维生素 C 含量分别在 10.8~31.2 g/100 g,处理 2 中,A 组合辣椒果实维生素 C 含量最高为 195.1 g/100 g,D 组合辣椒果实维生素含量次之,E 组合辣椒果实维生素 C 含量最低为 12.9 g/100 g,其他 3 个组合辣椒果实维生素 C 含量分别在 21.0~95.7 g/100 g;不同穗砧组合辣椒果实可溶性糖含量有差异,E、F 组合辣

表 7 不同砧穗组合辣椒品质比较
Table 7 Quality comparison of pepper with different anvil combinations

处理 treatment		维生素 C vitamin C /g/100 g	可溶性糖含量 soluble sugar content/g/100 g	可溶性蛋白含量 soluble protein /g/100 g	干物质含量 dry matter content /g/100 g
处理 1	A	24.7	0.37	0.66	5.28
	B	31.2	0.37	0.83	5.01
	C	10.8	0.34	0.69	5.56
	D	151.0	0.35	0.93	6.67
	E	10.8	0.44	0.64	7.75
	F	127.9	0.42	0.85	8.99
处理 2	A	195.1	0.30	0.72	6.47
	B	21.8	0.29	0.55	5.90
	C	21.0	0.34	0.62	6.61
	D	138.7	0.37	1.62	4.74
	E	12.9	0.32	0.88	4.28
	F	95.7	0.39	1.12	7.90

椒果实可溶性糖含量最高为 0.44 g/100 g、0.42 g/100 g,其他组合辣椒果实可溶性糖含量分别在 0.35~0.37 g/100 g,处理 2 中,D、F 组合辣椒果实可溶性糖含量为 0.37 g/100 g、0.39 g/100 g,其他组合辣椒果实可溶性糖含量分别在 0.29~0.34 g/100 g;不同穗砧组合辣椒果实可溶性蛋白含量差异显著,处理 1 中,D 组合辣椒果实可溶性蛋白含量最高为 0.93 g/100 g,B、F 组合辣椒果实可溶性蛋白含量次之,A、E 组合辣椒果实可溶性蛋白含量最低为 0.66 g/100 g、0.64 g/100 g,处理 2 中,D、F 组合辣椒果实可溶性蛋白含量最高为 1.62 g/100 g、1.12 g/100 g,其他组合辣椒果实可溶性蛋白含量分别在 0.55~0.88 g/100 g;不同穗砧组合辣椒果实干物质含量差异显著,处理 1 中,F 组合辣椒果实干物质含量最高为 8.99 g/100 g,E 组合辣椒果实干物质含量次之为 7.75 g/100,其他组合辣椒果实干物质含量分别在 5.01~6.67 g/100 g,处理 2 中,F 组合辣椒果实干物质含量最高为 7.9 g/100 g,A、C 组合辣椒果实干物质含量次之,其他组合辣椒干物质含量分别在 4.28~5.9 g/100 g。

3 讨论与结论

嫁接成活率是判定砧木与接穗亲和性的一个重要指标[8],本试验中,2 种砧木,6 种接穗共 12 个组合,嫁接辣椒成活率均较高,分别在 94%~100%,这与前人研究的结果基本一致。嫁接能促进植株生长,主要表现在加快根系生长,增强根的吸收能力,从而增强植株长势、生长量等方面[9],本试验中,处理 1,C 组合("神根"/"金惠 TN-1")株高最高为 95.6 cm,F 组合("神根"/"犇腾 2 号")辣椒茎粗最粗为 14.12 mm,B 组合("神根"/"卓越")辣椒开展度最宽为 72.8 cm;处理 2 中,D 组合("神威 4F1"/"金惠 13-E")株高、茎粗均最高,分别为 92.2 cm、13.98 mm,F 组合("神威 4F1"/"犇腾 2 号")辣椒开展度最宽为 74.0 cm;不同穗砧组合嫁接辣椒病毒病发病率有差异,处理 1 中,A 组合("神根"/"超越")、C 组合("神根"/"金惠 TN-1")辣椒病毒病发病率最高为 2%,处理 2 中,A 组合("神威 4F1"/"超越")、B 组合("神威 4F1"/"卓越")辣椒发病率最高为0.67%。

光合作用是作物产量形成的基础。光合作用主要取决于3个生理过程,即光合物CO_2的传导、光反应和暗反应。较强的CO_2传导能力,较高的光反应和暗反应活性是叶片提高光合速率的生理基础[10],本试验中,处理1中,D组合("神根"/"金惠13-E")、F组合("神根"/"犇腾2号")辣椒叶片净光合速率最大,分别为18.01 $\mu mmol \cdot m^{-2} \cdot s^{-1}$、19.07 $\mu mmol \cdot m^{-2} \cdot s^{-1}$;不同穗砧组合辣椒叶片荧光有差异,处理1中,B组合("神根"/"卓越")叶片荧光参数F0最大为569.11,Fm最大为2 735.78,E组合("神根"/"泰达")叶片荧光参数F0/Fm最大为0.26,A组合("神根"/"超越")叶片荧光参数Fv/F0最大为4.04,处理2中,C组合("神威4F1"/"金惠TN-1")叶片荧光参数Fm、Fv最大分别为2 401.75、1 865.50,F组合("神威4F1"/"犇腾2号")叶片荧光参数Fv/F0最小为2.91。

植物体内广泛存在能清除活性氧代谢的保护酶,其中SOD,POD和CAT是细胞抵御活性氧伤害的主要酶类,可以保护膜系统免受自由基的伤害,这3种酶与植物抗病性有重要关系[11]。处理1中,E组合("神根"/"泰达")叶片过氧化物酶含量最高为1999.15 $\mu g/g \cdot min$,叶片过氧化氢酶含量最高为0.084 $\mu g/g \cdot min$,F组合("神根"/"犇腾2号")叶片可溶性糖含量最高为1.65%,C组合("神根"/"金惠TN-1")可溶性蛋白含量最高为0.35%,处理2中,F组合("神威4F1"/"犇腾2号")叶片过氧化物酶、可溶性糖含量最高,分别为1 807.33 $\mu g/g \cdot min$、1.7%,C组合叶片可溶性蛋白含量最高为0.31%,A组合叶片过氧化氢酶含量最高为0.097 $\mu g/g \cdot min$。试验中,处理1,D组合叶片丙二醛含量最高为7.11 mmol/g,处理2,A组合、C组合叶片丙二醛含量最高,分别为6.09 mmol/g、7.54 mmol/g。

土壤酶作为土壤的重要组成部分,其活性的大小可较敏感地反映土壤中生化反应的方向和强度[12,13]。处理1中,A~F组合土壤过氧化氢酶活性分别在0.26~0.27 $ml \cdot g^{-1} \cdot 30\ min^{-1}$,A组合土壤蔗糖酶活性最强为 21.84 ml·g⁻¹·30 min⁻¹,E 组合土壤纤维素酶活性最强为 9.33 mg·g⁻¹·h⁻¹,B 组合土壤脲酶活性最强为 63.72 mg·g⁻¹·h⁻¹,处理 2 中,A~F 组合土壤过氧化氢酶活性分别在 0.26~0.28 ml·g⁻¹·30 min⁻¹;E 组合、F 组合土壤蔗糖酶活性最强,分别为 20.58 ml·g⁻¹·30 min⁻¹、17.66 ml·g⁻¹·30min⁻¹,C 组合土壤纤维素酶活性最强为 2.01 mg·g⁻¹·h⁻¹，处理 2 中,A 组合、B 组合土壤脲酶活性最强,分别为 66.89 mg·g⁻¹·h⁻¹、75.71 mg·g⁻¹·h⁻¹。

成功的嫁接组合不仅表现在结构上，更重要的是所产生的维管束桥能够执行生理功能，进行接穗和砧木间的物质交换[14]。因此,嫁接可能对果实品质产生一定影响。处理 1 中,D 组合辣椒果实维生素 C 含量最高 151.0 g/100 g, E、F 组合辣椒果实可溶性糖含量最高为 0.44 g/100 g、0.42 g/100 g,D 组合辣椒果实可溶性蛋白含量最高为 0.93 g/100 g,F 组合辣椒果实干物质含量最高为 8.99 g/100 g，处理 2 中,A 组合辣椒果实维生素 C 含量最高为 195.1 g/100 g,D、F 组合辣椒果实可溶性糖含量为 0.37 g/100 g、0.39 g/100 g, D、F 组合辣椒果实可溶性蛋白含量最高为 1.62 g/100 g、1.12 g/100 g,F 组合辣椒果实干物质含量最高为 7.9 g/100 g。

嫁接辣椒比自根辣椒具有更强的吸水吸肥能力,生长旺盛,从而有利于产量的提高[4,15,16],前人研究发现,5 种砧木和 7 种接穗共 35 种西瓜嫁接组合均提高了产量,平均增产达 30.1%[17]。本试验中,处理 1,B 组合辣椒单株产量、小区产量、折合总产量、产值均最高,分别为 2.57 kg/株、339.2 kg/35.1 m²、6 445.8 kg/667 m²、7 090.4 元/667 m²,处理 2,B 组合辣椒单株产量、小区产量、折合总产量、产值均最高,分别为 2.48 kg/株、327.4 kg/35.1 m²、6 221.5 kg/667 m²、6 843.7 元/667 m²。

综合分析:神根/卓越、神根 4F1/卓越 2 个穗砧组合表现优良,可以在生产中推广应用。

参考文献

[1] 塔国民.朝阳地区辣椒丰产栽培技术[J].现代农业,2017,5(4).

[2] 马玲,陶君.宁夏设施辣椒发展现状和增产措施探析[J].蔬菜,2017.9.

[3] 李丁仁,鲁长才.宁夏压砂地现状与可持续发展建议[J].宁夏农林科技,2011.01.001.

[4] 付玲.嫁接对辣椒生长发育及产量和品质的影响[D].山东农业大学,2016.06.

[5] 唐启义.DPS统计软件简介[J].中国医院统计,2009,16(01):99.

[6] 翟玉莹,余朝阁,韩絮,等.不同砧木嫁接对辣椒品质的影响[J].浙江农业科学,2016(8):1207-1209.

[7] 金丹,戴黎华.上海地区不同品种甜樱桃嫁接对成活率和生长的影响[J].上海农业科技,2015(6):76.

[8] 白小军,冯海萍,曲继松.不同砧木与接穗对茄子嫁接亲和性及产量的影响[J].宁夏农林科技,2014,55(08):1-3.

[9] 刘新华,曹春信,刘林,等.不同砧木对嫁接西瓜生长和果实品质的影响[J].浙江农业学报,2015,27(6):966-969.

[10] 张建恒,李宾兴,王斌.不同磷效率小麦品种光合碳同化和物质生产特性研究[J]中国农业科学,2006.11.006:0578-1752.

[11] 王吉伟.同砧木梨树嫁接愈合过程中相关酶活性的观察[J].林业科技通讯,2018(5):1671-4938.

[12] 尹承苗,相立.不同苹果砧木对连作土壤微生物及酶活性的影响[J].园艺学报,2016-0217:0513-353X.

[13] 高玉红,许娟.嫁接对西瓜根际土壤酶和叶片氮代谢酶活性及产量的影响[J].中国瓜菜,2016.(01):1673-2871.

[14] 史星雲,郭艳兰.不同酿酒葡萄砧穗组合硬枝嫁接亲和力研究[J].中国果树,2016(01):1000-8047.

[15] 刘新华,曹春信.不同砧木对嫁接西瓜生长和果实品质的影响[J].浙江农业学报,2015(06):1004-1524.

[16] 王金玉.砧木选择对嫁接西瓜生长及品质的影响[D].山东农业大学,2006.

[17] 俞更才.日光温室的西瓜嫁接效应研究[J].甘肃科学学报,2002(6):1004-0366.

不同作物秸秆腐解对连作辣椒生长及根际环境的影响

高晶霞[1]，高　昱[2]，牛勇琴[2]，吴雪梅[2]，谢　华[1]

（1.宁夏农林科学院种质资源研究所，宁夏银川　750002；
2.宁夏彭阳县蔬菜产业发展服务中心，宁夏彭阳　756500）

摘　要：【目的】植物的秸秆完全腐解后产生了氮、磷、钾等多种有机物及肥料，提高了土壤的微生物含量及活性，对种植作物的生长发育有着显著的促进作用。【方法】以辣椒为研究对象，通过试验研究及数据分析，重点研究玉米秸秆、万寿菊秸秆以及芹菜秸秆腐解对辣椒生长及土壤的影响。【结果】玉米、万寿菊及芹菜作物的秸秆腐物，提高了辣椒的株高、茎粗、生物量及壮苗指数，同时一定程度上增强了辣椒植株的光合作用及连作辣椒的叶绿素含量，另外作物的秸秆均能够改善土壤的pH值及电导率以及土壤的微生物结构，增强土壤的酶活性，不同作物的改善程度也有所不同。【结论】研究的成果对改善辣椒等经济作物连作的土壤活性及生产能力有重要的参考价值，可在农业生产中进行推广应用。

关键词：秸秆腐解；连作辣椒；土壤微生物；叶绿素

Effect of Decomposition of Different Crop Straws on Growth and Rhizosphere Environment of Continuous Cropping Pepper

Gao Jingxia[1], Gao Yu[2], Wu Xuemei[2], Niu Yongqin[2], Xie Hua[1]

(1. Ningxia Academy of Agriculture and Forestry Sciences, Yinchuan 750002, China; 2. Vegetable Industry Development Service Center of Pengyang County, Ningxia Hui Autonomous Region, Pingyang 756500, China)

Abstract: 【Objective】After the straw of different crops is completely decomposed, various organic substances and fertilizers such as nitrogen, phosphorus, and potassium are produced, which improves the soil microbial content and activity, and has a significant promotion effect on the growth and development of planted crops. 【Method】Taking pepper as a research object, through experimental research and data analysis, the effects of corn stalk, marigold stalk and celery stalk decomposition on pepper growth and soil were mainly studied.【Result】The results showed that the straw rot of corn, marigold, and celery crops increased the plant height, stem thickness, biomass, and seedling index of peppers, and enhanced the photosynthesis and continuous cropping of peppers to a certain extent. Chlorophyll content, and crop straws can improve soil pH and electrical conductivity, as well as soil microbial structure, enhance soil enzyme activity, and the degree of improvement varies among different crops. 【Conclusion】 The results of the study have important reference value for improving soil activity and productivity of continuous cropping of cash crops such as pepper, and can be popularized and applied in agricultural production.

Key words: straw decomposition; continuous cropping pepper; soil microorganisms; chlorophyll

在我国农业种植中,秸秆资源是一种重要的可再生资源,通过对秸秆进行粉碎或腐解处理可以产生较多的营养物质,有效提升土壤的活性,对于绿色可持续农业的发展具有重要的推动意义(于寒等,2018;Tian et al.,2019;陈昱等,2019)。目前连作障碍在较大程度上影响了农作物的种植产量及种植品质。特别是在设施环境下,这种现象最为普遍。设施栽培中大量施用氮肥、磷肥,钾肥,这种不科学的施肥导致土壤大量元素超标而微量元素匮乏,土壤营养元素不均衡,进而使土壤离子平衡和缓冲能力降低。因此研究连作障碍发生机制并通过科学环保的方法克服连作障碍,已成为设施农业发展中亟待解决的问题。

利用植物间的化感作用原理合理安排伴生或间套作不仅可以提高蔬菜产量和品质,而且还可以有效改善土壤生态环境,减轻连作障碍的发生。目前大量的研究显示(孙玲等,2019;Wu 等,2019;Zhao et al.,2019),农作物秸秆的腐解可以有效增加土壤中的微生物和其他养分的含量,一定程度上降低土壤的容重,有效调节土壤活性含量的平衡,进而降低辣椒等农作物对土壤的连作伤害(Xie et al.,2019;李换平等,2019)。学者对玉米类秸秆研究发现,玉米秸秆腐解中包含的微生物可以提升土壤的活性,有效改善烟叶种植土壤的微生物成分,提升烟叶产量并降低烟叶含梗率。也有学者对小麦秸秆进行研究,认为小麦秸秆的腐解物中包含了大量的氮素积累,增加了土壤中的氮肥的含量,提升了水稻种植的产量及效率(杨封科等,2019;Han et al.,2020;班允赫等,2019)。此外秸秆腐蚀中带来的微生物群落的改变,可以进一步提升土壤中微生物的活性,有学者研究发现,荠菜植物的秸秆腐解能够提升土壤中微生物物种的丰富程度指数,同时降低土壤中的真菌和嫌气菌的数量,增加土壤的酶活性。另外有学者对花科植物的秆茎腐物进行研究,发现花科植物能够消除土壤中的大量虫卵,降低土壤的活性,如果将这些花科植物的根茎进行二次腐解,可以有效调解土壤中的微生物菌体含量,有效调解辣椒等农作物连作对土壤的伤害(刘微等,2019;柴如山等,2019;赵海岚等,2019;付佑胜等,2019)。

综合目前秸秆腐解的文献研究可以发现,目前针对秸秆腐解对土壤及农作物改善的研究文献较多,但涉及土壤连作特性改善的文献较少,特别是研究多种秸秆腐解对辣椒等经济作物连作特性的影响,目前几乎没有文献涉及,为此,本研究以辣椒为研究对象,重点研究玉米秸秆、万寿菊秸秆及芹菜秸秆腐解对辣椒生长及土壤的影响,通过本文的研究,力求为降低辣椒等经济作物连作对土壤影响提供理论参考及实践指导。

1 材料与方法

1.1 试验材料及试剂

本研究选用的辣椒品种为"巨峰1号"(宁夏巨丰种苗有限公司提

供),供试土壤取自宁夏彭阳县新集乡辣椒连作 3 a 以上的土壤,土壤面积为 0.15 hm²,土壤的基本参数如下:pH 值为 8.37,碱解氮 78.2 mg·kg⁻¹,有效磷 18.3 mg·kg⁻¹,速效钾 406 mg·kg⁻¹,有机质 38.7 g·kg⁻¹,玉米、万寿菊及芹菜秸秆购买自当地农户,所用的肥料为经济作物专用的果蔬有机肥,主要购买自宁夏嘉禾源种苗有限公司。

1.2 试验设计

对于买入的玉米、万寿菊及芹菜秸秆首先进行腐解处理,将植物秆茎烘干后,用高速粉碎机将其粉碎,并将粉碎的秸秆与施有有机肥的土壤按照 1:50 的比例进行完全混合,并装入容器中,加入一定量的水,使得土壤的湿度保持在 50%左右,放置 7 d,每天都要监测容器重量,适当补充水分,盆外包裹黑色遮阳膜。随后进行自然腐解处理,腐解时间为 30 d,腐解均匀完全后装入固定容器中,进行备用。本次实验共进行 4 种方式处理,处理方式 1 为对照处理(CK),不进行任何处理,处理方式 2 为玉米秸秆的腐解处理(YM);处理方式 3 为万寿菊秸秆腐解处理(WS);处理方式 4 为芹菜秸秆腐解处理(QC),每小区面积 23.4 m²,重复 3 次,腐解 30 d 后定植辣椒苗,腐解后 5 d 内,进行浇水,并随机排列,同时在测试过程中,不进行药剂喷洒。

1.3 测定项目

1.3.1 生长指标

在腐解 30 d 后,测量辣椒植株的各项生长指标,其中辣椒的株高采用杜克 ls-p 激光测距仪进行测量,茎粗选用实验室的游标卡尺测量,植株生物量采用梅特勒托乐多 ME204/02 物理电子天平进行测量,同时分析辣椒植株根系指标,并给出辣椒植株的壮苗指数计算公式如下所示:

$$壮苗指数 = \left(\frac{茎粗}{株高} + \frac{根干重}{地上部分干重} \right) \times 总干重 \quad (1)$$

1.3.2 辣椒的叶绿素及光合作用指标测量

在定植 30 d 后,辣椒植株的叶绿素含量采用经典文献中的丙酮乙醇混合液法进行测量,辣椒植株的光合作用采用荧光法测量,随机选择5株种植30 d的辣椒植株,在每株上面选取位置一致的地方对辣椒植株的复叶进行展开,利用实验室购买的美国OPTI-SCIENCES OS-5p+便携式脉冲调制叶绿素荧光仪对相关参数进行测量,并在间隔5 min后进行数据采集。

1.3.3 土壤指标

在腐解30 d后,对于连作辣椒的土壤测量采用pH测量仪、细菌培养基以及比色法等方法,分别测量连作辣椒土壤的pH、电导率、土壤微生物以及酶活性等参数。

1.4 数据统计

本研究的试验数据采用EXCEL 2019以及SPSS 22.0进行数据分析,多重比较采用LSR法(Duncan's法),显著水平 $P<0.05$。

2 结果与分析

2.1 不同作物秸秆腐解处理对连作辣椒生长的影响

图1为30 d的辣椒植株的株高和茎粗生长情况,从图中可以发现,辣椒植株的株高和茎粗在经过不同作物秸秆腐解处理后,均高于不处理的对照情况,其中万寿菊植株(WS)的秸秆腐解作用效果最好,对应在各个时期的辣椒植株的株高和茎粗都是最高的,其次是玉米秸秆(YM)和芹菜秸秆

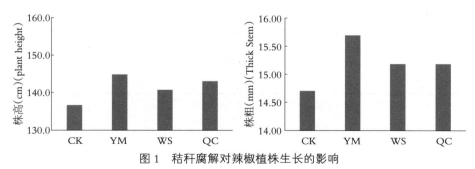

图1 秸秆腐解对辣椒植株生长的影响

Fig.1 Effects of straw decomposing on the growth of pepper plants

注:CK为对照组;YM 玉米秸秆腐解处理方式,WS 万寿菊秸秆腐解处理方式,QC 芹菜秸秆的处理方式

(QC),在各个时期也均高于不处理的对照情况。与对照组相比,辣椒在定植 30 d 的株高和茎粗分别增加了 5%、0.2%、0.3%以及 8.7%、7.6%、7.7%增幅较为明显,达到显著水平。

2.2 不同作物秸秆腐解处理对连作辣椒的发育状态的影响

不同秸秆腐解处理对连作辣椒发育状态的影响如表 1 所示,从表 1 中可以发现,相比对照组(CK),不同作物处理后的辣椒作物总重均高于对照组(CK),特别是 WS 的秸秆腐解处理,总重增长量显著增加,存在显著性差异。同时从表 1 中可以看到,YM 和 QC 处理后的辣椒作物总重、根重及茎重都比对照(CK)组增加,但增加量较小,WS 处理后的辣椒根干重相比对照组(CK)减小,但减少量并不显著,而 WS 处理后的辣椒茎干重相比对照组(CK)显著增加,呈现显著性差异,也即 WS 处理后对辣椒的茎部有明显的促进生长作用,CK 组、YM 组以及 WS 组的壮苗指数基本一致,无显著性差异,表明 YM、WS、QC 处理对辣椒的总体促进作用较小,并无显著影响。

表 1 不同处理对连作辣椒发育状态的影响
Table 1 Effects of different treatments on the development status of continuous cropping pepper

处理方式 Processing method	总干重/g Total dry weight	根干重/g Root dry weight	茎干重/g Stem dry weight	壮苗指数 Seedling Index
CK	0.103	0.048	0.052	0.11
YM	0.114	0.051	0.063	0.11
WS	0.183*	0.046	0.116*	0.103
QC	0.132	0.047	0.068	0.105

注:字母 * 表示相比对照组 CK 差异显著($p<0.05$),以下同上

2.3 不同作物秸秆腐解处理对连作辣椒根系形态的影响

从表 2 中的数据可以看到,相比对照组(CK),YM 组、WS 组及 QC 组均能显著促进辣椒根系的生长发育,辣椒根系的长度、根系表面积以及体积方面显著增长,与对照组(CK)存在显著性差异,其中 WS 组的处理效果最

好,辣椒植株的根系长度、根系表面积、根系体积以及根尖数相比对照组均增加了24.49%,46.62%,76.92%以及31.63%,对辣椒植株根系的整体生长状态发育效果最好,YM和QC组低于WS组的影响效果。

表2 不同处理方式对连作辣椒根系状态的影响

Table 2 Effects of different treatments on root status of continuous cropping pepper

处理方式 Processing method	根系长度/cm Root length	根系表面积/cm² Root surface area	根系体积/cm³ Root volume	根尖数 Number of apex
CK	176.38	29.66	0.39	273.80
YM	218.45*	38.61*	0.58*	323.43*
WS	231.12*	43.49*	0.69*	362.46*
QC	223.73*	39.28*	0.62*	351.36*

2.4 不同作物秸秆腐解处理对连作辣椒光合作用能力的影响

从表3中的数据可以看到,相比对照组(CK),YM组、WS组及QC组的处理方式均能提高辣椒植株的光合作用,对应的叶绿素荧光参数包含PSII最大量子产量Fv/Fm、PSII实际量子产量Y(II)、光化学淬灭系数qP等参数均有一定程度的提高,但相比对照组(CK)差异并不显著,另外相比对照组CK,YM组、WS组及QC组的非化学淬灭系数NQP有一定程度的下降,但差异也并不显著。

表3 不同处理方式对连作辣椒叶片光合作用能力的影响

Table 3 Effects of different treatments on photosynthetic capacity of continuous cropping pepper leaves

处理方式 Processing method	PSII最大量子产量 Fv/Fm Maximum quantum yield of PSII the amount	PSII实际量子产量 Y(II) PSII actual quantum yield the amount	光化学淬灭系数 qP Photochemical quenching coefficient	非化学淬灭系数 NQP Non-chemical quenching coefficient
CK	0.82	0.36	0.47	1.35
YM	0.85	0.38	0.51	1.27
WS	0.87	0.43	0.52	1.26
QC	0.83	0.41	0.5	1.23

2.5 不同作物秸秆腐解处理对辣椒叶绿素含量的影响

从表4中数据可以看到,YM组、WS组及QC组的处理方式对辣椒植株叶片叶绿素含量均有一定程度的提高,其中YM处理方式对辣椒植株的叶绿素含量提升最为明显,相比对照组(CK)呈现显著性差异,3种叶绿素分别相比对照组(CK)增加了29.19%、56.07%及30.16%,差异呈现显著状态,WS组及QC组与CK组相比,并无显著差异。

表4 不同处理方式对辣椒叶绿素含量的影响
Table 4 Effect of different treatments on chlorophyll content of pepper

处理方式 Processing method	叶绿素 A/(mg·g^{-1}) Chlorophyll A	叶绿素 B/(mg·g^{-1}) Chlorophyll B	叶绿素 A+B/(mg·g^{-1}) Chlorophyll A+B
CK	1.85	0.66	2.52
YM	2.39*	1.03*	3.28*
WS	2.13	0.78	2.83
QC	2.19	0.83	2.65

2.6 不同作物秸秆腐解处理对连作辣椒土壤酸碱度及电导率的影响

对辣椒种植土壤的酸碱度及电导率的影响如表5所示,从表中可以看到,进行YM、WS及QC处理后,土壤酸碱度pH值相比对照组(CK)都有提高,其中YM处理方式对土壤酸碱度提高显著,相比对照组(CK)增加了27.58%,呈现显著性差异,另外从表中可以看到,YM、WS及QC处理后对土

表5 不同处理方式对连作辣椒土壤酸碱度及电导率的影响
Table 5 Effects of different treatments on soil pH and electrical conductivity of continuous cropping pepper

处理方式 Processing method	土壤酸碱度 Soil pH	处理电导率/(mS·cm^{-1}) Handling conductivity
CK	5.55	1.54
YM	7.08*	0.86*
WS	6.91	1.03
QC	6.73	0.92

壤的处理电导率有所降低,其中 YM 组降低程度最多,相比对照组(CK),降低了 44.19%,影响较为显著,其他两组也存在一定程度下降。

2.7 不同作物秸秆腐解处理对种植辣椒土壤微生物的影响

对种植辣椒土壤微生物的影响如表 6 所示,从表 6 中可以看到,YM、WS 及 QC 处理方式对土壤微生物元素含量都有提高,其中 WS 处理方式提高的最为显著,显示了 WS 处理方式对土壤微生物的重要调节作用,另外从表 6 中可以看到,YM 和 QC 处理方式能够有效降低土壤中真菌的数量,但 WS 处理方式相比对照组(CK)最为显著,有效地降低了辣椒等经济作物连作对土壤的破坏。

表 6 不同处理方式对种植辣椒土壤微生物的影响
Table 6 Effects of different treatments on soil microorganisms in pepper planting

处理方式 Processing method	放线菌 /(10^5cfu·g^{-1}) Actinomycetes	细菌 /(10^6cfu·g^{-1}) Bacterial	真菌 /(10^3cfu·g^{-1}) Fungus	总菌数 /(10^7cfu·g^{-1}) Total bacteria
CK	125.38	112.63	61.16	12.53
YM	139.38	116.17	48.33	13.11
WS	212.03	175.67*	19.33*	19.69*
QC	153.06	128.17	46.26	15.12

2.8 不同作物秸秆腐解处理对辣椒种植土壤酶活性的影响

从表 7 中可以看到,YM、WS 及 QC 处理方式对土壤酶的活性有显著增强,除了对土壤脲酶有降解作用外,其他蔗糖酶、酸性磷酸酶及多酚氧化酶均相比对照组(CK)显著增加,特别是 YM 处理方式,对以上几种土壤酶的活性有显著增强,增长率分别为 132.69%、186.36%以及 198.73%,显示了玉米秸秆对土壤酶的显著改善作用,增强了辣椒植株连作土壤的总体活性,其他 2 种处理方式的影响效果相比 YM 较低。

表7 不同处理方式对辣椒种植土壤酶活性的影响
Table 7 Effects of different treatments on soil enzyme activities in pepper plantation

处理方式 Processing method	脲酶 /(mg·g^{-1}·h^{-1}) Urease	蔗糖酶 /(mg·g^{-1}·h^{-1}) Sucrase	酸性磷酸酶 /(mg·g^{-1}·h^{-1}) Acid phosphatase	多酚氧化酶 /(mg·g^{-1}·h^{-1}) Polyphenol oxidase
CK	0.925	1.627	0.132	0.079
YM	0.587*	3.786*	0.378*	0.236*
WS	0.528*	2.836*	0.259*	0.178*
QC	0.563*	3.168*	0.313*	0.201*

3 讨论

大量研究显示,不同作物的秸秆完全腐解后产生了氮、磷、钾等多种有机物及肥料,能够满足种植植物所需,这些有机物进入种植植物的土壤中,提高了土壤的微生物含量及活性,对种植作物的生长发育有着显著的促进作用,显著提高了农作物的产量。学者研究发现将常见蔬菜作物的秸秆腐解处理后放置在种植的甘蔗的土壤中,促进了甘蔗作物幼苗根茎的生长,同时改善了土壤的微生物活性剂含量,促进了甘蔗的土壤养分的吸收,提升了甘蔗的产量(Huang et al.,2020;Na et al.,2019;李永刚等,2017;严吴炜等,2017)。本研究也得到了如下结论,通过在辣椒连作土壤中加入玉米、万寿菊以及芹菜三种作物的秸秆,压碎处理后放入辣椒连作土壤中,提高了辣椒的株高、茎粗、生物量及壮苗指数,其中万寿菊秸秆的腐解效果最好,究其原因,主要是不同作物秸秆腐解处理对辣椒根系长度、体积、表面积以及根尖数均显著提高,促进了根系对养分的吸收,有效改善了辣椒连作土壤的根部环境,增强了辣椒作物吸收土壤肥料的能力,其中万寿菊秸秆腐解后对辣椒根系生长的促进作用最为显著。

叶绿素荧光是测定叶片光合作用的无损伤探针,其与光合色素之间存在很好的相关性,能够反映逆境因子对光合作用的影响。植物叶绿素等参数是衡量植物光合作用的主要参数,对于改善整体农业作业环境,增强有机物合成有显著作用,植物叶绿素等参数的测量通常采用荧光法。结合本研究的结果,目前玉米秸秆、芹菜秸秆以及万寿菊秸秆均能提高辣椒植株的光合作用,相关的光合作用等参数均有一定程度的提高,显示腐解处理能够增加辣椒叶片的光合作用利用率,但农作物秸秆对辣椒非化学淬灭系数(NQP)影响较小,猜测可能是辣椒自身叶绿素系统的胁迫效应及自我保护机制有关,以上研究成果与文献中给出的荠菜秸秆对种植茄子光合作用的提高类似,充分说明了秸秆腐物对土壤活性成分的变强,增强了光合作用影响[18,19]。另外结合本文的研究发现,玉米秸秆、芹菜秸秆及万寿菊秸秆均能提高连作辣椒的叶绿素含量,增强辣椒的根系活性,提升辣椒作物的生长能力。

在农作物种植过程中,适当的土壤pH值及较低的土壤电导率可以有效提升农作物的生长活力,在本次试验中,玉米秸秆、万寿菊秸秆及芹菜秸秆均能够改善土壤的pH值及电导率,促进了辣椒的生长。这与文献中给出的通过玉米秸秆的粉碎提高黄瓜等农作物土壤的活性,有效改善土壤的pH值及电导率有相似之处(乔天长等,2015;于寒等,2015)。主要原因可能与玉米秸秆腐解后改善了土壤的有机物质,活化了土壤养分,改善了土壤的各项性能。另外从本文的研究中发现,秸秆腐解方式释放进土壤的有机物质被土壤微生物利用,进而活化了土壤中的养分所致。土壤微生物是土壤生命活体中最活跃的有机体,其主要组成类群包括细菌、放线菌和真菌,在维持土壤微生态结构和功能中发挥重要作用。本研究显示,玉米秸秆及万寿菊秸秆处理对连作辣椒的土壤微生物的放线菌、细菌及总菌数都有提高,同时玉米秸秆、芹菜秸秆及万寿菊秸秆也能够改善土壤的酶活性,是土壤有益的催化剂,增强辣椒连作植株生理代谢及养分吸收能力。

4 结论

本研究以辣椒为研究对象,通过试验研究及数据分析,重点研究玉米秸秆、万寿菊秸秆以及芹菜秸秆腐解对辣椒生长及土壤的影响，研究结果显示，玉米、万寿菊及芹菜作物的秸秆腐物，提高了辣椒的株高、茎粗、生物量及壮苗指数，同时一定程度上增强了辣椒植株的光合作用及连作辣椒的叶绿素含量，另外作物的秸秆均能够改善土壤的pH值及电导率以及土壤的微生物结构，增强土壤的酶活性，不同作物的改善程度也有所不同。综上所述，本研究的成果对改善辣椒等经济作物连作的土壤活性及生产能力有重要的理论及实践参考价值，研究成果可以进一步在农业生产中进行推广。

参考文献

[1] 于寒,高春梅,谷岩.秸秆腐解液对玉米幼苗根系生长及生理特性的影响[J].分子植物育种,2018,16（23）:7795-7799.[Yu Han, Gao Chunmei, Gu Yan. Effect of straw decomposing solution on root growth and physiological characteristics of corn seedlings [J]. Molecular Plant Breeding, 2018, 16 (23): 7795-7799.]

[2] 陈昱,张福建,范淑英等.秸秆腐解物对豇豆连作土壤性质及幼苗生理指标的影响[J].核农学报,2019,33(07):1472-1479.[Chen Yu, Zhang Fujian, Fan Shuying, et al. Effects of straw decomposed matter on soil properties and physiological indexes of seedlings of cowpea continuous cropping[J]. Chinese Journal of Nuclear Agriculture, 2019, 33 (07): 1472-1479.]

[3] 孙玲,吴景贵,李建明等.纤维素降解细菌对玉米秸秆的降解效果[J].吉林农业大学学报,2019,41(04):402-407.[Sun Ling, Wu Jinggui, Li Jianming et al. Degradation effect of cellulose-degrading bacteria on corn straw[J]. Journal of Jilin Agricultural University, 2019, 41 (04): 402-407.]

[4] 李换平,张永仙,江解增等.小麦秸秆截段覆盖对设施蔬菜产量和土壤性质的影响[J].北方园艺,2019(17):14-20.[Li Huaiping, Zhang Yongxian, Jiang Jiezeng, et al. Effects of wheat straw covering on greenhouse vegetable yield and soil properties[J]. Northern Horticulture, 2019(17):14-20.]

[5] 杨封科,何宝林,张国平.膜下秸秆还田添加腐解剂对旱地土壤碳氮积累及土壤肥力性状的影响[J].草业学报,2019,28(09):67-76.[Yang Fengke, He Baolin, Zhang Guoping. Effects of Adding Decomposers on the Soil Carbon and Nitrogen Accumulation and Soil Fertility in Dry Land under Film Returning to the Field[J]. Acta

Practicum, 2019, 28 (09): 67-76.]

[6] 班允赫,李旭,李新宇.降解菌系和助腐剂对不同还田方式下水稻秸秆降解特征的影响[J].生态学杂志,2019,38(10):2982-2988. [Ban Yunhe, Li Xu, Li Xinyu. Effects of degrading bacteria and preservatives on degradation characteristics of rice straw under different returning methods [J]. Chinese Journal of Ecology, 2019, 38 (10): 2982-2988.]

[7] 刘微,张晓翔,彭辉.秸秆还田对玉米田病虫害影响的研究[J].农业科技通讯,2019(10):198-199. [Liu Wei, Zhang Xiaoxiang, Peng Hui. Study on the effect of returning straw to field on pests and diseases of corn field[J]. Agricultural Science and Technology Newsletter, 2019 (10): 198-199.]

[8] 柴如山,王擎运,叶新新.我国主要粮食作物秸秆还田替代化学氮肥潜力[J].农业环境科学学报,2019,38(11):2583-2593. [Chai Rushan, Wang Qingyun, Ye Xinxin. Potential of Returning Chemical Nitrogen Fertilizer to China's Major Grain Crop Straws [J]. Journal of Agro-Environment Science, 2019, 38 (11): 2583-2593.]

[9] 赵海岚,李冰,王昌全.秸秆腐解滤出液对淹水土壤镉形态变化的影响[J].环境科学与技术,2019,42(09):1-6. [Zhao Hailan, Li Bing, Wang Changquan. Effects of straw decomposed filtrate on the change of cadmium form in flooded soil[J]. Environmental Science and Technology, 2019, 42 (09): 1-6.]

[10] 付佑胜,刘伟中,张凯.麦秸秆高留茬条件下不同秸秆覆盖量对稻田杂草及水稻产量的影响[J].西南农业学报,2019,32(10):2313-2318. [Fu Yousheng, Liu Weizhong, Zhang Kai. Effects of Different Straw Covers on Weeds and Rice Yield in Paddy Field under the Condition of High Stubble of Wheat Straw[J]. Southwest Agricultural Journal, 2019, 32 (10): 2313-2318.]

[11] 李永刚,王丽艳,张思奇.玉米连作障碍主要因子对苗期玉米生长影响的初步分析[J].东北农业科学,2017,42(02):27-31. [Li Yonggang, Wang Liyan, Zhang Siqi. Preliminary analysis of the effects of major factors of continuous cropping obstacles on corn growth in seedling stage[J]. Northeast Agricultural Science, 2017, 42 (02): 27-31.]

[12] 严吴炜,朱丽丽,张路.土表覆盖水稻秸秆对大棚水蕹菜产量和土壤肥力的影响[J].中国蔬菜,2017(08):51-57. [Yan Wuwei, Zhu Lili, Zhang Lu. Effects of rice straw mulching on soil yield and soil fertility in greenhouse[J]. China Vegetables, 2017 (08): 51-57.]

[13] 乔天长,赵先龙,张丽芳.玉米秸秆腐解液对苗期根际土壤酶活性及根系活力的影响

[J].核农学报,2015,29(02):383-390. [Qiao Tianchang, Zhao Xianlong, Zhang Lifang. Effect of corn straw decomposing solution on soil enzyme activities and root activity in seedling rhizosphere[J]. Journal of Nuclear Agriculture, 2015, 29 (02): 383-390.]

[14] 于寒,谷岩,梁烜赫.玉米秸秆腐解规律及土壤微生物功能多样性研究[J].水土保持学报,2015,29(02):305-309. [Yu Han, Gu Yan, Liang Yanhe. Study on the Decomposition of Corn Straw and Functional Diversity of Soil Microbes[J]. Journal of Soil and Water Conservation, 2015, 29 (02): 305-309.]

[15] Xiaoping TIAN, Lei WANG, Yahong HOU. Responses of Soil Microbial Community Structure and Activity to Incorporation of Straws and Straw Biochars and Their Effects on Soil Respiration and Soil Organic Carbon Turnover[J]. Pedosphere,2019,29(4): 12-32.

[16] Junnan Wu, Yanfen Liao. Study on thermal decomposition kinetics model of sewage sludge and wheat based on multi distributed activation energy[J]. Energy,2019,185 (3):73-76.

[17] Hongyu Zhao,Yuhuan Li,Qiang Song. Catalytic reforming of volatiles from co-pyrolysis of lignite blended with corn straw over three different structures of iron ores [J]. Journal of Analytical and Applied Pyrolysis,2019,144(8):36-39.

[18] Tian Xie, Ruichao Wei, Zhi Wang. Comparative analysis of thermal oxidative decomposition and fire characteristics for different straw powders via thermogravimetry and cone calorimetry[J]. Process Safety and Environmental Protection,2020,134(8): 81-87.

[19] Ya Han, Shui-Hong Yao, Heng Jiang. Effects of mixing maize straw with soil and placement depths on decomposition rates and products at two cold sites in the mollisol region of China[J]. Soil & Tillage Research,2020,197(2):28-36.

[20] Junjie Huang, Ke Ma, Xingxuan Xia. Biochar and magnetite promote methanogenesis during anaerobic decomposition of rice straw[J]. Soil Biology and Biochemistry, 2020,143(3):98-102.

[21] Yin Na, Koide Roger T. The role of resource transfer in positive, non-additive litter decomposition[J]. PloS one,2019,14(11): 83-89.

拱棚辣椒水肥一体化技术试验研究

高晶霞[1], 吴雪梅[2], 牛勇琴[2], 赵云霞[1], 颜秀娟[1],

裴红霞[1], 王学梅[1], 谢 华[1]

（1. 宁夏农林科学院种质资源研究所, 银川　750002;
2. 宁夏回族自治区彭阳县蔬菜产业发展服务中心, 宁夏彭阳　756500）

摘　要:【目的】制定拱棚辣椒水肥一体化施肥制度。【方法】设置3个施肥量和3个灌水量,按生育期进行调整,完全随机设计试验,旨在研究不同灌水量与施肥量对拱棚辣椒土壤酶活性、土壤硝酸盐含量、果实品质等指标的影响。【结果】苗期—开花期灌水量和施肥量为201 m³/hm² 和 60 kg/hm²,开花—果实膨大期灌水量和施肥量 226.5 m³/hm² 和 119.7 kg/hm²,果实膨大期—拉秧期灌水量和施肥量为 300 m³/hm² 和 180.0 kg/hm²,辣椒植株根系活力最强,为 18.905 μg/h;土壤脲酶、蔗糖酶、过氧化氢酶、纤维素酶活性最强, 分别为 182.59、874.1、7.5、0.15 mg/(g·72 h);土壤硝态氮含量最低,为 5.91 mg/kg;辣椒根干质量、根冠比最大,为 20.2 g、0.22;辣椒可溶性蛋白含量、可溶性糖含量、维生素C含量最高,分别为 32.24 mg/g、37.9 mg/g、130.42 mg/g。【结论】苗期—开花期、开花—果实膨大期、果实膨大期—拉秧期3个时期,灌水量是 201 m³/hm²、226.5 m³/hm²、300 m³/hm²,施肥量是 60 kg/hm²、119.7 kg/hm²、180 kg/hm²,拱棚辣椒土壤酶活性、辣椒品质均比其他处理高,为制定拱棚辣椒水肥一体化施肥制度提供理论依据。

关键词:拱棚辣椒;水肥一体化;试验研究

Experimental Study on Water and Fertilizer Integration Technology of Capsicum in Arch Shed

Gao JingXia[1], Wu XueMei[2], Niu YongQin[2], Zhao YunXia[1], Yan XiuJuan[1], Pei HongXia[1], WangXueMei[1], Xie Hua[1]

(1. *Ningxia Academy of Agriculture and Forestry Plant Resources, Ningxia, Yinchuan, 750002*; 2. *The Ningxia Hui Autonomous Region, Pengyang County AgriculturalTechnology Extension and Service Center, Pengyang, Ningxia 756500*)

Abstract: 【purpose】Establish the integrated fertilization system of pepper water and fertilizer in arch shed. 【method】In this study, 3 fertilizer rates and 3 irrigation rates were set, The adjustment was made according to the growth period, Completely randomized design trial, The aim of this study was to investigate the effects of different irrigation and fertilizer rates on soil enzyme activity, soil nitrate content and fruit quality of pepper in arch shed. 【results】The amount of irrigation water and fertilizer applied from seedling stage to flowering stage was 201 m^3/hm^2 and 60 kg/hm^2, The irrigation water and fertilizer amount from flowering to fruit expansion was 300 m^3/hm^2 和 180.0 kg/hm^2, The irrigation water and fertilizer application amount from fruit expansion stage to seedling planting stage was 20 m^3/12.0 kg, Pepper plants have the strongest root system, 18.91ug/FW.h; Soil urease, sucrase, catalase, cellulase activity is the strongest, 182.59, 874.1, 7.5 and 0.15 mg/g·72 h, respectively; Soil nitrate nitrogen content is the lowest, 为 5.91 mg/kg; Pepper root dry weight, root cap ratio is the largest, 20.2g, 0.22; The soluble protein content, soluble sugar content and vitamin C content of pepper were the highest, 32.24 mg/g, 37.9 mg/g and 130.42 mg/g, respectively. 【conclusion】there are three stages from seedling stage to flowering stage, from flowering stage to fruit expansion stage, and from fruit expansion stage to seedling seedling stage, the irrigation were 201 m^3/hm^2、226.5 m^3/hm^2、300 m^3/hm^2, the fertilizer application rates were 60 kg/hm^2、119.7 kg/hm^2、180 kg/hm^2, respectively, The soil enzyme activity and quality of capsicum in arch shed were higher than those in other treatments and controls, which provided theoretical basis for the establishment of integrated fertilization system of water and fertilizer for capsicum in arch shed.

Key words: Arch shed pepper;Water and fertilizer integration;Experimental study

辣椒(*Capsicum annuum L.*)又名番椒,原产于拉丁美洲热带地区,茄科辣椒属,为一年生草本植物,在我国普遍栽培[1]。辣椒营养丰富,维生素 C 含量在蔬菜中居第一位[2]。辣椒是宁夏南部山区栽培面积较大的蔬菜之一,年平均气温 7.5℃,无霜期 158 d,降水量 442.7 mm,地貌类型复杂多样,且远离大城市,污染少,具有发展绿色蔬菜得天独厚的自然条件,土质、气温非常适宜辣椒生长,以塑料拱棚做春提前秋延后栽培,目前是该区域辣椒主要栽培模式,也是连片规模种植最大区,种植面积达到 1.333 多万 hm^2。但是宁夏南部山区塑料拱棚辣椒水肥管理方式全部为滴灌或膜下沟灌模式下的人工操作,为经验型模糊操作,引起的水资源利用率低、土传病害加重,肥料养分严重流失、环境污染加剧和产品品质下降等问题,生产上推广应用水肥一体化技术已成为必然[3,4]。研究[5-10]表明,灌水量较高时会降低蔬菜品质,使果实内可溶性糖、有机酸、可溶性固形物、可溶性蛋白和维生素 C 含量降低;而较低的灌水量对蔬菜的品质也有较大影响,水分胁迫会严重降低蔬菜的光合作用,导致作物品质降低[10,11];干旱条件会使蔬菜硝酸还原酶活性降低,从而加剧硝酸盐的累积[12]。陈秀香[13]研究得出灌水前土壤相对田间持水率为 70%~75%处理的加工樱桃番茄品质较好。张鲁鲁[14]研究表明,温室膜下滴灌甜瓜在初花期、开花—坐果、膨大期、成熟期灌水量分别为 40.62 mm、25.27 mm、63.54 mm、35.37 mm 时,甜瓜的总体品质较好。合理的水肥搭配才能提高蔬菜的品质。丁果[21]研究表明,影响樱桃番茄品质的最主要因素是水肥交互作用和肥料用量,中肥有利于樱桃番茄果实维生素 C 含量的提高,高肥有利于果实中可溶性糖的提高。本试验以单体漩涡式施肥罐、速溶高效复合滴灌肥、有机海藻肥等为物化产品，开展不同灌水量、施肥量的"拱棚辣椒水肥一体化试验研究"，分析辣椒土壤酶活性、土壤硝酸盐含量、果实品质等指标的变化，旨在提出拱棚辣椒水肥一体化施肥制度，为拱棚辣椒丰产栽培提供保障。

1 材料与方法

1.1 试验设计

试验地位于宁夏彭阳县新集乡沟口村拱棚辣椒核心示范基地，位于N35°45′~36°14′，E106°52′~106°21′。供试材料为辣椒"朗月407"，肥料为高氮型有机海藻肥、高钾型有机海藻肥、海藻广谱型（宁波费尔诺生物科技有限公司提供）。基肥施优质农家肥羊粪3 000 kg/667 m²，复合肥(15:15:15)30 kg/667 m²。试验设置3个灌水量与3个施肥量两个因子，其中灌水量(G1/G2/G3)设置3个水平(苗期至开花期分别为0.35 m³、0.41 m³、0.47 m³/23.4 m²/10 d；开花坐果期至果实膨大期分别为0.41 m³、0.47 m³、0.53 m³/23.4 m²/10 d；果实膨大期至拉秧期分别为0.47 m³、0.58 m³、0.70 m³/23.4 m²/10 d，施肥量(S1/S2/S3)设置3个水平(苗期至开花期：高氮型有机海藻肥分别为0.09 kg、0.14 kg、0.19 kg/23.4 m²；开花坐果期至果实膨大期：高氮型有机海藻肥+海藻广谱分别为0.18 kg、0.28 kg、0.38 kg/23.4 m²；果实膨大期至拉秧期高钾型有机海藻肥+水溶肥高钾型+海藻广谱型分别为0.27 kg、0.42 kg、0.57 kg/23.4 m²)。共9个处理组合。定植时间为2018年4月16日，完全收获期为2018年9月30日。小区面积23.4 m²，辣椒100株，各小区间设有隔离行。辣椒栽植行距65 cm，株距35 cm。辣椒定植前，先将滴灌带铺放在垄上，使滴灌带上的滴头与辣椒植株根部相对并保持10 cm左右的间隔，再覆塑料薄膜，以后进行膜下滴灌、灌水施肥，每小区均安装开关以调节灌水及施肥量，安装水表精确记录每次灌水量。田间栽培管理一致，试验处理具体设置见表1。

1.2 调查项目

根系活力采用TTC法测定。土壤体积质量采用环刀法。土壤含水率采

表1 试验设计
Table 1 Experimental design

处理 Treatment	苗期至开花期 Seedling stage to flowering stage /(m³/kg/667 m²)	开花坐果期至果实膨大期 Flowering and fruiting stage to fruit swelling stage /(m³/kg/667 m²)	果实膨大期至拉秧期 Fruit swelling stage to seedling stage /(m³/kg/667 m²)
G1S1	10.0/2.56	11.6/5.13	13.4/7.7
G1S2	10.0/4.0	11.6/7.98	13.4/12.0
G1S3	10.0/5.4	11.6/10.8	13.4/16.2
G2S1	11.6/2.56	13.4/5.13	16.5/7.7
G2S2	11.6/4.0	13.4/7.98	16.5/12.0
G2S3	11.6/5.4	13.4/10.8	16.5/16.2
G3S1	13.4/2.56	15.1/5.13	20.0/7.7
G3S2	13.4/4.0	15.1/7.98	20.0/12.0
G3S3	13.4/5.4	15.1/10.8	20.0/16.2

用烘干法测定,计算公式为土壤含水率%=(原土质量-烘干土质量)/烘干土质量×100%。过氧化氢酶采用$KMnO_4$滴定法;脲酶采用靛酚蓝比色法;蔗糖酶采用$Na_2S_2O_3$滴定法,纤维素酶采用DNS法。土壤硝态氮、铵态氮采用Smartchem全自动化学分析仪测定。维生素C用2,6-二氯酚靛酚法测定;蛋白质含量用碱滴定法测定;可溶性糖含量用蒽酮比色法测定。

辣椒生育后期,选取5株长势均匀的植株,齐地面剪去地上部分,称其鲜质量。同时以对应植株为中心,挖取留在土壤中的根系。用吸水纸吸干根表面的水分后,立刻称其鲜质量。采用烘干称质量法测定冠干质量(即地上部分)和根干质量。辣椒根冠比=根干质量/冠干质量。

1.3 数据处理

数据用EXCEL和SPSS11.0软件进行单因素方差分析,多重比较采用LSR法(Duncan's法),显著水平$P<0.05$。

2 结果与分析

2.1 不同处理对拱棚辣椒根系活力的影响

从图1可以看出,不同处理对拱棚辣椒根系活力有差异,G3S2辣椒根系活力最强,为18.905 μg/FW·h,G2S3辣椒根系活力次之,为12.47 μg/FW·h,G2S2辣椒根系活力最弱,为4.7063 μg/FW·h,其他处理及对照(CK)辣椒根系活力分别在5.421~9.3467 μg/FW·h之间,差异不显著。

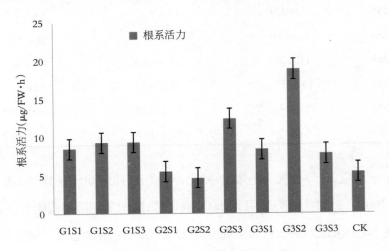

图1 不同处理对拱棚辣椒根系活力的影响

Fig. 1 Effects of different treatments on root activity of gongpeng pepper

2.2 不同处理对拱棚辣椒土壤容重、含水率的影响

从图2~3可知,在辣椒结果盛期,采集土壤(深度30 cm)进行土壤容重和土壤含水率的调查,各处理土壤容重及土壤含水率有差异,G1S1处理土壤容重最大,为1.22 g/cm³,G1S1、G3S1土壤容重次之,均为1.16 g/cm³,G2S2土壤容重最小,为0.96 g/cm³,其他处理及对照(CK)差异不显著,分别在1.05~1.12 g/cm³;G2S3土壤含水率最高,为24.65%,G3S2土壤含水率次之,为22.0%,G3S3土壤含水率最低,为9.42%,其他处理及对照土壤含水率分别在20.13%~21.61%。

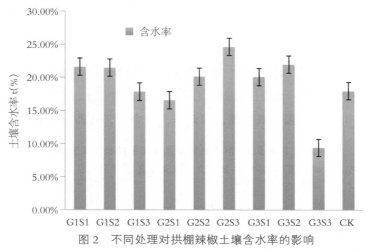

图 2 不同处理对拱棚辣椒土壤含水率的影响

Fig. 2 Effects of different treatments on soil moisture content of pepper in arch shed

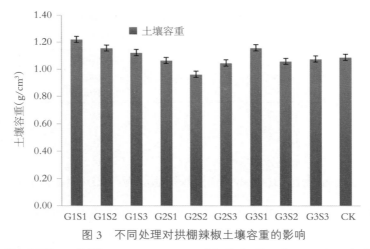

图 3 不同处理对拱棚辣椒土壤容重的影响

Fig. 3 Effects of different treatments on soil bulk density of pepper in arch shed

2.3 不同处理对连作辣椒土壤酶活性的影响

从表 2 可知,不同处理对连作辣椒土壤酶活性有差异,土壤脲酶是表征土壤氮素转化的关键酶。G3S2 土壤脲酶活性最强,为 182.59 mg/g·24 h,G2S3 土壤脲酶活性次之,为 169.58 mg/g·24 h,对照(CK)土壤酶活性最弱,为 120.36 mg/g·24 h,其他处理土壤脲酶活性分别在 126.85~159.13 mg/g·24 h;土壤蔗糖酶是影响土壤碳代谢的关键酶。G3S2 土壤蔗糖酶活性最强,为

874.10 mg/g·24 h，G2S3土壤蔗糖酶次之，为862.70 mg/g·24 h，对照（CK）土壤蔗糖酶活性最弱，为389.21 mg/g·24 h，其他处理土壤蔗糖酶分别在408.20~827.79 mg/g·24 h；过氧化氢酶是一种分解土壤中过氧化氢进而减小其对植物毒害作用的氧化还原酶，可以用来表征土壤的生化活性。G3S2土壤过氧化氢酶活性最强，为7.50 mg/g·min，G1S1、G2S3、G3S1土壤酶活性次之，分别为7.43 mg/g·min、7.43 mg/g·min、7.47 mg/g·min，对照土壤过氧化氢酶活性最弱，为7.02 mg/g·min；纤维素酶是碳循环的重要酶。G3S2土壤纤维素酶活性最强，为0.15 mg/g·72 h，G2S3土壤纤维素酶活性次之，为0.12 mg/g·72 h，对照（CK）土壤纤维素酶酶活性最弱，为0.04 mg/g·72 h，其他处理土壤纤维素酶活性分别在0.04~0.11 mg/g·72 h。

表2 不同处理对辣椒连作土壤酶活性的影响
Table 2 Effects of different treatments on soil enzyme activity in continuous cropping of pepper

处理 treatment	脲酶 urease /(mg/g·24 h)	蔗糖酶 sucrase /(mg/g·24 h)	过氧化氢酶 catalase /(mg/g·min)	纤维素酶 cellulase /(mg/g·72 h)
G1S1	159.13	545.81	7.43	0.04
G1S2	132.65	408.00	7.35	0.05
G1S3	134.26	511.12	7.35	0.06
G2S1	136.52	758.87	7.37	0.08
G2S2	157.97	827.79	7.35	0.11
G2S3	169.57	862.71	7.43	0.12
G3S1	126.85	693.78	7.47	0.05
G3S2	182.59	874.10	7.50	0.15
G3S3	152.17	530.49	7.17	0.08
CK	120.36	389.21	7.02	0.04

注：各列在P<0.05水平达到显著，以下相同

2.4 不同处理对拱棚辣椒土壤硝态氮、铵态氮的影响

从表3可以看出，不同处理对拱棚辣椒土壤硝态氮、铵态氮有影响，

表3 不同处理对辣椒连作土壤硝态氮、铵态氮的影响
Table 3 Effects of different treatments on nitrate and ammonium nitrogen in pepper continuous cropping soil

处理 treatment	硝态氮 nitrate nitrogen/(mg/kg)	铵态氮 ammonium nitrogen/(mg/kg)
G1S1	13.5	8.14
G1S2	22.8	6.40
G1S3	12.3	7.20
G2S1	18.1	9.97
G2S2	10.7	16.1
G2S3	20.1	7.46
G3S1	15.2	10.9
G3S2	5.91	8.69
G3S3	7.16	8.57
CK	19.8	8.05

G3S2硝态氮含量最低为5.91 mg/kg，G3S3硝态氮含量次之，为7.16 mg/kg，G1S2硝态氮含量最高，为22.8 mg/kg，其他处理及对照(CK)硝态氮含量分别在7.16~19.8 mg/kg；G2S2铵态氮含量最高为16.1 mg/kg，G3S1铵态氮含量次之，为10.9 mg/kg，G1S2铵态氮含量最低为6.4 mg/kg，其他处理及对照铵态氮含量分别在7.2~9.97 mg/kg。

2.5 不同处理对拱棚辣椒干物质分配的影响

从表4可以看出，辣椒盛花期至盛果期，不同处理对辣椒地上部、地下部干物质分配有影响，G3S2、对照(CK)、G3S1根干重最大，分别为20.2 g、20.0 g、19.7 g，G1S2、G2S1根干重最小，均为16.4 g，其他处理根干重分别在17.0~18.9 g；G1S3、G2S3茎干重最大，分别为75 g、70 g，G2S1、G3S2茎干重最小，分别为35.0 g、40 g，其他处理和对照(CK)茎干重分别在45~60 g；各处理及对照(CK)叶干重差异不显著，分别在40~48 g；G3S2、G2S1、G3S1根冠比最大，分别为0.21、0.22、0.22，G1S2根冠比最小为0.14，其他处理及对照

表4 不同处理对拱棚辣椒干物质分配的影响

Table 4 Effects of different treatments on the dry matter distribution of tunnel pepper

处理 treatment	根干重 dry weight of root /g	茎干重 stem dry weight /g	叶干重 leaf dry weight /g	根冠比 root shoot ratio
G1S1	18.6	55.0	45.0	0.19
G1S2	16.4	55.0	45.0	0.14
G1S3	18.9	75.0	45.0	0.17
G2S1	16.4	35.0	40.0	0.22
G2S2	16.9	60.0	40.0	0.17
G2S3	18.4	70.0	45.0	0.16
G3S1	19.7	45.0	45.0	0.22
G3S2	20.2	40.0	40.0	0.21
G3S3	17.0	55.0	45.0	0.17
CK	20.0	55.0	48.0	0.19

(CK)分别在0.16~0.19。

2.6 不同处理对拱棚辣椒品质的影响

从表5可知,辣椒盛果期,不同处理对辣椒品质有影响,G3S2辣椒可溶性蛋白含量最高,为32.24 mg/g,G2S3、G1S1可溶性蛋白含量次之,分别为31.32 mg/g、31.20 mg/g,G1S3可溶性蛋白含量最低,为26.57 mg/g,其他处理及对照(CK)可溶性蛋白含量差异不显著;G3S2辣椒可溶性糖含量37.9 mg/g,G1S1辣椒可溶性糖含量次之,为35.27 mg/g,其他处理及对照(CK)可溶性糖含量差异不显著;G3S2辣椒维生素C含量最高,为130.42 mg/g,G2S3维生素C含量次之,为91.31 mg/g,对照(CK)维生素C含量最低,其他处理及差异不显著,分别在32.21~50.73 mg/g。

表5 不同处理对拱棚辣椒品质的影响

Table 5 Influence of different treatments on quality of tunnel pepper

处理 treatment	可溶性蛋白含量 soluble protein content/(mg/g)	可溶性糖含量 contents of soluble sugar/(mg/g)	维生素C vitamin C /(mg/g)
G1S1	31.20	35.27	50.73
G1S2	29.73	31.24	40.83
G1S3	26.57	28.36	32.21
G2S1	29.85	32.69	40.33
G2S2	27.31	29.69	42.71
G2S3	31.32	34.16	91.32
G3S1	27.71	29.41	33.57
G3S2	32.24	37.90	131.42
G3S3	29.45	30.75	34.60
CK	28.34	29.44	30.57

3 讨论

水肥一体化技术是设施蔬菜节本增效的有效措施,前人研究,保护地黄瓜水肥一体化技术与大水冲施进行对比试验,结果表明:节水220 m³/667 m²,节省化肥58%,节省投入51.4元/667 m²;水肥一体化技术一次性投资约2 810元/667 m²,毛管使用寿命一般可达3年以上,支管和其他设备寿命以10年计算,每年投入652元[11-13]。滴灌施肥技术对大棚甜椒产量与土壤硝酸盐的影响,试验结果表明:15 cm和100 cm土层土壤溶液中硝态氮和无机态氮在甜椒整个生育期内保持稳定是滴灌施肥、节肥高产的主要原因。大棚土壤和地下水的无机氮素污染物质主要是硝态氮。滴灌处理100 cm土层土壤溶液中的硝态氮在整个甜椒生育期内显著低于常规施肥沟灌处理,滴灌施肥技术对减轻土壤和地下水硝酸盐污染是十分有效的措施之一[14-16]。灌溉施肥技术对温室辣椒干物质积累及叶片光合特性的影响试验结果表明:2种滴灌施肥处理下的辣椒植株较沟灌施肥处理在盛花期和盛果期有较高的地上部干物质积累量[17-20]。

4 结论

本研究探讨了塑料拱棚辣椒水肥一体化栽培，在不同灌水量和施肥量条件下，苗期至开花期，开花期至果实膨大期，果实膨大期至拉秧期3个时期，G3S2处理灌水量是 201 m³/hm²、226.5 m³/hm²、300 m³/hm²，施肥量是 60 kg/hm²、119.7 kg/hm²、180 kg/hm²，20 m³/12.0 kg/667 m² 辣椒根系活力最强，为 18.91 ug/FW.h；土壤脲酶、蔗糖酶、过氧化氢酶、纤维素酶活性最强，分别为 182.59 mg/g·72 h、874.1 mg/g·72 h、7.5 mg/g·72 h、0.15 mg/g·72 h；硝态氮含量最低，为 5.91 mg/kg；辣椒根干重、根冠比最大，为 20.2 g、0.22 g；辣椒可溶性蛋白含量、可溶性糖含量、维生素 C 含量最高，分别为 32.24 mg/g、37.9 mg/g、130.42 mg/g，为制定拱棚辣椒水肥一体化施肥制度、拱棚辣椒高产优产提供理论依据。

参考文献

[1] 虞娜,张玉龙,黄毅,等.温室滴灌施肥条件下水肥耦合对番茄产量影响的研究[J].土壤通报,2003,34(3):179-183.

[2] 姚静,邹志荣,杨猛,等.日光温室水肥耦合对甜瓜产量影响研究初探[J].西北植物学报,2004,24(5):890-894.

[3] 贺超兴,张志斌,刘富中,等.日光温室水钾氮耦合效应对番茄产量的影响[J].中国蔬菜,2001(1):31-33.

[4] 刘祖贵,段爱旺,吴海卿,等.水肥调配施用对温室滴灌番茄产量及水分利用效率的影响[J].中国农村水利水电,2003(1):10-12.

[5] 齐红岩,李天来,曲春秋,等.亏缺灌溉对设施栽培番茄物质分配及果实品质的影响[J].中国蔬菜,2004(2):10-12.

[6] 王军,陈双臣,邹志荣.肥料增效剂对大棚番茄产量、品质的影响[J].陕西农业科学,2004(2):33-35+69.

[7] 赵宏儒,刘建英,张彦萍,等.反季节基质无土栽培甜瓜技术[J].华北农学报,2005

[8] 刘建英,赵宏儒,张丽清,等.保护地黄瓜水、肥一体化高效栽培技术[J].华北农学报,2005(S1):206-208.

[9] F Wang,S Kang,T Du,et al. Determination of comprehensive quali-ty index for tomato and its response to different irrigation treatments[J]. Agricultural Water Management,2011,98:1228-1238.

[10] Feng Wang,Shao zhong Kang,Tai sheng Du,et al.Determination of compre hensive quality index for tomato and itsresponseto diffe rentirrig ationtre atments[J].Agricultural Water Management,2011.

[11] PATANèC,COSENTINOSL.Effectsofsoil water deficitony ield and quality of processing tomato under amedit erranean climate[J]. Agricultural Water Management,2010.

[12] 陈秀香,马富裕,方志刚,等.土壤水分含量对加工番茄产量和品质影响的研究[J].节水灌溉,2006(04):1-4.

[13] 张鲁鲁,蔡焕杰,王健.膜下滴灌对温室甜瓜水分利用效率及品质影响[J].节水灌溉,2011(04):7-10.

[14] 隋方功,王运华,长友诚,樗木直也,乌尼木仁,稻永醇二.滴灌施肥技术对大棚甜椒产量与土壤硝酸盐的影响[J].华中农业大学学报,2001(04):358-362.

[15] SALOKHEVM,BABELMS,TANTAUHJ.Waterre quirement of dripirrig ated to matoes growning reen house in tropical environment[J]. AGRICUL TURAL WATER MANAGE MENT,2005,71(3):225-242.

[16] 李建明,潘铜华,王玲慧,等.水肥耦合对番茄光合、产量及水分利用效率的影响[J].农业工程学报,2014,30(10):82-90.

[17] 杨慧,曹红霞,刘世和,等.水氮耦合对温室番茄光合特性与产量的影响[J].灌溉排水学报,2014,33(Z1):58-62

[18] 丁果.温室蔬菜滴灌灌溉施肥水肥耦合效应的研究[D].内蒙古农业大学,2005.